Gene structure and expression

Other structure and cooperation

Gene structure and expression

JOHN D. HAWKINS

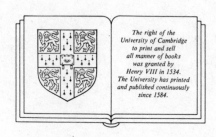

The right of the
University of Cambridge
to print and sell
all manner of books
was granted by
Henry VIII in 1534.
The University has printed
and published continuously
since 1584.

CAMBRIDGE UNIVERSITY PRESS

Cambridge

London   New York   New Rochelle

Melbourne   Sydney

Published by the Press Syndicate of the University of Cambridge
The Pitt Building, Trumpington Street, Cambridge CB2 1RP
32 East 57th Street, New York, NY 10022, USA
10 Stamford Road, Oakleigh, Melbourne 3166, Australia

First published 1985

Printed in Great Britain by the University Press, Cambridge

Library of Congress catalogue card number: 84–17459

British Library Cataloguing in Publication Data

Hawkins, J. D.
Gene structure and expression.
1. Deoxyribonucleic acid
574.87′3282   QP624

ISBN 0 521 25824 3 hard covers
ISBN 0 521 27726 4 paperback

Contents

List of abbreviations   ix

Introduction   xi

1 DNA   1
1.1   The genetic material   1
1.2   DNA is a polar molecule   1
1.3   DNA generally exists as a double helix   2
1.4   DNA molecules are very long but can be twisted into
      compact forms   5
1.5   Replication is semi-conservative   8
1.6   The gene or cistron is the functional unit of DNA   10
1.7   Mutations can arise in various ways   11

2 Ribonucleic acid   14
2.1   Expression of the information in DNA is mediated by RNA   14
2.2   Transcription is a major stage of gene expression   14
2.3   The four major classes of RNA   16
2.4   The genetic code   21
2.5   Translation is a later stage of gene expression   23

3 Methodology   28
3.1   Introduction   28
3.2   mRNA isolation   29
3.3   Separation of nucleic acids   30
3.4   Reverse transcriptase   30
3.5   Nucleases   31
3.6   Restriction endonucleases   33
3.7   Restriction fragment length polymorphisms   37
3.8   Hybridisation of nucleic acids   38
3.9   The determination of base sequence in DNA   40

4   Vectors used in work with recombinant DNA   47

4.1   Plasmids   47
4.2   Bacteriophages   52
4.3   Viruses   55

5   Prokaryotic gene organisation and expression   56

5.1   Replication   56
5.2   Transcription   59
5.3   Some RNAs are processed after transcription   62
5.4   Transposable genetic elements   63

6   The operon concept   66

6.1   Genes for sets of metabolically related enzymes are transcribed
      as one long message   66
6.2   The *lac* operon   70
6.3   The *gal* operon   72
6.4   The *ara* operon   74
6.5   The *hut* operons   75
6.6   The *mal* regulon   76
6.7   The *trp* operon: control by attenuation   77
6.8   Pyrimidine biosynthesis   80
6.9   Arginine biosynthesis   81
6.10 Ribosomal proteins   82
6.11 The stringent response   84

7   Eukaryotic gene organisation and expression   86

7.1   DNA is in the nucleus in discrete linear chromosomes   86
7.2   Nuclear DNA is associated with proteins   87
7.3   Histones associate in a regular fashion with RNA to form
      nucleosomes   88
7.4   Replication of DNA is a mystery   90
7.5   Transcription   90
7.6   Enhancers   92
7.7   Many mRNA molecules have a cap and a tail added
      post-transcriptionally   93
7.8   The coding sequence of many genes is interrupted by non-coding
      sequences   95
7.9   Introns are transcribed into RNA and then removed in the
      nucleus   96
7.10 Post-translational modifications may be required to produce
      functional proteins   99

8   Repeated sequences and oncogenesis   101

8.1   Histone genes   101

8.2	rRNA and tRNA genes	101
8.3	Repeated sequences	102
8.4	Retroviruses and oncogenic viruses	105
8.5	Chromosomal alterations in cancer	109
9	**Haemoglobin**	**111**
9.1	Genes for globins are found in two clusters	111
9.2	Thalassaemias	113
9.3	Other mutations	116
9.4	Prenatal diagnosis of anaemias	117
10	**Proteins of the immune system**	**118**
10.1	Immunoglobulins consist of H and L chains	118
10.2	L chain genes	120
10.3	H chain genes	122
10.4	DNA processing is employed during the course of the immune response	125
10.5	Enhancers	126
10.6	Allelic exclusion	127
10.7	The major histocompatibility complex	127
10.8	Class I genes	129
10.9	Class II genes	130
10.10	Complement genes	131
11	**Hormone genes**	**132**
11.1	Insulin	132
11.2	Growth hormone family	134
11.3	Polyproteins	135
11.4	Glycoprotein hormones	141
12	**The mitochondrial genome**	**143**
12.1	Yeast mitochondrial genome	143
12.2	Mammalian mitochondrial genome	147
12.3	Mitochondrial genome of higher plants	150
13	**The control and plasticity of the genome**	**151**
13.1	Sequences 5′ to genes control their expression	151
13.2	Control in prokaryotes	151
13.3	Control in eukaryotes	153
13.4	DNA methylation	154
13.5	DNase sensitivity	157
13.6	The plasticity of the genome	158
13.7	Evolution	159
13.8	Future developments	162

Reading lists: 163
chapter 1   163
chapter 2   163
chapter 3   164
chapter 4   164
chapter 5   165
chapter 6   165
chapter 7   165
chapter 8   166
chapter 9   167
chapter 10   167
chapter 11   167
chapter 12   168
chapter 13   168

Index   169

List of abbreviations

In general, standard biochemical abbreviations are used throughout this book, particularly in figures and tables. The following is a list of those that are used without further explanation.

Amino acids:

Alanine Ala
Arginine Arg
Asparagine Asn
Aspartate Asp
Cysteine Cys
Glutamine Gln
Glutamate Glu
Glycine Gly
Histidine His
Isoleucine Ile
Leucine Leu
Lysine Lys
Methionine Met
Phenylalanine Phe
Proline Pro
Serine Ser
Threonine Thr
Tryptophan Trp
Tyrosine Tyr
Valine Val
Termination Ter
 (at end of protein sequences)

Purine and pyrimidine bases and other symbols used in writing DNA or RNA sequences. It should be clear from the context whether bases or nucleotides are intended:

Adenine A
Guanine G
Cytosine C
Uracil U
Pseudo-uridine Ψ
Thymine T
Pyrimidine Y
Purine R
Any nucleotide N
Any base B
Phosphate radical P
Ribose R
Deoxyribose dR
Methyl group m

Introduction

There has been an explosive growth in our detailed knowledge of genetics at the molecular level over the last few years, and it is likely that accretion of new knowledge will occur at an ever increasing rate. It is therefore very difficult even for the specialist to keep abreast of all the latest ideas which rapidly progress from hypothesis to theory to accepted dogma. In the time that it takes to write a comprehensive text book it is inevitable that new ideas will be generated and many problems in the field elucidated so that such a book will certainly be out of date before the writing is finished, let alone published. Even during the writing of this small book, over the course of a little more than a year, much new information has come to light so that were it to be re-written in the next few months, appreciable differences would appear. It does not therefore claim to be a complete guide to the subject under review; nevertheless it attempts to present ideas that are reasonably well established and at the same time to cover a fairly wide field, albeit mostly not in great depth. The selection of topics as examples of our knowledge is somewhat arbitrary and conditioned by the author's own interests and expertise.

I believe that it should be a useful book for medical students who wish to become familiar with recent ideas and techniques in molecular biology to help in understanding further advances when they arrive. It will also be of use to honours and graduate students in genetics, biochemistry and those who would not necessarily regard the topics discussed here as their major interests in these subjects. It assumes a working knowledge of biochemistry that a first or second year university or polytechnic student should have acquired in a fairly elementary course in that subject. This basic material is already excellently covered in such books as Lubert Stryer's *Biochemistry* and Albert Lehninger's *Biochemistry*, as well as a host of others.

Reading Lists for each Chapter are to be found at the back of this book. They are mostly made up of review articles in which references to original work can be found. In several cases parallel reviews covering more or less the same ground have been cited.

I am most grateful to Dr Fay Bendall of Cambridge University Press who encouraged me to write this book in the first place, widening my horizons and giving me a good deal of pleasure in the process: to Dr Audrey O. Smith whose patience and skill in pointing out errors and inconsistencies in the text have helped to clarify it. Needless to say, any errors that do remain are my own responsibility. I am also grateful to many colleagues at St Bartholomew's Hospital Medical College, particularly Dr Clem Lewis for stimulating discussions which I have frequently found helpful in clarifying my ideas. Last, but not least, I am grateful for my wife's forebearance in the face of a neglectful husband who has been preoccupied for many months with the genome rather than with its ramifications as revealed in family life.

June 1984

1
DNA

1.1   The genetic material

The classic experiments of Avery in 1944 demonstrated that DNA (deoxyribonucleic acid) is the material that can pass genetic information from one bacterium to another. He showed that strain-specific properties of related bacteria could be transferred by DNA that was free of proteins and other substances. DNA is a polymeric molecule built up from only four similar but distinct monomers – nucleotides which are the 5'-phosphates of deoxyguanosine (dGMP), deoxyadenosine (dAMP), deoxycytidine (dCMP), and thymidine (TMP) (Fig. 1.1). In DNA these are joined by phosphodiester linkages between the 3'- and 5'-positions of successive deoxyribose moieties. The initial letters of the bases in the nucleotides are used as abbreviations when writing out sequences in DNA.

1.2   DNA is a polar molecule

One end of a DNA molecule has a phosphoryl radical on the C-5' of its terminal nucleotide, while the other end possesses a free -OH on C-3' of its nucleotide. Thus a polynucleotide possesses *polarity* in an analagous way to the more familiar polarity of proteins with free -NH$_2$ and -COOH groups at each end. This means, for example, that the tetranucleotides TCGA and AGCT are different chemical entities with distinct properties, even though they behave in a very similar way in many respects (Fig. 1.2). By convention, sequences of DNA are written with the nucleotide containing the free phosphoryl radical at the left. Sequences to the left of a given nucleotide are said to be on the 5'-side (often called upstream), and those to the right are said to be on the 3'-side (often called downstream). The symbols N, R and Y are used to denote any nucleotide, a purine nucleotide, and a pyrimidine nucleotide respectively.

1.3   DNA generally exists as a double helix

DNA generally exists in double strands because of the propensity of the bases for hydrogen bonding to each other in a highly specific way (Fig. 1.1). A bonds with T, and G with C, though very occasional mismatches or alternative bonding can occur. Thus a double-stranded DNA will always contain equal molar proportions of A and T and of G and C though the content of A (or T) and G (or C) varies widely in DNA from different sources.

Fig. 1.1. The four deoxyribonucleotides that make up DNA, showing how the bases form hydrogen bonds.

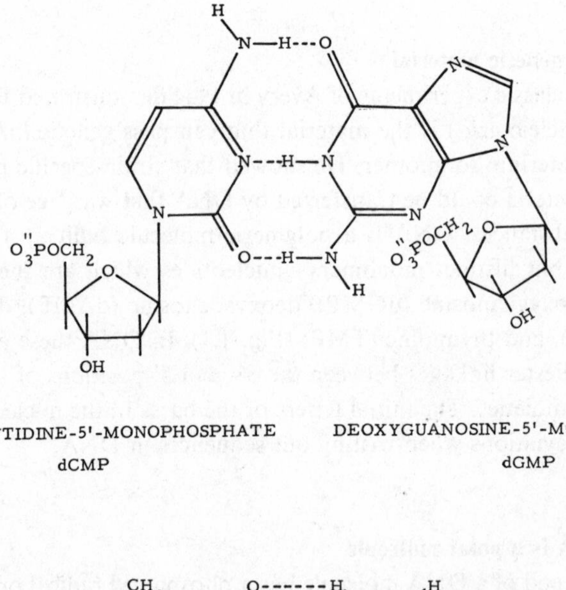

DEOXYCYTIDINE-5'-MONOPHOSPHATE   DEOXYGUANOSINE-5'-MONOPHOSPHATE
dCMP   dGMP

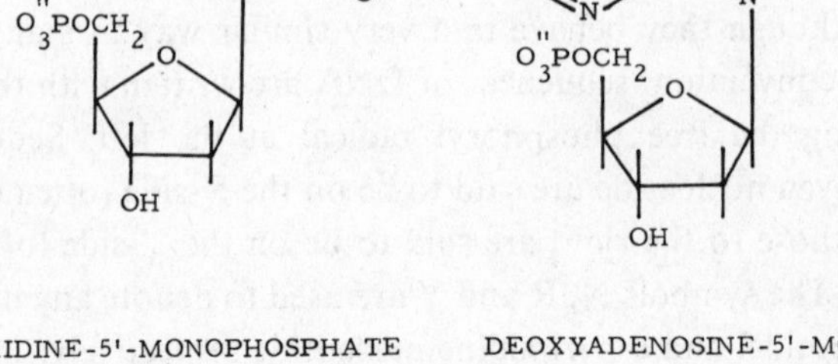

THYMIDINE-5'-MONOPHOSPHATE   DEOXYADENOSINE-5'-MONOPHOSPHATE
dTMP   dAMP

The basic unit in a DNA molecule is the pair of nucleotides hydrogen bonded to each other, which is generally known as a base-pair (bp; with kbp used as an abbreviation for 1000 bp).

This double-stranded molecule takes up a helical conformation in which the continuous deoxyribose-phosphate strands twine round the outside of the *helix* with the base pairs (A and T or G and C) in the interior (Fig. 1.3). Since A and T associate together with only two hydrogen bonds, while the G-C pair has three hydrogen bonds, the former pairing is less stable than the latter. This has important consequences for the stability of different regions of the double helix. In regions that are rich in A and T residues

Fig. 1.2. Polarity in DNA. The two tetranucleotides TCGA (left) and AGCT (right) are different, even though they have the bases in the same order.

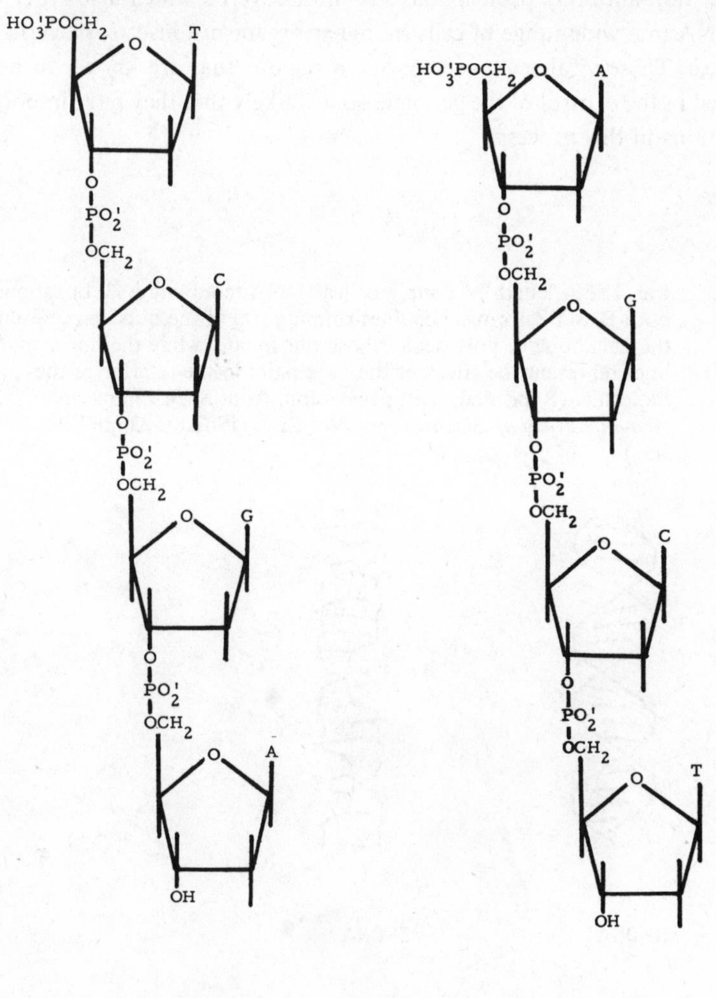

the helix can be more easily destabilised and unwound than in G-C rich regions.

The polarity of the two DNA strands is *anti-parallel* – that is to say, one runs in the 5′ to 3′ direction while the complementary strand runs the opposite way. The helix can adopt several conformations. The commonest form (B-DNA) has a pitch of just over 10 residues per turn, and is right-handed when viewed end on. Another form, known as Z-DNA, can arise under certain conditions when there are alternating purine and pyrimidine residues in the sequence. This is a left-handed helix and has a pitch of 11.5 residues per turn (Fig. 1.3). This form is believed to have some biological significance, since short stretches of alternating purine and pyrimidine residues occur at numerous sites in many DNAs. They can be detected by the binding of antibodies which specifically recognise this form of DNA. Other non-antibody proteins have been discovered which also react with Z-DNA in a wide range of cells in higher organisms, insects, bacteria and viruses. These viral sequences occur in regions that are known to be involved in the control of the genome, so it is likely that they have important functions in this process.

Fig. 1.3. A length of double helical DNA, containing 20 bp, showing both B and Z forms. The lines running round the outside represent the backbone of poly deoxyribose phosphate, while the horizontal lines represent the edges of the base pairs in the interior of the molecule. (Reprinted, with permission, from S. B. Zimmerman, *Annual Review of Biochemistry*, Vol **51** © 1982 by Annual Reviews Inc.)

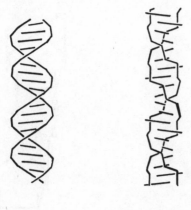

B—DNA   Z—DNA

Table 1.1. *Chromosome numbers and DNA content of cells of a representative set of species*

Species	Number of chromosomes	DNA content kilobase pairs
Bacillus subtilis	1	2×10^3
Escherichia coli	1	3.8×10^3
Saccharomyces cerevisiae	34	14×10^3
Drosophila melanogaster	8	2×10^5
Sea urchin	52	1.6×10^6
Frog	26	45×10^6
Chicken	78	2.1×10^6
Mouse	40	4.7×10^6
Human	46	5.6×10^6
Maize	20	30×10^6

All figures are for diploid cells, except for bacteria.

1.4   DNA molecules are very long but can be twisted into compact forms

DNA molecules are extremely long and can be visualised by the electron microscope. A double helix of 10^6 base pairs is 0.34 mm long and only 2 nm in diameter. In prokaryotes the DNA is circular so that there are no free 3'- and 5'-ends, and all the chromosomal DNA is in a single molecule. In eukaryotes the DNA in the chromosomes exists as linear molecules, and different species possess different numbers of chromosomes. The amount of DNA in cells of different species varies very widely, generally with an increase in the DNA content as species become more complex (Table 1.1).

With one exception all eukaryotic chromosomes are paired, with one partner coming from each parent. The exception is the sex chromosome. Females carry two X chromosomes, while males carry an X chromosome inherited from their mother and a Y chromosome from their father. This is the case in mammals and many other orders, but other methods of sex determination do occur. The two chromosomes of a pair are said to be *homologous* since they will nearly always be identical in their organisation and frequently in the genes they carry. However, since there are many mutant genes in a population a pair of homologous chromosomes may carry different genes at particular loci. These are known as *alleles*.

A mutant gene encoding a defective product can generally be *complemented* by a 'good' copy of the gene on the homologous chromosome, but if there is a defect on the single copy of the X chromosome that a male carries it cannot be complemented in this way. Thus there are many sex-linked inherited diseases which are carried by females, but expressed only in males. These disorders can actually occur in females but the chances of a female inheriting two defective genes are very low.

Since all somatic cells contain a homologous pair of each of the chromosomes they are known as *diploid*. The gametes – sperm and ova – which only contain one member of each pair of chromosomes are known as *haploid* cells. The contribution of one parent to the genetic make-up of the offspring is known as the *haplotype*.

Individual chromosomes are morphologically distinguishable when they are suitably stained. For purposes of identification they have been given numbers in numerical order, starting with the largest one.

In the cell both linear and circular molecules are found in much more compact forms. The helix is coiled on itself several times (like the element in an electric light bulb) so that the overall length is greatly reduced at the expense of an increase in diameter. This conformation is stabilised by proteins in eukaryotic cells (see Chapter 7, sections 2, 3). This kind of structure is very suitable for packaging DNA into a minimum of space, but when the DNA, or strictly speaking, portions of it, becomes functional some uncoiling of this structure must occur accompanied by temporary separation of the two helical strands.

In a circular DNA molecule containing 4000 bp (such as might occur as a bacterial plasmid – Chapter 4.2) the double helix is in the B-form. As this has a pitch of 10 residues per turn there should be 400 turns. In practice such DNA is found to have only about 380 turns because the helix is untwisted to a certain extent. This is known as negative *supercoiling*, and is an important feature of the structure of DNA in bacterial cells. It results in a puckered form of the molecule which gives it a more compact structure, and also places considerable torsional strain on it. An analogy can be made by twisting a rubber band held firmly at two diametrically opposite positions. A molecule in which there is no supercoiling is said to be *relaxed*, and there is a dynamic balance between relaxed and supercoiled forms of DNA as a result of the action of two classes of enzymes called *topoisomerases I* and *II* which catalyse the production of one form from the other.

When a DNA molecule is supercoiled it migrates more rapidly on electrophoresis than when it is relaxed. A family of otherwise identical DNA molecules with different degrees of supercoiling can be made visible as a ladder of bands by this technique (Fig. 1.4). Supercoiled molecules appear more compact than relaxed ones when viewed in the electron microscope.

Formation of the supercoiled form in the cell is brought about by an enzyme called DNA *gyrase* (a class II topoisomerase) and, because it is a strained structure, there is an energy requirement for this reaction, which

Fig. 1.4. The effect of supercoiling on the electrophoretic mobility of Simian virus 40 DNA. The DNA was treated with a class II topoisomerase and then electrophoresed. The thick band at the bottom represents fully relaxed DNA, while molecules with increasing degrees of supercoiling appear as bands of increasing mobility. The arrow shows the direction of electrophoretic migration. (Reproduced from W. Keller, *Proc. Natl. Acad. Sci. USA* (1975), **72**, 2550.)

Fig. 1.5. The action of DNA gyrase in forming a negative supercoil in a circular molecule of DNA. By convention, when the upper strand crosses above the lower strand from left to right the supercoiling is said to be positive. Negative supercoiling is the converse of this.

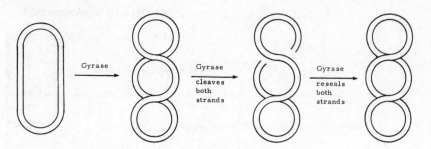

is met by the concomitant hydrolysis of ATP. The reaction proceeds by breaking both strands of the DNA so that another part of the molecule can be passed through the break. The broken strands are then re-sealed (Fig. 1.5). The reverse reaction, in which supercoiled DNA is converted to the relaxed form, is catalysed by a class I topoisomerase and requires no input of energy since a less-strained molecule is being produced. This reaction involves nicking one strand of the DNA, when the 5'-phosphate at the break point becomes bound to a tyrosyl residue on the enzyme. This strand also has a free 3'-OH group and this end is rotated round the other strand which is intact. The nicked strand is then re-sealed (Fig. 1.6).

Another way of relieving the tension in supercoiled DNA is by local disruption of the double helix to produce *single-stranded* regions. These can be stabilised by various proteins which bind to single-stranded DNA (known variously as single-stranded binding proteins or helix-destabilising proteins). This opening up of the double helix is an essential step for both replication of DNA (see next section) and its transcription by RNA polymerase (see Chapter 2.2), since each of these processes involves the use of single-stranded DNA as a template. Thus the supercoiling of DNA provides a favourable situation in which the operation of these two vitally important processes is facilitated. Supercoiling of DNA also promotes the flipping of B-DNA to the Z-form when the base sequence is suitable.

1.5   Replication is semi-conservative

When a cell divides each of the daughter cells contains a full complement of DNA, identical to that of the parent (except during the production of gametes in eukaryotes). Thus the DNA must be precisely *replicated*. This is done by the separation of the two strands, followed by pairing of deoxyribonucleoside triphosphates through specific hydrogen bonding with the bases in each strand. These triphosphates are joined together (ligated) by the enzyme DNA polymerase with the release of inorganic pyrophosphate (Fig. 1.7). Synthesis always proceeds from the

Fig. 1.6. Relaxation of supercoiled DNA by topoisomerase I.

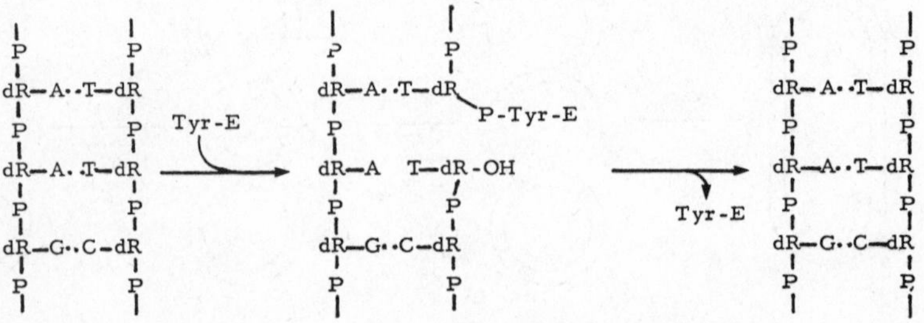

Fig. 1.7. Replication fork, showing strand separation and incorporation of deoxyribonucleoside triphosphates. Continuous synthesis takes place on the lower strand. Synthesis on the upper strand, which is discontinuous, does not start so near the replication fork.

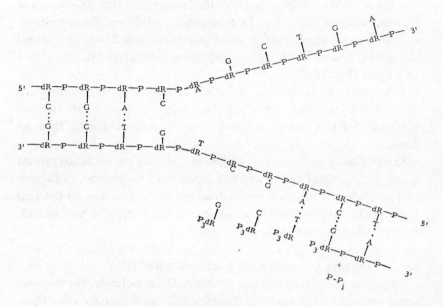

Fig. 1.8. Continuous and discontinuous synthesis of two strands of DNA during replication. Note that the scale is much smaller than in Fig. 1.7.

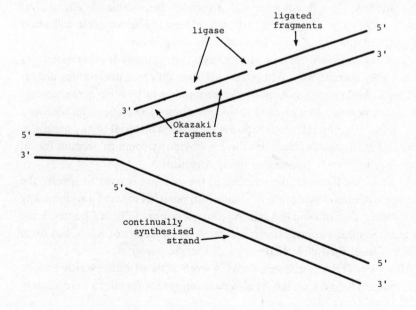

3'-end of the growing chain. In practice the two parental strands do not separate completely, but are opened up at what is known as the *replication fork* which is moved along as the process proceeds.

Because of the opposite polarity of the two parental strands, one strand is in the incorrect orientation for continuous synthesis of the new strand. On this strand, comparatively short lengths of new DNA are formed (Okazaki fragments, named after their discoverer) which are then joined by a ligase (Fig. 1.8).

This mode of replication is called *semi-conservative* because one of the DNA strands is conserved. This means that each daughter cell possesses one new strand of DNA and one derived directly from its parent.

This outline of replication may sound very simple, but the actual process is extremely complex and by no means completely understood. In *Escherichia coli*, where it has been most intensively studied, at least 14 different polypeptides, some of which are enzymes, are known to be involved. Further details will be found in Chapter 5.

1.6   The gene or cistron is the functional unit of DNA

Replication ultimately involves the whole DNA molecule, but the functional unit in DNA is much smaller than this – perhaps only a few thousand or even hundred base pairs. These units correspond to the original concept of *genes* postulated by earlier geneticists who associated them with various visible (*phenotypic*) characters which could be inherited in predictable ways. A more recent term with a very similar meaning is *cistron*, but this is not widely used, and we still talk of genes. The complete collection of genes in one organism is referred to as the *genome*.

Genes contain information which directs the synthesis of other molecules in a highly specific way. Although only four different nucleotides are involved in building a DNA molecule the number of possible arrangements is extremely large. In a piece of DNA containing 100 nucleotides there are 4^{100} (1.7×10^{60}) possible sequences. As the genetic material of any organism contains millions of nucleotides there is obviously room to account for all the amazingly wide diversity of living organisms.

Most of the information encoded in the genome is used to specify the sequence of amino acids that are made into proteins to serve a wide variety of essential functions in the cell. As discussed more fully in Chapter 2, the expression of this information is mediated by molecules of RNA, but some RNA is itself used directly for various specific purposes.

However, DNA sequences flanking both ends of those which encode information for making RNAs also have important functions in regulating the activity of genes. In prokaryotes such factors as the availability of metabolites will determine which genes are active at any particular time. In metazoans there is specialisation of cells of various types so that some genes in certain organs are permanently inactive ('switched off'), while others may be activated or not depending on particular local environmental signals.

There also appear to be stretches of DNA to which no definite function can be assigned, particularly in eukaryotes. These are sometimes considered to provide a pool of DNA from which new genes can arise by mutations of various kinds.

1.7 Mutations can arise in various ways

The fact that offspring generally resemble their parents very closely means that the constitution of DNA must be fairly stable. However, changes in the base sequence can and do occur, though at a rate which is very slow in relation to human life span. If this were not so, evolution could never have occurred. These changes are known as *mutations* and occur as a result of several different mechanisms.

A *point mutation* arises when there is a change in a single nucleotide in a sequence coding for a particular protein or RNA. They can occur as a result of a chemical reaction affecting one of the functional groups on a base (e.g. deamination of cytosine gives rise to uracil, see Fig. 1.9). This particular change can result by the action of nitrous acid on DNA. Such an agent altering the structure of the bases in DNA is called a *mutagen*. Some pollutants of our environment are believed to act in this way. Mutations resulting in a change of one pyrimidine to another or one purine to another are known as *transitions*: changes of a purine to pyrimidine or vice versa are known as *transversions*.

Electromagnetic radiation of various wavelengths (e.g. ultraviolet or X-rays) is also potentially mutagenic by causing the formation of highly reactive free radicals. These can cause chemical changes in the bases and

Fig. 1.9. Deamination of a cytosine residue to a uracil residue.

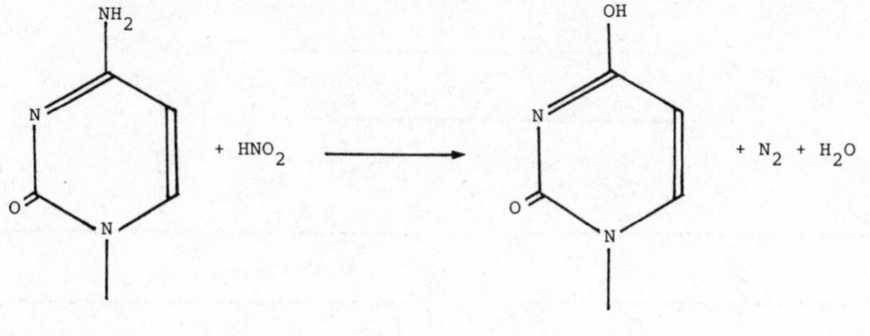

break DNA chains. Although enzymes can repair them they do not always function quite correctly. This may lead to either deletion or insertion of one or more nucleotides.

Finally, more drastic changes in the DNA may be brought about by the *translocation* of whole segments from their original site to a new one, either within the same chromosome or on a different one. A special case of this is the phenomenon of *crossing over* which was postulated as a result of classical genetic experiments long before the actual mechanism was known. Crossing over involves reciprocal changes between two homologous chromosomes. The consequence is that two heritable characteristics which were formerly expressed simultaneously will no longer appear together.

Gene conversion is the name given to another important transformation of DNA in which two genes interact so that part of the nucleotide sequence of one is incorporated into the other one. Both genes retain their integrity and location, but a non-reciprocal change in the structure of one of them occurs. Gene conversion probably occurs mostly at meiosis or mitosis, and results from a misalignment of sequences that are normally not paired but which are sufficiently homologous to allow base-pairing to occur (Fig. 1.10). Such an intermediate can give rise either to a cross-over or to transfer of DNA in one direction only (i.e. gene conversion). Gene conversion involves breaks in single strands of DNA, whereas in crossing over both strands must be broken. Gene conversion can occur between genes on different chromosomes (both homologous and non-homologous), or between genes located on the same chromosome. It is probably more common in the latter situation, especially where there are families of re-iterated genes of very similar structure, such as the globin genes (Chapter 9); V-region genes of immunoglobulins (Chapter 10); and some genes of the major histocompatibility complex (Chapter 10).

By studying differences in the amino acid sequence of analogous proteins in two different species (e.g. haemoglobins or cytochromes), or, better

Fig. 1.10. Gene conversion and crossing over. ABC and abc are reasonably homologous regions of either one chromosome or two different ones.

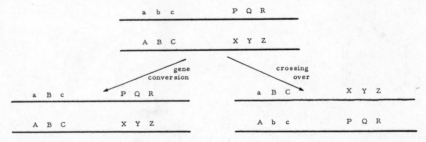

still, the DNA sequences of their genes, it is possible to make an estimate of the natural mutation rate if other evidence for the time at which the two species diverged is available. The best estimates suggest that on average a mutation occurs about once in 10^6 cell divisions. This sounds a high rate, bearing in mind that there are 10^{14} cells in an adult human, but only those mutations that actually occur in the gametes will be heritable. It is also important to appreciate that many mutations are silent – that is to say they have no observable effect, either because they have occurred in parts of the DNA that do not encode vital information, or even because they have no effect on the encoded information (see Chapter 2.4).

2
Ribonucleic acid

2.1   Expression of the information in DNA is mediated by RNA

Genetic expression always involves the synthesis of another nucleic acid polymer called ribonucleic acid (RNA). RNA is composed of four nucleotide monomers – the monophosphates of adenosine (AMP), guanosine (GMP), cytidine (CMP), and uridine (UMP) (Fig. 2.1). These all contain the sugar ribose instead of deoxyribose found in DNA, and the individual nucleotides are linked together through 3'-, 5'-phosphodiester bridges, just as in DNA. Another major difference from DNA is the presence of uracil rather than thymine as one of the bases. These two compounds differ by the presence of a methyl group in thymine, which is absent from uracil. RNA is generally a single-stranded molecule though its bases can pair by hydrogen bonding to give hairpin-like or stem structures (Fig. 2.2). Guanine occasionally pairs with uracil, though this pairing is less stable than the more usual A-U and G-C pairs. RNA molecules are much smaller than DNA molecules, and only comparatively short stretches of DNA are used to direct the synthesis of individual RNAs.

2.2   Transcription is a major stage of gene expression

RNA is synthesized on a template of DNA by a process called *transcription*, in which the strands of the DNA must first separate in the region where it is going to start. As transcription proceeds the DNA strands separate in front of the growing RNA chain. Transcription is formally rather similar to replication, except that only one of the DNA strands is used as a template. Analogously, the immediate precursors of RNA are the nucleoside triphosphates (ATP, GTP, CTP, UTP) whose bases are hydrogen bonded to the complementary bases on the DNA. The enzyme involved is DNA-dependent RNA polymerase. It catalyses the formation of phosphodiester links between the nucleotides with the

14

Fig. 2.1. The four common ribonucleoside monophosphates.

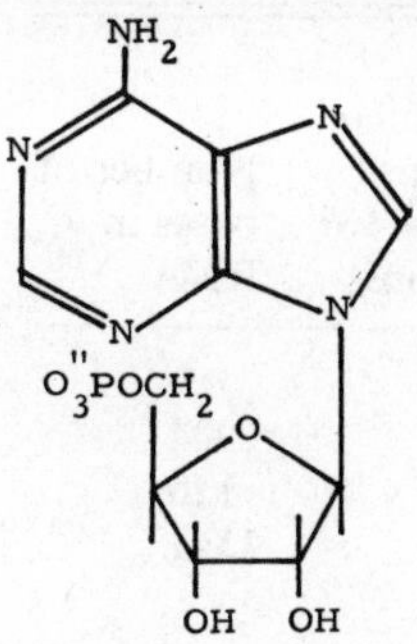

ADENOSINE-5'-MONOPHOSPHATE
ADENYLATE (AMP)

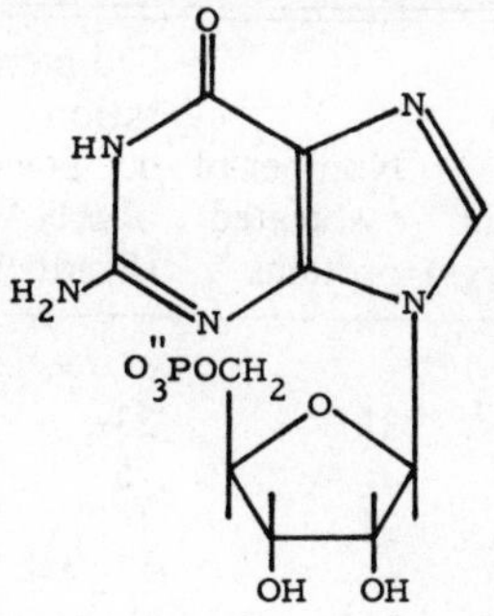

GUANOSINE-5'-MONOPHOSPHATE
GUANYLATE (GMP)

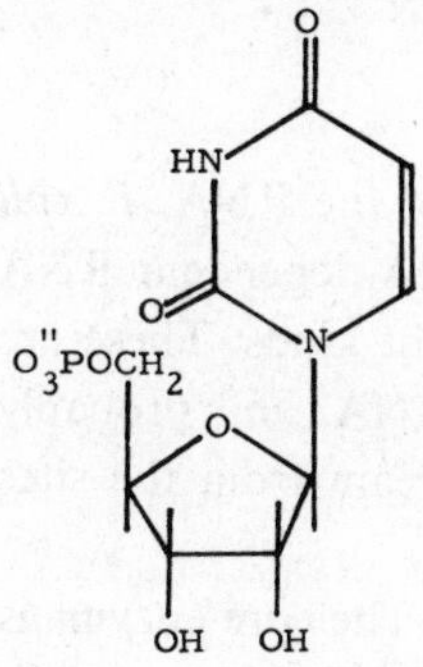

URIDINE-5'-MONOPHOSPHATE
URIDYLATE (UMP)

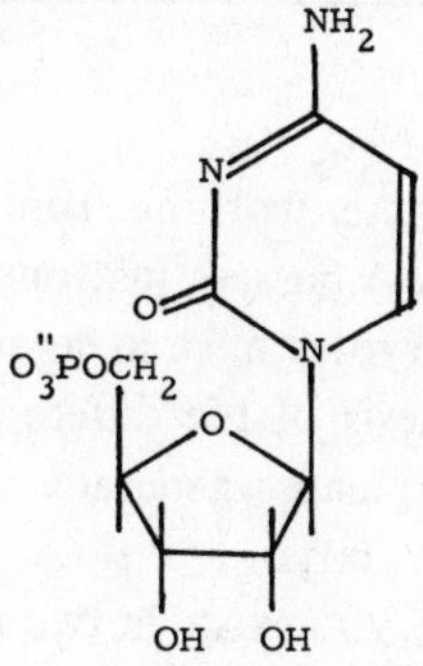

CYTIDINE-5'-MONOPHOSPHATE
CYTIDYLATE (CMP)

Fig. 2.2. A stem and loop structure in RNA. This particular sequence is the terminator structure of the *trp* operon in *E. coli* (Chapter 6.7).

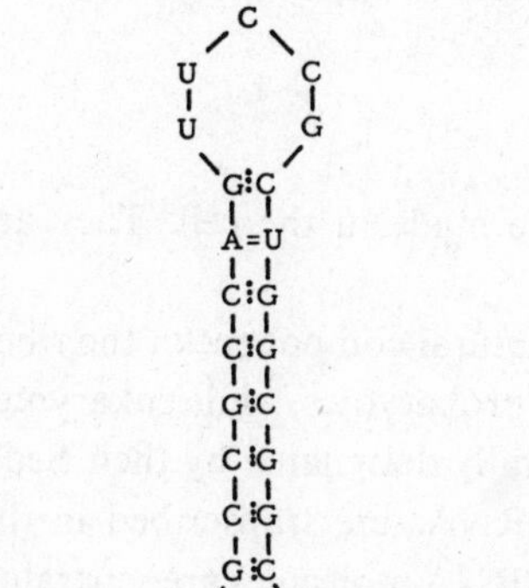

Table 2.1. *Composition and some properties of ribosomes and their subunits*

Subunit	Sedimentation coefficient (Svedbergs)	Number of associated proteins	Sedimentation coefficient of RNA (Svedbergs)	Molecular weight of RNA × 10⁵ (Daltons)	Number of bases in RNA
PROKARYOTES					
Large	50	31	23	11	2904
			5	0.4	120
Small	30	21	16	6	1541
EUKARYOTES					
Large	60	45	26* 28†	17	4000–5000
			5.8	0.5	158
			5	0.4	120
Small	40	33	18	7	1800

* Yeast; † Vertebrates.

release of pyrophosphate, working from the 5'-end of the RNA. *E. coli* (and other prokaryotes) possess just one type of DNA-dependent RNA polymerase, but eukaryotes have at least four different kinds. These are involved in the synthesis of the different kinds of RNA, and probably recognise different sequences generally situated upstream from the sites at which transcription starts.

The enzyme from *E. coli* exists in two major forms. The core enzyme is a tetramer of composition $\alpha_2\beta\beta'$ while the holoenzyme contains an additional polypeptide called sigma (σ). σ is required for accurate initiation of transcription, and dissociates from the holoenzyme during its passage along the DNA strand that is being transcribed. An additional polypeptide called rho (ρ) is sometimes required for termination of transcription.

Special features of transcription are considered further in Chapters 5 and 7.

2.3   The four major classes of RNA

Many different species of RNA are made in the cell. They are generally grouped into four classes.

1. *Ribosomal RNA (rRNA)* is a major structural component of the ribosomes. Three separate molecules are found in prokaryotes, while eukaryotes possess four distinct rRNAs. They are generally designated by their Sedimentation Coefficients (Table 2.1). These rRNAs are transcribed in the form of larger molecules in which the actual rRNA molecules are separated from each other by spacer RNA which has to be removed by endonucleases

(Fig. 2.3). In *E. coli* some of the spacers and also the 3'-flanking sequences may contain tRNA molecules. In both prokaryotes and eukaryotes the genes for these pre-rRNAs are repeated. In *E. coli* there are seven copies, while in some amphibia there are up to about 500 copies which are tandemly linked with untranscribed spacer DNA between the rRNA genes. During oogenesis, when the demand for ribosomes increases greatly, these genes are selectively amplified to form about two million copies which are present as extrachromosomal circular DNA.

The *ribosomes* themselves are composed of two subunits, designated large and small. In addition to their RNAs they also contain many proteins. Some details are set out in Table 2.1. The rRNAs in the ribosomes assume characteristic conformations with appreciable hydrogen bonding between bases in the individual molecules.

2. *Transfer RNAs* (*tRNAs*) are a family of smaller RNAs (3–4s), each containing about 80 nucleotides. There are about 40–60 different species in one cell. Like the rRNAs, tRNAs are first made as longer molecules which may have extra nucleotides at one or both ends and also in the interior of the molecule. These are removed in a specific manner by ribonucleases (RNases). The mature molecules contain a number of bases which are not generally found in other types of RNA (Fig. 2.4). These are produced by enzymic modification of the four types of base in the primary transcript.

They possess common structural features with considerable secondary

Fig. 2.3. Eukaryotic precursor rRNA, showing the stages in its cleavage by ribonucleases to the mature molecules.

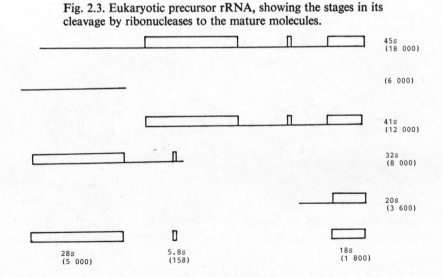

		45s (18 000)
		(6 000)
		41s (12 000)
		32s (8 000)
		20s (3 600)
28s (5 000)	5.8s (158)	18s (1 800)

structure, containing three base-paired stems and three or four unpaired loops. They can be represented in two dimensions by a 'clover-leaf' type of structure. In three dimensions they are L-shaped molecules (Fig. 2.5). Each one contains a specific sequence of three bases – the *anticodon* (see Fig. 2.5) – which is situated in the middle of an unpaired loop at one end of the molecule. All tRNA molecules contain the sequence CCA at the 3'-terminus which is at the opposite end to the anticodon. A specific amino acid can be attached to the 2'- or 3'-position of the adenylate residue in this sequence by means of a reaction catalysed by the appropriate amino

Fig. 2.4. Some bases found in tRNA as the result of chemical modifications of the four most commonly occurring bases.

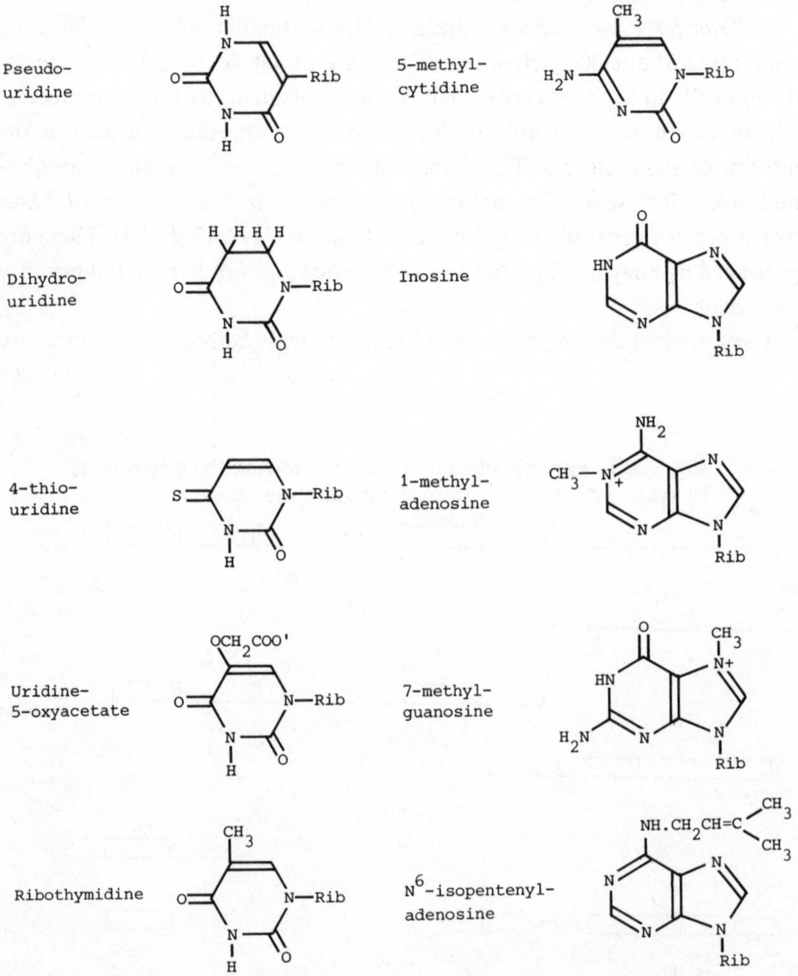

Fig. 2.5. Above: Yeast tRNA[Phe] drawn in its clover-leaf form (two-dimensional). D, AC and T are the dihydrouracil, anticodon and thymine loops respectively. The anticodon is the three bases in a horizontal line at the bottom of the figure.

Below: A representation of the same molecule as it appears in three dimensions. (Reproduced from S.-H. Kim, *Adv. Enzymol.* (1978), **46**, 279, by kind permission of John Wiley & Sons, New York, © 1978.)

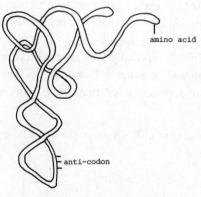

Table 2.2. *Small RNAs*

Symbol	Number of nucleotides	Site of occurrence	Function
U1	165	Nucleus	Splicing of axons
U2	188	Nucleus	?
U3	210–214	Nucleolus	?
U4	142–146	Nucleus	Polyadenylation
U5	116–118	Nucleus	?
U6	108	Nucleus	?
La 4.5	94	Nucleus	?
7s	295	Endoplasmic reticulum	Cleavage of leader peptides

acyl-tRNA synthetase. When combined with an amino acid a tRNA is said to be 'charged' (Fig. 2.6).

3. *Messenger RNA (mRNA)* is the name given to the transcripts that carry the specific information for the sequence of amino acids in proteins. There are a large number of mRNA species of widely varying sizes. They generally possess little secondary structure. In prokaryotes, several mRNAs are commonly transcribed in tandem as one large molecule with very short non-translated spacers between them (see Chapter 6). In eukaryotes primary transcripts synthesised by RNA polymerase are frequently much larger molecules which are processed by excision of parts of the molecule to yield mature mRNAs (see Chapter 7.9). A large proportion of the RNA in the nucleus is in the form of the primary or partly processed transcripts, and is known as *heterogeneous nuclear RNA (hnRNA)*. mRNAs contain flanking sequences of bases at their 5'- and 3'-ends in addition to the actual coding sequences that specify the amino acid sequence of proteins.

mRNAs are generally not very long-lived, as they are fairly readily degraded by RNases. This is especially true in prokaryotes where their lives are of the order of only a few minutes. In eukaryotes they are normally more stable, with lives ranging from a few minutes to many hours. rRNAs

Fig. 2.6. The charging of a tRNA molecule with its cognate amino acid.

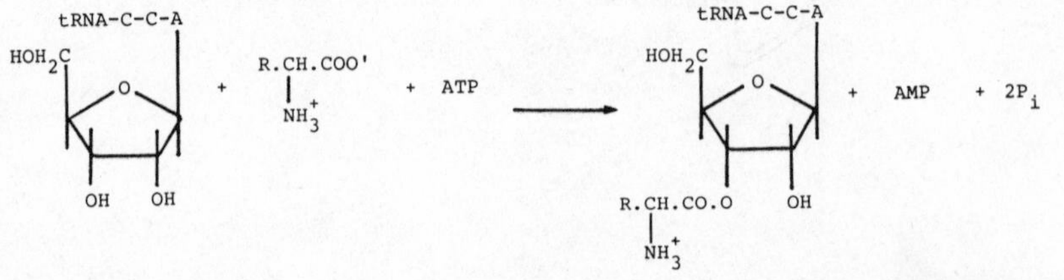

Table 2.3. *The genetic code*

First base (5'-end)	Second base				Third base (3'-end)
	U	C	A	G	
U	Phe	Ser	Tyr	Cys	U
	Phe	Ser	Tyr	Cys	C
	Leu	Ser	Ter	Ter	A
	Leu	Ser	Ter	Trp	G
C	Leu	Pro	His	Arg	U
	Leu	Pro	His	Arg	C
	Leu	Pro	Gln	Arg	A
	Leu	Pro	Gln	Arg	G
A	Ile	Thr	Asn	Ser	U
	Ile	Thr	Asn	Ser	C
	Ile	Thr	Lys	Arg	A
	Met	Thr	Lys	Arg	G
G	Val	Ala	Asp	Gly	U
	Val	Ala	Asp	Gly	C
	Val	Ala	Glu	Gly	A
	Val	Ala	Glu	Gly	G

Amino acids are designated by their standard three-letter abbreviations.
Ter is a termination codon.

and tRNAs, being much less rapidly degraded, are often referred to as *stable* RNAs.

4. A number of small RNA molecules have been discovered which mostly have no known function (Table 2.2).

2.4   The genetic code

The information encoded in the DNA that is to direct the synthesis of specific proteins is in the form of a nearly universal genetic code in which a sequence of three bases in a nucleic acid codes for a single amino acid residue in a protein. Such a triplet of three bases is known as a *codon*, and 64 codons can arise by permutating the four bases in triplets. Since only 20 different amino acids are incorporated into proteins, either many codons are not used, or some amino acids can be coded for by more than one codon. The latter is the true situation, and the code is therefore said to be degenerate. Three codons are also used for stop signals (*termination codons*). These are sometimes known as nonsense codons, and have been given the fanciful names amber, ochre and opal.

Codons are carried on mRNAs and are therefore also found on the DNA strand that is not transcribed (with T replacing U). The complementary triplets are known as anti-codons, and are found on the transcribed strand of DNA (known as the sense strand), and also in tRNAs. When writing the DNA sequence of protein-coding genes, it is usual to write the sequence of the non-coding strand, as this will contain codons in the correct order so that the amino acid sequence of the protein can be read off directly.

Inspection of the genetic code (Table 2.3) shows that the first two bases of the codon have the greatest effect in specifying the amino acid to be incorporated into a protein at a given position. One consequence of the degeneracy of the code is that many point mutations are 'silent', i.e. not expressed since, for example, mutation of the codon UUC to UUU still leads to the incorporation of phenylalanine. It is also found that the 5'-base of the anticodon in tRNA does not always pair exclusively with the complementary 3'- one on a codon. The base hypoxanthine is sometimes found in the 5'-position of anticodons, and this can pair with C, U, or A in the 3'-position of the codon. This lack of specificity was first proposed by Crick in what is known as the *wobble hypothesis*. A consequence is that some species of tRNA can recognise more than one codon, thus leading to economy in the production of these molecules so that less than the theoretical number of 61 tRNAs are needed to read all the codons on an mRNA.

The codon AUG for methionine is very nearly always used as an initiation codon for the first amino acid to be incorporated into a polypeptide chain. Successive triplets are read to give the specified protein sequence until a termination codon is reached. Thus, point mutations in which a single base is changed will lead to the incorporation of a single different amino acid (unless they are silent) or, in rarer cases, to the premature termination of a polypeptide chain or to the reading through of a termination codon. However, in mutations resulting in deletions or insertions of one or two bases, the reading frame will be thrown out of register from that point onwards, so a completely different protein will be synthesised (Fig. 2.7).

Fig. 2.7. A single nucleotide deletion in the gene for Haemoglobin$_{Wayne}$ throws the reading frame out of register, and leads to the synthesis of a longer α-chain. The codons of the mRNAs of the normal and mutated α chains are shown, together with the amino acids they specify, written in the standard three-letter code. The deleted nucleotide is boxed.

Hb   Thr Ser Lys Tyr Arg Ter
        ACN UC[U] AAA UAC CGU UAA GCU GGA GCC UCG GUA G

Hb$_{Wayne}$ ACN UCA AAU ACC GUU AAG CUG GAG CCU CGG UAG
        Thr Ser Asn Thr Val Lys Leu Glu Pro Arg Ter

Very exceptionally, in some bacterial viruses with small genomes, two or even three reading frames are used simultaneously so that portions of a long mRNA can code for parts of two or three proteins (Fig. 2.8). Intuitively, it seems surprising that two functional proteins can arise in this way, but it obviously leads to significant economy in the amount of coding DNA that is required.

Iso-accepting tRNAs (different tRNAs that can be charged with the same amino acid although they bear different anticodons) are not all present in a cell in the same amount, and the less-abundant ones are used less frequently. There is certainly good evidence that codon usage is not random. In both yeast and *E. coli*, mRNAs that code for proteins of low abundance contain a higher than usual proportion of 'rare' codons. Thus, to quote a fairly extreme case, the yeast genes coding for three proteins which are abundantly expressed (phosphoglyceraldehyde dehydrogenase alcohol dehydrogenase, enolase) use only 27 out of the possible 61 codons specifying amino acid residues more than 98 % of the time.

2.5 Translation is a later stage of gene expression

Gene expression involves translation of the triplet base code carried on mRNA into the sequence of amino acids in a protein. It is customary to divide it into three phases occurring sequentially.

The *initiation codon* is very nearly always AUG. This is the only codon for methionine, but there are two different tRNAs each containing the anticodon for this (CAU). In practice, the one that is always used for initiation in prokaryotes (tRNA$_i^{fMet}$) is charged with *N*-formyl-methionine (Fig. 2.9). A different tRNA (tRNA$_m^{Met}$) is used for inserting methionine into internal positions in a peptide chain. In eukaryotes there are again two tRNAs for methionine, but the methionine that is used for initiation is not formylated. Since the majority of mature proteins do not contain methionine as the N-terminal amino acid, the initiating methionine residue is often cleaved from the protein post-translationally.

Fig. 2.8. Overlapping genes in the bacteriophage X174. The beginning and end of the coding portion of the gene for protein E, and the end of the coding portion of the gene for protein D are shown, together with the amino acids specified, written in the three-letter code.

```
Met  Val Arg Tyr                        Ter
ATG GTA CGC TGG - - - protein E - - - G TGA TGT AA

TAT GGT ACG CTG G - - - protein D - - -GTG ATG TAA
Tyr  Gly Thr Leu                        Val Met  Ter
```

Initiation probably starts with the binding of a charged $tRNA_i^{Met}$ (eukaryotes) or RNA_i^{fMet} (prokaryotes) to the initiation codon on the mRNA in the presence of soluble protein initiation factors. This complex then binds to the small subunit of the ribosome in a reaction involving further initiation factors and GTP. The mRNA binds by base pairing of complementary sequences in its 5'-flanking region and on the 16s

Fig. 2.9. The charging of $tRNA_i^{Met}$ and formylation of the methionyl residue prior to initiation of protein synthesis in prokaryotes. PP_i = inorganic pyrophosphate; THF = tetrahydrofolate.

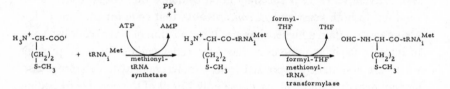

Fig. 2.10. Initiation of protein synthesis in eukaryotes. Initiation in prokaryotes is very similar in outline, but uses formyl-methionyl-$tRNA_i^{Met}$ and the ribosomal subunits are 30s and 50s with the formation of a 70s complex. IFs = initiation factors.

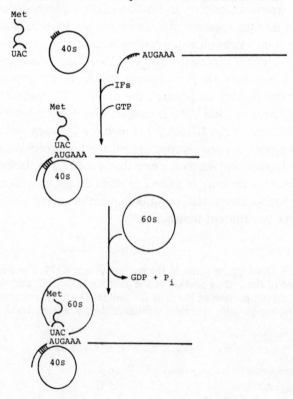

(prokaryotes) or 18s (eukaryotes) rRNA (see Chapter 5.2). Finally the large ribosomal subunit is bound, GTP is hydrolysed to GDP and P_i and the initiation factors dissociate from the complex (Fig. 2.11). The precise order and details of these reactions are not definitely known.

The second phase of translation is known as *elongation* (Fig. 2.10). This involves a second charged tRNA molecule pairing with the next available

Fig. 2.11. Elongation step of protein synthesis. Note the use of elongation factors (EFs), and the hydrolysis of GTP to provide energy for the reaction to proceed.

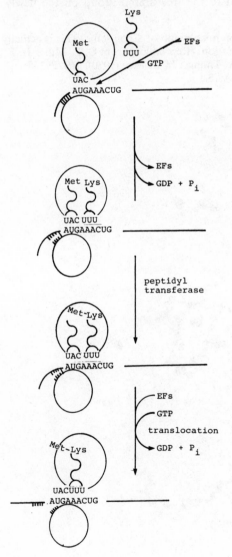

codon on the mRNA so that its amino acid residue lies on the A site on the ribosome next to the initiator methionine residue. This residue is then transferred onto the amino group of the incoming amino acyl residue. Again, soluble proteins (elogation factors) are required, and also GTP which is hydrolysed to GDP and P_i in the process, so providing the necessary energy. The uncharged tRNA dissociates from the ribosome, and the charged tRNA (now bearing a dipeptide) and the mRNA move relative to the ribosome so that a new charged tRNA ($tRNA_3$) can enter by base pairing with the next codon on the mRNA. The dipeptide already on $tRNA_2$ is transferred to the free amino group on the newly arrived

Fig. 2.12. Electron micrograph of a polysome. This is actually from a bacterial preparation. (Reproduced from O. L. Miller Jr, B. A. Hamkalo & C. A. Thomas Jr, *Science* (1970), **169**, 392–5. © 1970 by the AAAS.)

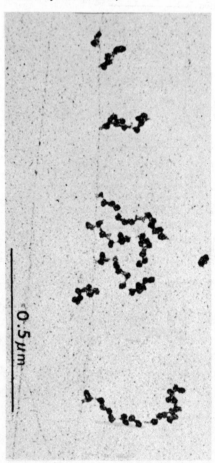

0.5 μm

amino acyl residue, and the polypeptide chain is built up sequentially by many repetitions of this process. Amino acid residues are incorporated into growing polypeptide chains at a rate of about 100 residues per second.

The *termination* phase of translation occurs when a termination codon is reached. Now there will be no corresponding tRNA and the peptide chain that has been built up is hydrolysed off the last tRNA which has been used. Soluble protein termination factors are involved which associate briefly with the ribosome. With the dissociation of the last, now uncharged, tRNA from the complex the ribosomal subunits also dissociate from the mRNA and from each other to join the pool of subunits which are available for reassociation in the initiation complex.

Several ribosomes are commonly found spaced out along an mRNA molecule separated from each other by about 80–100 bases. The mRNA coding for globin, a protein of about 145 amino acid residues, has a coding sequence of about 450 bases and could accommodate up to about five ribosomes. Such a structure is called a *polysome* (Fig. 2.12).

Thus, a single very complex process is used for the synthesis of proteins under the direction of specific mRNAs coding for particular proteins. Fuller details can be found in any good up-to-date textbook of biochemistry.

3
Methodology

3.1 Introduction

The explosive growth in our knowledge of the structure and function of the genome that has occurred in the last decade or so has only been possible thanks to the development of suitable techniques. In this chapter these are described in outline, as their proper appreciation should lead to a better understanding of the work that has been done. It is also necessary to master the jargon that is used in this field in order to understand current work.

Much of this work has centred on what has come to be known as 'genetic engineering' in which fragments of DNA are excised from their natural chromosomal sites, and incorporated into larger pieces of DNA which are capable of autonomous replication in cells of various kinds. Under the correct conditions this allows amplification of the original DNA fragments so that they can be obtained in large enough quantities and in a pure enough state for structural and other studies.

DNA for these purposes can be obtained in four ways:

(1) By physical breakage of larger DNA molecules by stirring a suspension at very high speed at 0 °C. This yields randomly sheared molecules.

(2) By enzymic synthesis on an RNA template by the enzyme reverse transcriptase (see 3.4) to produce what is known as *complementary DNA (cDNA)*.

(3) By hydrolytic cleavage at specific sites by the action of enzymes called restriction endonucleases (see 3.6).

(4) By chemical synthesis. This method is obviously confined to fragments smaller than those produced by other methods, which yield pieces of DNA up to about 50 000 kilobases (50 kb) in length.

DNA made in one of these ways is ligated to the DNA of one of several kinds of independently replicating vectors which will grow when introduced into cells of an appropriate sort (Chapter 4). This approach yields clones of cells infected with a particular type of DNA, and hence the technique is often referred to as cloning of DNA. The artificially produced DNA may be called *recombinant DNA*, since DNA will have been combined from two or more sources to produce the vector that is used in this type of work.

Early work showed that eukaryotic DNA could be incorporated into plasmids growing in bacteria (very frequently *E. coli*) and fears were expressed that if such bacteria escaped from the laboratory they might represent health hazards because they would be expressing eukaryotic (including human) genes in the wrong places. These fears have proved to be unfounded, largely because mutant strains of bacteria are used to grow the recombinant vectors. Growth of these strains requires certain nutrients which are not normally found in large enough concentrations for growth to take place outside a strictly controlled laboratory environment.

3.2   mRNA isolation

mRNAs are most easily isolated and purified from eukaryotic cells, especially from any tissue that is actively translating them. The simplest cases are found in tissues which synthesize only a very limited number of proteins, with one main product. This occurs in reticulocytes, synthesising predominantly haemoglobin, plasmacytomas synthesising predominantly immunoglobulins, and various cultured cell lines synthesising particular hormones. The microsomal fraction is separated from such tissues or culture, and contaminating proteins and lipids removed from the RNA by suitable extraction techniques. Most eukaryotic mRNAs possess a long tail of poly-A residues up to 200 nucleotides in length, so the crude RNA is passed down a column containing either poly-U or poly-T linked to a solid support. This binds poly-A containing RNA while other types (mainly tRNA and rRNA) pass through. The poly-A containing DNA is eluted with UMP or TMP. In the case of immunoglobulin mRNA there will be two species – one for the L-chain and one for the H-chain (Chapter 10.1). Because the H chain mRNA is about twice the size of the L chain mRNA they can be readily separated by electrophoresis on agarose, when the L-chain mRNA moves much faster.

When a tissue is producing many different proteins it is more difficult to separate one particular mRNA. However it can generally be achieved by immunoprecipitation. Antibodies to the pure protein specified by the desired mRNA are added to the crude microsomal fraction and only those ribosomes synthesising it will be precipitated. The mRNA can then be separated as described earlier.

Purification of prokaryotic mRNAs is more difficult since they contain no distinctive structural features like a poly-A tail. They are also much more readily hydrolysed by endogenous RNases in vivo. Isolation of prokaryotic genes is therefore generally accomplished by hydrolysing the total DNA with restriction endonucleases and isolating fragments in the 8–18 kb size range. These can be cloned into suitable vectors and after growing up the culture individual cells are isolated and grown on. Those that produce the product of the desired mRNA are selected by means of a sensitive test for that product.

3.3   Separation of nucleic acids

The size of both RNA and DNA molecules or fragments can be determined by measuring their electrophoretic mobility in gels of poly-acrylamide or agarose. The mobility of any sample depends on its size and also on the composition of the gel. The higher the concentration and the degree of cross-linking of a polyacrylamide gel, the slower will be the rate of migration of a nucleic acid molecule. It is not generally possible to separate fragments outside a fairly small range of sizes on one gel. Samples containing fragments spanning a wider range of sizes are generally separated on two (or even more) gels made from different concentrations of polyacrylamide. It is important that the molecules should be single-stranded as double-stranded DNA behaves differently from single-stranded species. Gels therefore usually contain denaturants such as high concentrations of urea or formamide, or else the DNA samples can be denatured first by treatment with gloxal or dimethyl sulphoxide. In a given gel the rate of migration of a nucleic acid is proportional to the logarithm of its chain length. It is usual to run a series of standards of known chain lengths in parallel with the experimental samples. These can be prepared by digestion of a plasmid or phage DNA with a restriction enzyme which will generate fragments of known sizes. Nucleic acids can be detected on gels by staining with ethidium bromide which has an intense fluorescence excited by ultraviolet radiation when it is complexed with nucleic acids.

3.4   Reverse transcriptase

This is an enzyme that was originally detected in viruses whose genome consists of RNA, rather than DNA. It catalyses the synthesis from the four deoxyribonucleoside triphosphates of a single strand of DNA complementary to an RNA strand. Thus it transcribes information in a direction opposite to that employed in the better known transcription

Table 3.1. *General properties of nucleases*

Type of enzyme	Particular characteristics
Ribonuclease	Hydrolyses RNA
Deoxyribonuclease	Hydrolyses DNA
Endonucleases	Attack internal phosphodiester bonds
Exonucleases	Attack the terminal phosphodiester bond, either at the 5'- or 3'-end of the chain

Table 3.2. *Base specificities of ribonucleases*

Name	Site of cleavage	Source
A1	Up N and Cp N	Mammalian pancreas
Phy M	Up N and Ap N	*Physarum polycephalum*
T1	Gp N	*Aspergillus oryzae*
U2	Ap N	*Ustilago sphaerogena*

These enzymes are all endonucleases, yielding 3'-phosphate and 5'-hydroxyl groups on the cleavage products.

of DNA by RNA polymerase – hence the name reverse transcriptase. There is good evidence that the enzyme is found in eukaryotic cells as well as in RNA viruses.

It is widely used in constructing molecules known as complementary DNAs (cDNAs) by using an RNA molecule as a template. A single-stranded DNA will result, and this can be converted to the double-stranded DNA by the action of DNA polymerase in the usual way. When these cDNAs are incorporated into suitable vectors (see Chapter 4) they can direct the synthesis of large quantities of DNA for further study.

3.5   Nucleases

Nucleases hydrolyse the phosphodiester bonds in nucleic acids and have important roles in processing RNA and DNA in the cell. Some are used in biochemical and genetic research. Table 3.1 sets out some general properties of these enzymes.

Several RNases only hydrolyse phosphodiester bonds next to a particular nucleotide bearing a particular base (Table 3.2). These can be useful in structural studies on RNA. Others seem to recognise tertiary structure and are used in the processing of primary transcripts. RNase III of *E. coli* recognises and cleaves the double-stranded stem of the precursor to rRNAs (Chapter 5.3).

RNase P generates the 5'-terminus of mature tRNAs by endonucleolytic hydrolysis of precursors. An interesting feature of its structure is that it consists of a faily small protein plus a very much larger RNA, known as M1 RNA. It is now known that in the presence of unphysiologically high concentrations of Mg^{2+} the RNA alone, in the absence of the protein, can catalyse the processing of tRNA precursors. However, the addition of the protein increases the rate of this reaction. The enzyme is widely distributed in prokaryotes, and a similar enzyme is found in eukaryotes.

The 3'-ends of tRNA precursors are trimmed by an exonuclease called RNase D. It can probably hydrolyse a phosphodiester bond on the 5'-side of any nucleotide.

The endonuclease RNase H acts on hybrid RNA–DNA chains and is believed to be important in trimming off ribonucleotides that are covalently linked to DNA when they are used as primers in DNA replication (Chapter 5.1).

There is a large number of restriction endonucleases (see next section) that recognise particular sequences in DNA.

At least six exonucleolytic DNases occur in *E. coli*. They all work from the 3'-end of the DNA, but some can also work in the other direction. Some act only on single-stranded DNA, while others hydrolyse only double-stranded DNA. Exonuclease III, which belongs to the latter class, will also hydrolyse RNA if it is complexed to a complementary strand of DNA, leaving the DNA intact.

Finally there are several nucleases that will hydrolyse both DNA and RNA, provided that they are single-stranded. These enzymes are generally regarded as endonucleases but they are capable of degrading their substrates completely to 5'-nucleotides. They have been widely used in various investigations when it is desirable to digest away portions of DNA or RNA sequences that are not base-paired to a complementary strand (Fig. 3.1). The best known of these enzymes are the S1 nuclease from *Aspergillus oryzae* and Mung Bean nuclease. The former is generally preferred in most experimental situations, since the latter does have a limited action on double-stranded DNAs.

A very important technique which makes use of S1 nuclease is in mapping the start site of transcription on DNAs. A restriction fragment of DNA containing the presumed site is prepared and is allowed to direct the synthesis of some RNA by adding the four ribonucleoside triphosphates plus RNA polymerase. Subsequent treatment with S1 nuclease digests away that part of the DNA sequence that was not used as a template for the transcription. If the RNA–DNA hybrid is now denatured the DNA can be separated and sequenced to establish the 3'end at which transcription began (Fig. 3.1).

DNA polymerases I and III both have exonucleolytic activity on DNA.

This action of the polymerase I is used in excision of damaged nucleotides (e.g. thymidine dimers formed by ultraviolet radiation) in the repair of DNA, prior to reformation of the correct sequence. This activity of DNA polymerase III is important for proof-reading of newly synthesised DNA (Chapter 5.1).

3.6   Restriction endonucleases

These are bacterial enzymes possessing an endonuclease activity which is directed to a specific sequence of bases in double-stranded DNA. In nature they serve to protect bacteria from the possible incorporation of foreign DNA into their genomes by digesting such material. The bacterium's own DNA is protected by being methylated on A or C residues which renders it unavailable for digestion by its own enzymes. The term 'restriction' arose because it was originally found that certain bacteriophages would not grow on certain bacterial strains – hence they were said to be restricted. Investigation of this phenomenon revealed that it was due to the action of this class of enzymes.

Over 350 restriction enzymes are known, although only about 85 distinct sequences are recognised. This is because, in many cases, enzymes from different organisms recognise the same sequence. Such enzymes are known as *isoschizomers*. They are named by three-letter abbreviations derived from the name of the species in which they occur, sometimes followed by a further letter or Roman numeral to differentiate enzymes from the same species, e.g. Eco RI is an enzyme derived from *E. coli* strain R. The base

Fig. 3.1. Detection of start sites of transcription by S_1 nuclease mapping.

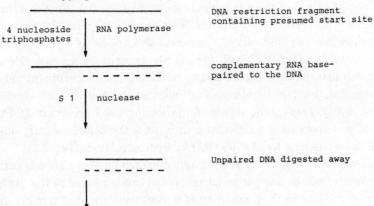

4 nucleoside triphosphates | RNA polymerase

DNA restriction fragment containing presumed start site

complementary RNA base-paired to the DNA

S 1 | nuclease

Unpaired DNA digested away

Denature to separate strands and electrophorese to determine precise length of protected DNA fragment

Table 3.3. *Some widely used restriction endonucleases*

Symbol	Organism and strain	Sequence recognised
EcoRI	*Escherichia coli R*	G|A A T T C C T T A A|G
Hpa I	*Haemophilus parainfluenzae*	G T T|A A C C A A|T T G
Hpa II[a]	*Haemophilus parainfluenzae*	C|C̊ G G G G C̊|C
Hae III	*Haemophilus aegypticus*	G G|C C C C|G G
Msp I[a]	*Moraxella species*	C|C G G G G C|C
Bam HI	*Bacillus amyloliquefaciens H*	G|G A T C C C C T A G|G
Pst I	*Providencia stuartii*	C T G C A|G G|A C G T C
Hinc II	*Haemophilus influenzae* R_c	G T Y|R A C C A R|Y T G
Ava II	*Anabaena variabilis*	G|G ^T_A C C C C ^A_T G|G
Sau 96I	*Staphylococcus aureus* PS96	G|G N C C C C N G|G

(*a*) These enzymes are isoschizomers (see text).
* This sequence is recognised and cleaved even if the C is methylated.
Note that in some cases (Hinc II, Ava II and Sau 96I) there may not be absolute specificity for one or more bases in the recognition sequence.

specificity of several of these enzymes is shown in Table 3.2. Note that the sequences in the two strands of the DNA that are recognised by the enzymes possess a two-fold axis of symmetry, also known as a *palindrome*. Some enzymes make cuts which are exactly opposite in the two DNA strands, so that the cut ends are said to be 'blunt': more usually the cuts are staggered, leaving a few unpaired bases at the end of each strand ('sticky ends'). This can be very useful in joining one fragment of DNA to another derived from a different source, since the complementary unpaired bases will tend to stick together by hydrogen bonds (Fig. 3.2).

Genomic DNA is likely to contain only a limited number of bonds that can be hydrolysed by any particular restriction endonuclease, so that such treatment will lead to the production of a comparatively small number of fragments. Many of these may be of suitable lengths (up to about 30 kb) for fractionation and purification by gel electrophoresis. So-called 'shotgun'

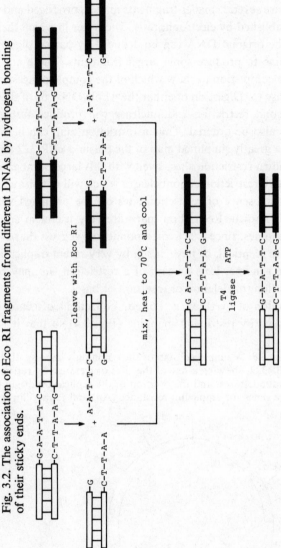

Fig. 3.2. The association of Eco RI fragments from different DNAs by hydrogen bonding of their sticky ends.

libraries of DNA fragments from the genome of both eukaryotes and bacteria have been made by cloning random mixtures of such fragments into suitable vectors.

If a large DNA fragment is digested to completion with a restriction endonuclease several smaller fragments may be produced and their lengths can be established by electrophoresis. The order in which these fragments occur in the original DNA can be deduced by partial digestion with the same enzyme to produce some larger fragments. These are isolated and digested to completion to show which of the complete digestion fragments they give rise to. Digestion of either the whole DNA or of these fragments with a second restriction endonuclease will produce further fragments which can also be ordered. Thus a restriction map can be built up. Fig. 3.3 shows a greatly simplified map of the plasmid pBR 322 on which there are 500 known restriction sites. Even with this large number of sites, there are at least 20 restriction endonucleases which will not cleave this plasmid. Presence or absence of restriction sites can be predicted when the base sequence is known. Restriction maps are very useful in comparing portions of genomes, since exact correspondences suggest that two fragments of DNA are identical, or have arisen by very recent duplication of a portion of the genome. Loss or gain of a restriction site may result from a point mutation, or deletion or insertion of bases.

The position of overlaps in shot-gun collections of genomic DNA can be established by restriction mapping (Fig. 3.4) so that it is possible to

Fig. 3.3. A simplified map of the commonly used synthetic plasmid pBR322, showing a few of the sites of cleavage by restriction endonucleases, and the position of the replication origin (ORI), and the genes for ampicillin resistance (Ap) and tetracyclin resistance (Tc).

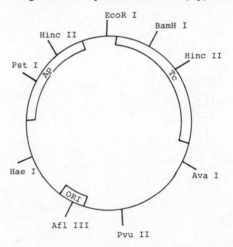

order correctly much larger pieces of DNA than can be cloned in one piece. This approach to building up maps of relatively long lengths of DNA is known as 'chromosome walking'.

3.7 Restriction fragment length polymorphisms

If the total human genomic DNA is digested to completion with a restriction endonuclease an extremely large number of fragments will be produced. These can be separated by electrophoresis and individual ones shown up by Southern blotting (see next section), using a probe made from any desired portion of the genome. For example, after digestion with Hind III, restriction fragments of 2.7, 3.5, 7.2 and 8.0 kb can be detected with a probe from the γ genes of the β-globin cluster (Chapter 9). When DNA from many individuals suffering from sickle cell anaemia is examined in this way the bands of 2.7 and 7.2 kb are no longer apparent. This is because of mutations which have caused the loss of two sites at which the Hind III enzyme can act. Since these sites are only present in the genome of certain individuals they are polymorphic.

Polymorphisms detected in this way are known as Restriction Fragment Length Polymorphisms (RFLPs). They can arise from point mutations leading to either loss or gain of a site at which a restriction endonuclease acts (Fig. 3.5). In addition, deletions or insertions can alter the length of fragments between two restriction endonuclease sites. Restriction fragment length polymorphism is being increasingly used for the diagnosis of genetic disorders, especially when it is known that a foetus is at risk of being

Fig. 3.4. Ordering of overlapping genomic fragments by restriction mapping. The horizontal lines represent cloned fragments of DNA from the region of the mouse genome encoding the fourth component of complement (Chapter 10.10). The vertical lines show the sites at which Bam HI acts. Other restriction enzymes were also used to confirm these results.

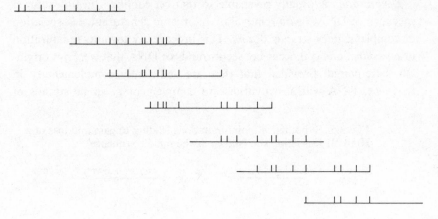

genetically disabled if it has inherited defective genes from both parents. Note that these polymorphisms are not necessarily in that sequence of the genome which codes for the harmful aberrant product (as in the example cited above). Provided they are closely and stably linked to such a mutant sequence they can be used to detect its presence. Foetal cells can be obtained by amniocentesis, but this technique only yields rather few cells so that they may have to be cultured for several weeks to produce enough DNA to make the test. A newer technique, which is still being evaluated for safety, involves removing some trophoblastic villi which yield larger amounts of DNA sufficient for performing the test.

RFLPs have recently been reported that are linked to the inheritance of phenylketonuria and to Huntington's chorea, and doubtless it will be possible to detect many other inherited diseases by this technique in the future.

3.8   Hybridisation of nucleic acids

Double-stranded DNA can be denatured by various procedures which break the hydrogen bonds linking the complementary base pairs. The simplest way of doing this is to heat the DNA in aqueous solution. The dissociation of the two strands results in an increased absorbance at 260 nm, where the nucleic acids have an absorbance peak. This provides a simple way of monitoring the denaturation and hence the separation of the strands in the molecule. This occurs over a fairly small temperature range, frequently around 60–80 °C. This temperature is called the melting temperature (T_m). Since triple-bonded G-C pairs are more stable than double-bonded A-T pairs, the greater the proportion of G-C in a DNA sample, the higher its T_m is likely to be. Other agents that denature DNA are high salt concentrations, alkali, and various organic solvents, of which formamide and dimethyl sulphoxide are often used.

Denaturation is usually reversible so that on cooling a heat-denatured specimen of DNA, or on removal of an organic denaturant, reassociation of complementary strands occurs. This process is known as renaturation or *annealing*, and can occur between strands of DNA that were not originally base-paired, provided that they are sufficiently complementary in sequence. RNA will also hybridise to complementary single strands of

Fig. 3.5. Hypothetical point mutations leading to gain and loss of a Hind III restriction site (shown as the middle sequence).

AAGTTT	POINT MUTATION	AẢGCTT	POINT MUTATION	AGGCTT
TTCAAA	──────────────▶	TTCGAẠ	──────────────▶	TCCGAA

DNA in an analogous way. Complexes between DNA and RNA are known as heteroduplexes, to distinguish them from homoduplexes formed from two complementary single strands of DNA. They are more stable than homoduplexes, so RNA–DNA heteroduplexes will form preferentially in a mixture of denatured double-stranded DNA and RNA.

Since DNA adsorbs strongly to nitrocellulose, DNA solutions can be spotted on to nitrocellulose paper and denatured and fixed by heating. Alternatively, prints can be made from gels on which DNA fragments have been electrophoresed to separate them on the basis of their sizes. Specific sequences in the DNA are detected by hybridisation to probes of radioactively labelled DNA or RNA at an elevated temperature in the presence of formamide. On cooling, the probe will anneal to any DNA on the original blot containing the complementary sequences, and excess uncomplexed probe is washed away, or removed by digestion with a single-strand-specific nuclease. The binding of the probe can be detected by autoradiography, or by direct counting of the filter. This technique is known as Southern blotting, after the name of its originator. A complementary technique of Northern blotting hybridises single-stranded DNA to RNA which has been blotted onto nitrocellulose paper.

The precise conditions under which hybridisation is carried out may affect the extent of hybridisation with similar but not identical sequences. Stringent conditions, using higher temperatures and higher concentrations of formamide only allow hybridisation of perfectly complementary sequences or, at best, those with only few mismatches. Conditions of lower stringency will permit the formation of hybrids from sequences containing appreciable numbers of mismatching bases. This can often be useful in detecting related, though not identical, sequences either in different parts of one genome or in homologous sections of the genome from different species.

It is now quite feasible to synthesise DNA oligomers of up to 20 nucleotides in length. Such synthetic oligomers at least 12 nucleotides long can be used as probes to detect complementary DNA or RNA sequences by hybridisation on nitrocellulose paper. Oligomers can be designed which match the probable mRNA sequence coding for short portions of proteins whose amino acid sequences are known, and they have been used as probes in Southern blots to detect the genes for these proteins.

Homoduplex formation can also be used to detect insertions or deletions in DNA from mutant organisms. DNA from both the wild type and mutant is prepared and they are mixed and denatured. On renaturation some of the single-stranded mutant and wild type DNA will combine, but a loop of unpaired bases will be present on one or other strand, depending on whether a deletion or an insertion has occurred. These unpaired lengths of DNA can be visualised by examining the renatured mixture in an electron microscope, when the unpaired DNA appears as a bubble or blister.

Heteroduplexes can also be examined in the electron microscope. This procedure has been widely used to visualise eukaryotic genes and map the position of introns in them (see Chapter 7.8). mRNA lacks these introns, so if it is hybridised to genomic DNA, introns will be seen as bubbles of single-stranded DNA (Fig. 3.6). If size markers are included in the mixture spread for microscopy, it is possible to measure the length of each intron and hence calculate the number of bases that it contains. Such measurements are usually accurate to about ± 50 bases.

3.9   The determination of base sequence in DNA

The determination of the base sequence in DNA is now a relatively easy procedure, thanks to the development of two methods both of which produce a series of DNA molecules differing in length by one nucleotide that can be separated by electrophoresis.

Fig. 3.6. R loops formed by reaction of cloned DNA coding for mouse IgC (Chapter 10.3) with the corresponding mRNA. On the left the actual electron micrograph. On the right is an interpretive drawing. A and B are double strands of DNA outside the region being studied. The dotted line is mRNA. V is the region coding for the variable part of the Ig chain. R1, R2 and R3 are loops formed when the mRNA anneals to one strand of DNA. I1 and I2 are introns where the two strands of DNA have re-associated.

(Reprinted by permission from H. Sakano *et al.*, *Nature* (1979), **277**, 627. © 1979 Macmillan Journals Limited.)

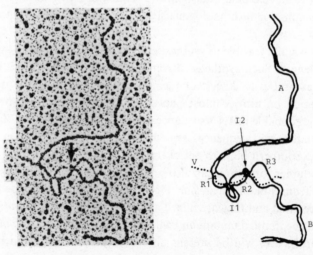

In the method devised by Gilbert and Maxam, specific chemical reactions are performed which cleave the DNA at certain base residues (Fig. 3.7). Reaction with dimethyl sulphate methylates guanine residues in such a way that they will be displaced by reaction with piperidine. This also eliminates the sugar residue to which the guanine was attached, and breaks the DNA chain at this point. Since the initial methylation reaction has an equal chance of occurring at any guanine residue, a series of fragments will be produced, broken at each of these residues in the sample, provided that the reagents are not in excess. Heating the DNA in acid solution

Fig. 3.7. Selective cleavage of DNA at a guanosine residue by reaction with dimethyl sulphate and piperidine. Guanosine and adenosine residues are both removed by treatment with acid prior to reaction with piperidine, while thymidine and cytidine residues are lost from the DNA chain by reaction with hydrazine.

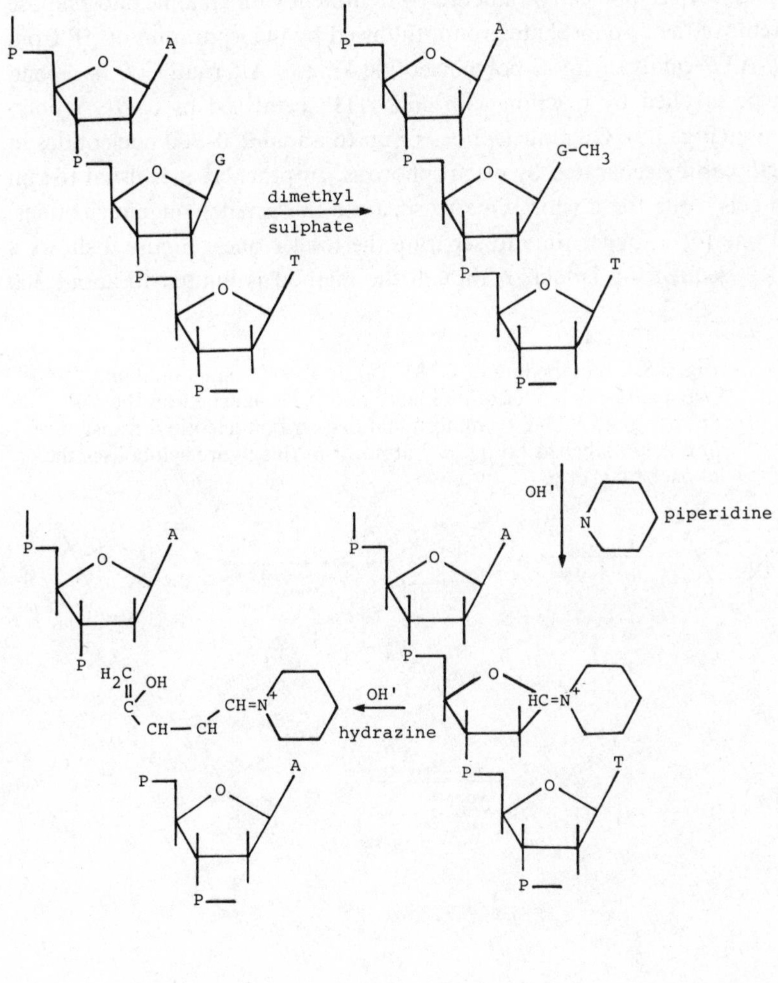

specifically removes both purine bases. Subsequent treatment with piperidine eliminates the deoxyribose and cleaves the chain at both adenine and guanine residues. Thymine and cytosine residues both react with hydrazine, and subsequent reaction with piperidine cleaves the chain at these two residues. Finally, if the reaction with hydrazine is carried out in the presence of a high concentration of salt, only cytosine residues react and subsequently piperidine cleaves at these residues only.

Thus, four simultaneous reaction mixtures are set up, and after the reactions have gone to completion each mixture is electrophoresed on an acrylamide gel. This system separates polymers according to their sizes, so a series of bands (a 'ladder') appears on each gel and the sequence of bases can be read off by observing which treatment caused the breaks in the original DNA. In order to detect the oligonucleotides on the gel, the DNA to be sequenced is first made highly radioactive. This can be done in two ways. Its 5′-end can be labelled by treatment with alkaline phosphatase to remove the 5′-phosphate group, followed by incorporation of ^{32}P from γ^{32}P-ATP catalysed by a polynucleotide kinase. Alternatively, its 3′-end can be labelled by reaction with α^{32}P-ATP, catalysed by a DNA polymerase (Fig. 3.8). Oligonucleotides of up to about 250–300 nucleotides in length can be separated by electrophoresis. In practice, it is usual to run two gels – one for a short time to separate the smaller oligonucleotides, and one for a longer time to separate the longer ones. Figure 9 shows a typical sequencing ladder. Although the method is limited to about 300

Fig. 3.8. End labelling of DNA: (a) at 5′-end, using alkaline phosphates, polynucleotide kinase and ATP, labelled on the γ-P atom; (b) at 3′-end, using terminal deoxyribonucleotidyl transferase and ATP, labelled on the α-P atom. P in this figure symbolises the phosphate group.

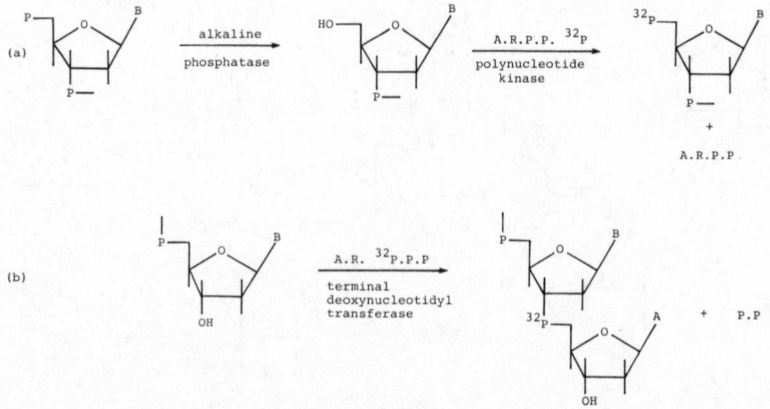

nucleotides, very much longer sequences can be determined by first cleaving the DNA by restriction endonucleases to give overlapping fragments, and then sequencing each one.

The second widely used method of sequence determination was developed by Sanger and his colleagues. This involves the enzymic synthesis of stretches of DNA on a template of the DNA whose sequence is being

Fig. 3.9. Sequence ladders used in DNA sequencing by the technique of Maxam & Gilbert.

Left. Ladder for determination of sequence of part of the vasopressin-neurophysin II gene in normal rats.
Right. Ladder for determination of sequence for part of the vasopressin-neurophysin II gene in rats of the Brattleboro strain with diabetes insipidus (D.I.). (Chapter 10.3). The deduced sequences are shown at the bottom. The arrow in the right-hand ladder shows where the deleted G is missing. (Reprinted by permission from H. Schmale & D. Richter, *Nature* (1984), **308**, 706. © 1984. Macmillan Journals Limited.)

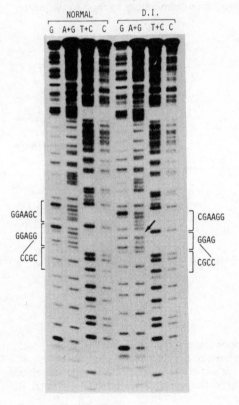

NORMAL   - G G A A G C G G A G G C C G C -
D.I.   - G G A A G C   G A G G C C G C -

sought. Newly synthesised fragments are terminated by a nucleotide analogue which blocks further addition of nucleotide residues. This is usually a dideoxynucleoside triphosphate (Fig. 3.10). Because this lacks a 3′-OH group, it cannot accept the next nucleotide in the sequence.

A series of genetically engineered derivatives of the bacteriophage M13 (see Chapter 4.3) which contains a single stranded circular genome, have been constructed with contiguous sites for a number of restriction endonucleases (Fig. 3.11). This DNA can be opened at one of these sites and ligated to the piece of DNA whose sequence is to be determined. Next, a synthetic oligonucleotide primer whose sequence is complementary to about 15 bases on the 3′-side of the restriction sites is annealed to the phage. Four parallel reaction mixtures are set up, each containing some of this plus three of the deoxyribonucleoside triphosphate plus the fourth one (usually dATP) highly labelled with ^{32}P in the α position. Each one of the

Fig. 3.10. Sanger's dideoxy chain terminating method for sequence determination.

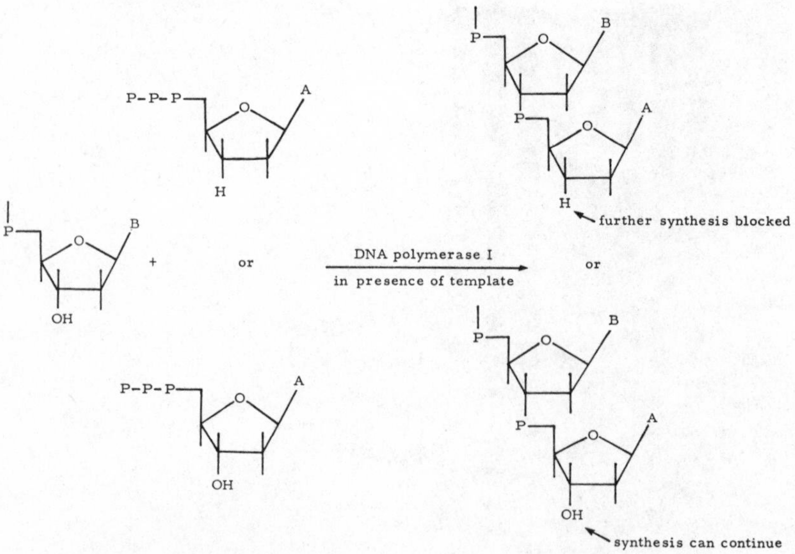

Fig. 3.11. Sequence of nucleotides introduced into the phage M13mp8, which is often used in Sanger's sequencing method. Several restriction sites are available, and a 17-nucleotide stretch has a complementary oligonucleotide which is available as a primer.

Sma 1 Hinc II Primer
—GAATTCCCGGGGATCCGTCGACCTGCAGCCAAGCTTGGCACTGGCCGTCGTTTTAC—
EcoR I BamH I Pst I Hind III

four mixtures receives one of the four dideoxyribonucleoside triphosphates and also a proteolytic fragment of DNA polymerase I (Klenow fragment) which lacks an exonucleolytic activity of the enzyme, but retains the polymerase activity. This will catalyse the synthesis of a new DNA strand complementary to that which has been inserted into the M13 vector. Four sets of DNA fragments will be synthesised, with all those in any one set terminating at a particular base residue. They are separated as a ladder by electrophoresis, just as in the Maxam & Gilbert method.

The key to the success of both methods lies in the fact that the sites of specific cleavage or chain termination are completely random. This is achieved by ensuring that the chain-breaking or chain-terminating reagents are not in excess.

Figure 3.12 shows the fragments that would be generated from a hypothetical short length of DNA by both methods.

The largest DNA that has been sequenced so far is that of the Epstein Barr virus, with just over 107000 nucleotides, but this number will doubtless be exceeded in the near future. The ease and simplicity of these elegant methods mean that it is generally easier to sequence DNA than protein.

Fig. 3.12. Determination of the sequence of a simple oligonucleotide by (a) Sanger's method, and (b) Maxam & Gilbert's method, showing the products that will be obtained by the two procedures.

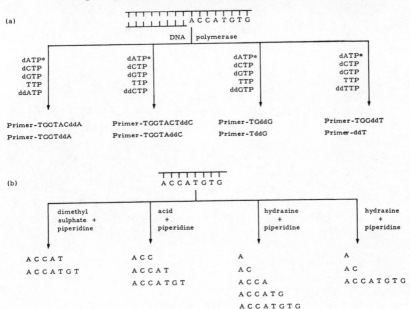

Thus, the sequences of many proteins have been inferred from either the sequence of the gene encoding them or that of the cDNAs made from their mRNAs. A number of protein sequences have been determined in this way before the proteins themselves have been isolated. It was recently estimated that it costs about $5 a nucleotide to establish the sequence of a DNA fragment.

4

Vectors used in work with recombinant DNA

DNA fragments or cDNAs that have been produced as described in the last chapter can be propagated by cloning them in suitable hosts. This involves ligating the DNA covalently to a suitable vector, which can then be manipulated so as to multiply in an appropriate type of host cell. Vectors include plasmids and bacteriophages, both of which will grow and replicate in bacteria, and viruses which can be grown in eukaryotic cells.

4.1   Plasmids

Plasmids consist of double-stranded DNA, found in bacteria, but generally not associated with the bacterial chromosome. They are stably inherited and utilise a specific region of their genome (the replication origin) for autonomous replication. Under natural conditions they will usually replicate to give 20–30 copies per cell, but if, for example, the bacteria are grown on a medium containing chloramphenicol yields of up to 1000 or more copies per cell can be obtained. They generally exist as covalently closed circular DNA molecules, smaller than the bacterial chromosome by 2–3 orders of magnitude (i.e. they contain 10^3–10^5 base pairs). Under suitable conditions their DNA can be broken open to yield a linear molecule which passes from one bacterium to another in a process known as transfection. They can be isolated in a relatively pure state from a bacterial lysate by centrifuging in a density gradient.

The DNA of plasmids is transcribed to give mRNAs which direct the synthesis of corresponding proteins. Some plasmids (R factors) carry genes for antibiotic resistance, specifying an enzyme that will detoxify the antibiotic by various chemical reactions, or interfere with its uptake (Table 4.1). Other types of plasmid carry genes for the synthesis of proteins known as bacteriocins that are lethal to other bacteria not carrying the same type

Table 4.1. *Mechanisms by which plasmids confer resistance to some antibiotics*

Antibiotic	Mechanism
Chloramphenicol	acetylation by chloramphenicol-acetyl transferase (CAT)
Tetracyclin	interferes with uptake
Ampicillin	hydrolysis by β-lactamase
Kanamycin	induction of transferase

Table 4.2. *Some properties of commonly used plasmids*

Plasmid	Size (bp)	Genetic markers[a]	Restriction sites[b]
pBR 322	4362	Ap^R, Tc^R	EcoRI Hind III (Tc) Pst I (Ap)
ColE1	4200	colicin production	EcoRI (CP) Sma I (CP)
pCR1	7400	colicin immunity, Km^R	EcoRI Hind III (Km)
pACYC 184	4000	Cm^R, Tc^R	EcoRI (Cm) BamHI (Tc)
pJC 74[c]	16000	Ap^R	EcoRI Bgl II Pst (Ap)
pJC 79[c]	6000	Ap^R, Tc^R	EcoRI Pst I (Ap) Hind III (Tc)

(a) Ap^R = ampicillin resistant; Tc^R = tetracyclin resistant; Km^R = kanamycin resistant; Cm^R = chloramphenicol resistant.
(b) When the restriction enzyme is followed by a symbol for an antibiotic, this indicates that it cuts within the gene specifying resistance to that antibiotic.
(c) Cosmid (see Chapter 4.3).

of plasmid. The best known ones are the colicins which are made by plasmids such as ColE1 in *E. coli*.

Many naturally occurring plasmids have been isolated and modified in various ways by genetic engineering. Genes can be introduced whose products can serve as markers; portions of their DNA can be deleted so as to make them smaller; various control elements, such as promoters (see Chapter 5.2) that respond to the presence of certain metabolites in the medium can also be incorporated. It is obviously essential that the region around the origin of replication should not be altered by these procedures. Some details of a few plasmids are shown in Table 4.2.

Plasmids can be cleaved by restriction endonucleases. Figure 3.3 shows a greatly simplified restriction map of the plasmid pBR 322 on which there are 500 known restriction sites. Even with this large number of sites, there are at least 20 restriction endonucleases that will not cleave this plasmid.

Recombinant DNA can be prepared by cleavage at one of these restriction sites. If this leaves short lengths of unpaired bases – 'sticky ends' –

and if a fragment of DNA derived from some other source (*passenger DNA*) by treatment with the same enzyme is mixed with the cleaved plasmid, some of these sticky ends of the plasmid will associate by hydrogen bonding with the restriction fragment. The passenger DNA and the plasmid are joined covalently by DNA ligase from either *E. coli* or the bacteriophage T4 to form a new recombinant DNA molecule (Fig. 4.1). An alternative method is to cleave both plasmid DNA and the passenger DNA to be cloned with a restriction endonuclease (e.g. Hae III) which yields 'blunt-ended' cuts. These can also be joined together by the action of the T4 ligase, though not by the enzyme from *E. coli*. In both these methods undesired ligations (i.e. passenger to passenger or vector to vector) may be introduced.

Blunt-end ligations are also used to introduce sites at which specified restriction endonucleases can act at a later time. This is achieved by ligating synthetic oligonucleotides, known as linkers, to DNA fragments (Fig. 4.2). Some of these linkers are fairly complex, and may contain sites for several restriction endonucleases.

Fig. 4.1. Construction of a cloned vector in the plasmid pBR322. The passenger DNA would have been isolated after cleavage of a larger DNA by the restriction endonuclease BamH I. Ap[R] and Tc[R] are the sites of the genes for ampicillin resistance and tetracyclin resistance respectively. The latter gene will have been inactivated by the cloning of the passenger DNA.

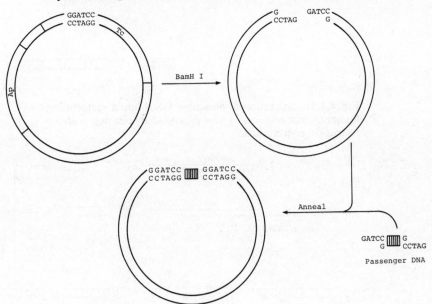

A further method for linking two blunt-ended pieces of DNA together is to add homopolymers to their 3'-ends with the enzyme terminal deoxy-nucleotidyl-transferase (Fig. 4.3). Chains of up to several hundred nucleotides can be added in this way; however reaction conditions are chosen such that a much smaller number are incorporated. The bases in the nucleotides added to the two DNAs are chosen as the complementary pair (e.g. G to DNA and C to the other). The two DNAs are annealed, and a stable enough hybrid is formed so that if one of the DNA fragments is a cleaved plasmid, the hybrid can be used directly for transfection. Any gaps in the hybrid are filled in by the host's own synthetic enzymes. Since there will be no circularisation of cleaved plasmids with homopolymer tails, only the hybrid molecules are infectious. Thus all bacteria containing the plasmid will also contain the inserted DNA. However, the infectivity

Fig. 4.2. Introduction of restriction sites into DNA by the use of synthetic linker oligonucleotides.

Fig. 4.3. Incorporation of passenger DNA into a vector using homopolymer addition to sites generated by cleavage with a restriction endonuclease.

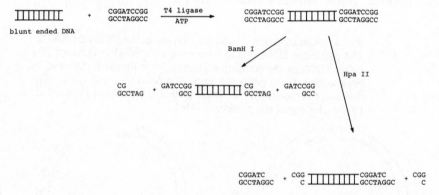

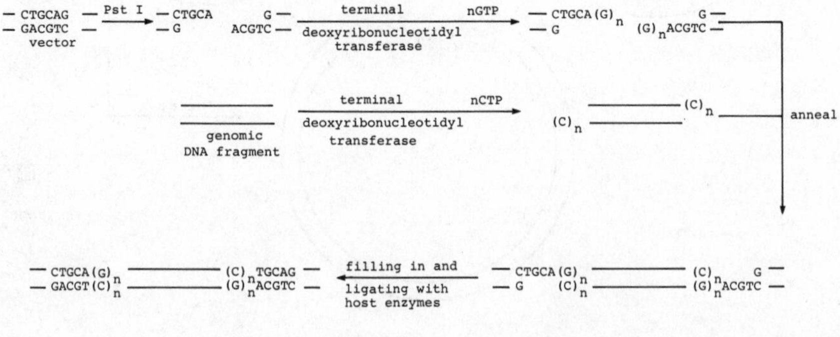

of these annealed forms is generally lower than that of covalently closed forms.

Plasmids containing passenger DNA can be cloned into suitable bacteria. This process is known as *transfection*, and is usually achieved by mixing suspensions of the plasmid and bacterial host in the presence of a relatively high concentration of calcium (e.g. 20–50 mM). Not all the plasmids will contain the passenger DNA, so when the bacterial culture has grown, it must be screened for those that do. Firstly, the bacteria are grown in a medium containing an antibiotic against which the plasmid confers resistance. All the bacteria that have not been transfected will be killed. Many of the survivors containing the plasmid will not contain passenger DNA. If this has been inserted at a site in the middle of a second gene specifying resistance to a different antibiotic, this resistance will have been lost and this property can be used to select the cells that do contain the passenger DNA.

Other methods may be used to screen the cells that have been through some of these procedures to select those that actually contain the desired DNA sequences in the plasmids. One method is to spread a growth of the bacteria on a nitrocellulose filter supported on a nutrient agar plate and allow the separate colonies to grow to about 1 mm in diameter. A replicate set of colonies is made. The nitrocullulose filter is then removed and washed. The cells on it are lysed by a suitable detergent and the filter dried at 80 °C to fix the DNA to the nitrocellulose. The filter is then soaked in a solution containing 50 % formamide and an isotopically labelled RNA probe which is known to hybridise to the desired DNA sequence. After washing and drying, autoradiography will reveal any clones that contain the desired DNA. These colonies are then picked off the replicate plates and grown on to provide any desired quantity of the DNA.

Finally, the DNA is recovered from the plasmids, now amplified, by lysing the bacteria, usually with lysozyme followed by a detergent, and separating the plasmid DNA from the chromosomal DNA by centrifuging on a density gradient of caesium chloride on which the plasmid DNA sediments faster than the bacterial DNA. Digestion of the plasmid DNA with a suitable restriction endonuclease generates the desired fragments which can be separated from other DNA fragments by electrophoresis.

There is also a yeast plasmid, referred to as the 2 µm plasmid (63 kbp in length), and this can be manipulated in similar ways to bacterial plasmids to provide a vector that will grow and replicate in yeast cells. Some hybrids of bacterial and yeast plasmids are actually capable of growth in either organism (shuttle vectors), and these are valuable tools for the study of gene expression.

4.2   Bacteriophages

Alternative vectors for amplifying selected DNA sequences are derived from bacteriophages (frequently referred to just as phages). These are usually considered as viruses whose specific hosts are bacteria. Their genome of DNA (occasionally RNA) encodes a number of associated proteins. Phages will only replicate in a bacterial cell of the correct species or strain. They may eventually kill an infected cell when a sufficient number of new phage particles have been synthesised. This causes lysis of that cell so that they are released and can infect other cells. Some phages (temperate) can infect a cell and replicate without causing it to lyse. They are then said to lysogenise the cell. Others (virulent) will always replicate and lyse the host cell.

Phages have some advantages over plasmids as vectors for the production of recombinant DNA:

(1) it is comparatively easy to screen very large numbers of phage particles by nucleic acid hybridisation for a given DNA sequence;
(2) they can infect bacteria more efficiently than plasmids;
(3) large numbers of DNA fragments derived from a single large genome can be packaged, replicated and stored as a 'library' containing very many DNA sequences; and
(4) on the whole, larger DNA fragments can be cloned into phages than into plasmids.

Two phages and their derivatives have been most widely used in work with recombinant DNA. Phage M13 is a virulent phage containing its DNA as a single-stranded circular molecule. This is about 6.4 kb in length. A number of derivatives have been made which are particularly useful for Sanger's method for the determination of the base sequence in DNA (Chapter 3.8).

Phage λ (lambda) is a temperate phage containing its DNA as a double-stranded linear molecule. This is slightly less than 50 kb long (just over 1 % of the length of the *E. coli* chromosome). In the lysogenic mode its DNA integrates into a specific site on that chromosome. Its complete sequence has just been determined. Many derivatives of this phage have been made by deliberate manipulations. About a third of the genome can be removed without impairing its ability to replicate. This can be replaced by foreign DNA before infection of a host with the modified phage. In order to simplify this process, unique sites for restriction endonucleases should be present in the dispensable region of the genome. Wild type λ has five cleavage sites for Eco RI, and appropriate mutants have been selected which have lost three or four of these, leaving only one or two in the non-essential portion of the genome. Thus cleavage at these sites makes it possible to introduce passenger DNA in the same manner as into plasmids. The altered phage DNA must then be introduced into bacteria in which it is to replicate. Naked DNA will enter bacterial cells by the process of transfection in the presence of relatively high concentrations of Ca^{2+}, but the uptake of phage DNA is very inefficient. Greatly improved efficiency of infection results from packaging the DNA into the protein components of the phage head so as to make a complete bacteriophage. This is usually done by mixing it with two strains of the phage, each of which lacks one of the proteins that makes up the head so that they complement each other. The complete phage will now assemble itself and can be used to infect a suitable bacterial culture. Packaging of the DNA can only be achieved if its length is between 75 % and 109 % of the wild type phage genome. In practice, the maximum size of passenger DNA that can be used is about 15 kb. This is well above the size of many known genes, but because of the large amount of non-coding DNA in most genomes it is not likely to encompass more than a few complete genes.

After infection, the DNA circularises by the action of the host's own DNA ligase. This is a necessary step before the phage DNA becomes integrated into the bacterial chromosome (Fig. 4.4). In this form the phage DNA replicates in the lysogenic mode at the same time as the bacterial chromosome. Under appropriate conditions the phage can be induced into the lytic phase, when it becomes excised from the bacterial chromosome, replicates autonomously, and will eventually lyse the cell releasing many new phage particles.

At each end of the double-stranded linear DNA of λ is a short unpaired sequence of 12 nucleotides. Since these are complementary to each other (Fig. 4.4) they can form base pairs, and are known as cohesive ends (*cos*). When λ enters the lytic phase during its natural life cycle, these *cos* sites

Fig. 4.4. Integration and excision of λ from *E. coli* chromosome.

c and c′ are cos sites: 5′-AGGTCGCCGCCC-3′
3′-TCCAGCGGCGGG-5′

b′, b and p′, p are homologous sequences:
5′-GCTTTTTTATACTAA-3′
3′-CGAAAAAATATGATT-5′.

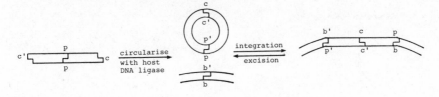

pair with each other and build up long chains (concatamers) of λDNA. During the packaging of the DNA into heads of the phage these are hydrolysed by an endonuclease called λ-terminase which recognises *cos* sites that are 38–52 kbp apart, so that DNA chains of this size are produced. The packaging system described above only requires that the DNA contains the *cos* sites at a suitable distance apart: it is indifferent to the rest of the DNA sequence. Advantage has been taken of this to construct vectors called *cosmids* which are specialised plasmids into which these *cos* sites have been engineered. A series of such cosmids have been constructed, each containing different restriction sites and a variety of antibiotic resistance genes which can be used for selection of the bacteria containing the desired cosmid. For use as vectors the cosmids are first cleaved by a specific restriction endonuclease and ligated to passenger DNA. During this ligation conditions are chosen so that formation of concatamers is favoured and occurs in high yields (Fig. 4.5). Because of the size limitation for packaging DNA into a λ head, these concatamers need to have *cos* sites about 38–52 kb apart. Since a minimum of about 4–6 kbp of plasmid

Fig. 4.5. Construction of recombinant DNA with a cosmid. ● = *cos* site. Note that concatamers of many different structures will be formed. Those of type A can be acted upon by λ terminase, and will be packaged efficiently. Those of types B, C and D (and others not shown) will not. On B the *cos* sites are too close together, while C lacks them. (Modified from J. Collins *Methods in Enzymology* (1979), **68**, 310, Fig. 1; reprinted by permission of John Wiley & Sons, Inc., © 1979.)

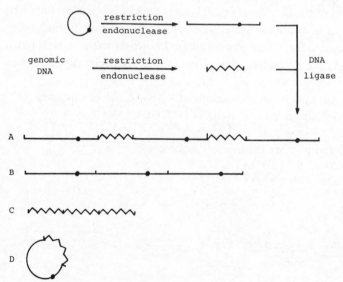

DNA is essential for the construction of cosmids, inserts of 30–45 kbp of DNA can be accepted.

After packaging and infection of a suitable host the cosmid behaves as an autonomously replicating plasmid. The advantages of using cosmids instead of ordinary plasmids are: (1) virtually all the DNA that is packaged contains the passenger DNA so that screening of the progeny for hybrid clones is unnecessary; and (2) there tends to be selection for large passenger DNA, whereas ordinary plasmids show a bias towards small fragments. Cosmids are smaller vectors than the λ vectors so that this system has the potential for packaging larger amounts of passenger DNA than the λ vectors.

4.3   Viruses

Viruses specific for eukaryotic cells can also be used as cloning vectors. The Simian virus 40 (SV40), a member of the group of papovaviruses, has been used in this way. It has a small genome of about 5 kbp. Half of this is transcribed early in its life cycle, and half late: the late gene region codes for viral structural proteins and is not all essential for viral replication. Portions of this can be cut out by restriction endonucleases and passenger DNA inserted. Cultured monkey cells can be either transfected directly with the recombinant DNA or infected with reconstructed viral particles containing the viral proteins as well as the recombinant DNA.

SV40, either as such, or carrying passenger DNA, can transform some rodent cell lines by a process in which the DNA is integrated into the host chromosome in a stable fashion. Other viruses such as the polyoma viruses, which are related to SV40, and the adenoviruses can also be used as vectors in a variety of appropriate mammalian host cells.

Experiments using these eukaryotic vectors are likely to be of more value than those using plasmids propagated in bacteria in studying the factors that are required for expression of genes in eukaryotic cells.

5

Prokaryotic gene organisation and expression

Prokaryotes have a single chromosome consisting of circular double-stranded DNA. The size may vary considerably, e.g. the *E. coli* chromosome contains about 3.8×10^6 bp, while that of *Bacillus subtilis* is about half that size (2×10^6 bp), and that of *Salmonella typhimurium* 10.5×10^6 bp. Phages and plasmids have chromosomes which are up to three orders of magnitude smaller than these. If the *E. coli* chromosome were in a linear extended form it would be about 1 mm long. In fact the DNA is a fairly compact molecule due to supercoiling (Chapter 1). This means that it must be opened up to allow access for enzymes involved in replication and transcription. This is a complex process, and requires several proteins including a helix destabilising protein and enzymes known as DNA gyrases and topoisomerases (Chapter 1.4). The helix destabilising protein (also known as single-strand binding protein) can bind to the single-stranded lengths of DNA produced by the action of topoisomerases. This helps to keep them in open structures which are more accessible to the enzymes that catalyse reactions on a single strand of DNA. Other proteins that are probably involved in this sort of process have been identified by genetic means, but their actual roles are not known. They are named after the gene that codes for them, whose chromosomal location is often known (e.g. *dnaB*, *dnaC*, etc.).

5.1   Replication

DNA replication begins at a definite site on the *E. coli* chromosome called oriC. A stretch of just over 1000 nucleotides in this region has been sequenced, and shorter sequences round the corresponding site in several other prokaryotic chromosomes are also known. A minimum of 245 bp is required to function effectively as a replication origin when this portion of the genome is engineered into plasmids. However, a longer sequence of at least 440 bp is necessary for bi-directional replication of both strands of the DNA. A consensus sequence can be derived from the known sequences of the different species in which short completely conserved stretches of up to 12 bp alternate irregularly with slightly longer unconserved sequences of up to 16 bp. These latter may serve as spacers between the binding sites of the various proteins involved in the initiation of replication. Several of the conserved sequences are inverted repeats which could form stem structures which might be involved in interaction with proteins.

DNA replication requires obligatory initiation ('priming') with a short length of RNA which must be synthesised first. There is some uncertainty about which enzyme catalyses this step. RNA polymerase is an obvious candidate, and there is some evidence that it may be involved. Two promoters at which RNA transcription could be initiated have been identified in the oriC region, and these run in opposite directions. Also the initiation of replication is inhibited by rifampicin, a known antibiotic inhibitor of RNA polymerase. However an enzyme known as DNA primase (the product of the *dnaG* gene) also functions as a DNA-dependent RNA polymerase, and this has been implicated in supplying the primer strand of RNA that is required before the DNA polymerase can start to function. There are certainly sites in the oriC sequence at which this enzyme can bind.

DNA synthesis is a processive process, that is to say, once it is initiated it proceeds steadily along the length of the chromosome from oriC. The actual site at which synthesis is occurring at any moment, where the strands are separated, is known as the *replication fork*. Since the process occurs in both directions starting from oriC, there will be two replication forks at any one time.

The synthesis of new strands of DNA is catalysed by a DNA-dependent DNA polymerase. Three such enzymes are known. DNA polymerases I and II (products of the genes *polA* and *polB* respectively) are not needed for DNA replication, since mutants lacking both these enzymic activities are viable. Polymerase I is probably involved in the repair of damaged pre-existing DNA, and also in the replication of certain plasmids, such as ColE1. The function of DNA polymerase II is unknown.

DNA polymerase III is the enzyme that is involved in replication. It is an oligomeric protein containing at least three and possibly as many as seven subunits. It catalyses the formation of phosphodiester bonds between the α-phosphoryl radical of deoxyribonucleoside triphosphates and the 3'-hydroxyl group of an oligo- or poly-nucleotide that is already base-paired in a double-stranded structure (Fig.1.7). The enzyme can incorporate about 800 nucleotides per second, so that the whole *E. coli* chromosome is replicated in about 40 minutes. Since replication is bi-directional $(3.8 \times 10^6)/$ 2 bp are copied as each replication fork moves round half the chromosome. Occasionally mispairing takes place, usually due to the occurrence of rare tautomeric forms of the bases. This is likely to occur at a rate of about 1 in 10^4 pairings, so the potential error in a single round of chromosomal replication is actually quite high (300–400 mispairings). This would lead to an unacceptably high rate of mutation if the 'errors' were not corrected.

Fig. 5.1. Multiple replication on one prokaryotic chromosome. RF are the replication forks where new DNA synthesis is occurring. RF′ are the replication forks where the second round of DNA synthesis occurs before the first round is completed. —— original strands of DNA, – – – DNA strands synthesised on first round of replication, ... DNA strands synthesised on second round of replication.

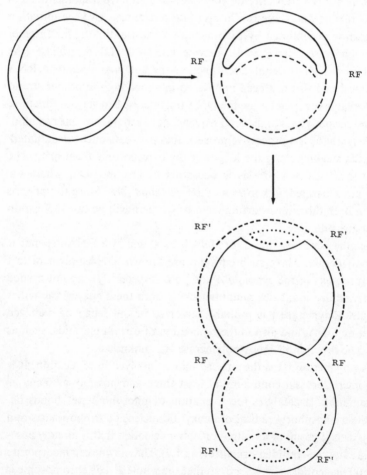

DNA polymerase also possesses a $3' \rightarrow 5'$ single-stranded exonuclease activity which provides a 'proof-reading' function. After incorporation of these unusual tautomeric forms of bases they rapidly revert to the more abundant forms which cannot base-pair, and these unpaired nucleotides are excised by this activity of the polymerase. The overall error rate is estimated to be about 1 in 10^{10} nucleotides.

As the replication fork opens up, one strand will be in the correct orientation $(3' \rightarrow 5')$ for the continuous synthesis of the complementary strand in the $5' \rightarrow 3'$ direction catalysed by the polymerase. This is known as the *leading strand*. The other $(5' \rightarrow 3')$ strand (known as the *lagging strand*) is wrongly oriented to provide a template for the continuous action of the polymerase. On this strand DNA synthesis is still in the $5' \rightarrow 3'$ direction, but it is discontinuous, occurring in lengths of about 1000 nucleotides at a time. RNA primer for the replication of this strand is likely to be provided by the action of DNA primase, rather than RNA polymerase which is probably used on the leading strand. The fragments of DNA produced from the lagging strand have actually been detected, and are known as Okazaki fragments, after their discoverer. A ribonuclease (probably ribonuclease H) which specifically hydrolyses RNA in covalently joined RNA–DNA hybrid molecules removes the RNA primer oligonucleotides, and the missing bases are filled in by DNA polymerase I. Finally an enzyme called DNA ligase joins the new DNA fragments together (Fig. 1.8).

When growth is occurring at very high rates and cells are dividing more rapidly than once every 40 minutes, multiple replications can take place from the oriC on the same chromosome (Fig. 5.1).

5.2 Transcription

Transcription requires the enzyme RNA polymerase. This is a large oligomeric enzyme which catalyses the formation of phosphodiester bonds between ribonucleoside triphosphates on an RNA template with the release of inorganic pyrophosphate (Fig. 5.2). The so-called core enzyme consists of three subunits, α, β and β' in the ratio of $2:1:1$. Other polypeptides may be associated with the core enzyme temporarily during the initiation and termination of transcription. These subunits are called sigma (σ) and rho (ρ) respectively. In purified systems the core enzyme binds indiscriminately to DNA and then proceeds to synthesise an RNA strand. However, in vivo, initiation of RNA synthesis starts at specific sites just before the beginning of genes specifying functional units. σ promotes this specificity of binding which allows initiation to occur correctly. Sites for this are signalled by two sequences in the DNA just upstream from the transcription start site, which are known as *promoters*. One, centred at 10 nucleotides before the site at which transcription starts, is known as the Pribnow box, after its discoverer, and has the consensus sequence TATAAT: the second is about 25 bases further upstream and has the consensus sequence TTGACA. This is sometimes referred to as the recognition site, where the initial interaction between the RNA polymerase and promoter is believed to occur. A crucial feature of this region of DNA is the space between the Pribnow box and the recognition site. For optimal rates of transcription this should be exactly 17 bases long, but the actual sequence here seems to be of less importance. A pomotor containing these two consensus sequences separated by exactly 17 nucleotides has recently been synthesised and shown to be more effective than several naturally occurring promoters. This optimal promoter sequence has so far not been shown to occur naturally. The sequence of the two conserved promoter sites is often somewhat different from the consensus, yet they still function reasonably well. Point mutations in these sites can lead to either increase or decrease in promoter efficiency. The precise sequence at these sites may be important in determining the amounts of transcripts that are normally made in a cell, with genes containing weak promoters being transcribed less frequently than those with strong promoters.

Other species of bacteria probably use different promoter sequences. For example, *B. subtilis* possesses several distinct σ factors that specifically recognise these different sequences.

Fig. 5.2. Transcription, showing the site at which nucleotides are being added to the growing RNA chain. To the left, the newly synthesised RNA is dissociating from the sense strand of the DNA, which is about to reassociate with its complementary anti-sense strand. To the right, the two DNA strands are separating under the influence of RNA polymerase. The incoming nucleoside triphosphates lose pyrophosphate on being incorporated into the growing RNA chain. The region covered by RNA polymerase where the two DNA strands are separated is much larger than that shown here.

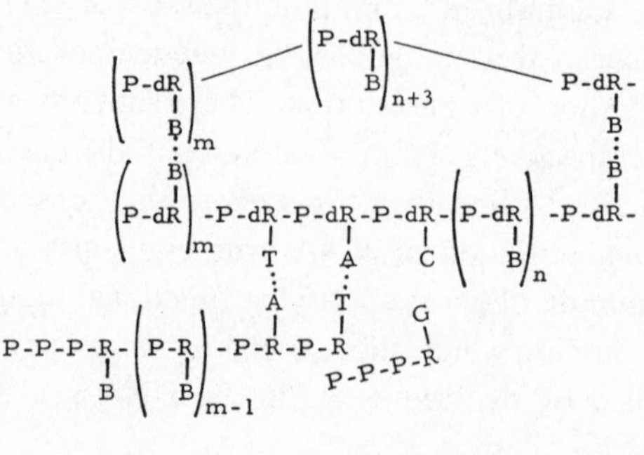

When RNA polymerase containing the σ subunit (referred to as the holoenzyme) binds to the promoter it is said to form a closed complex with the DNA base-paired in a double-stranded configuration. Its first action is to separate the two DNA strands, forming an open promoter complex. It will now start transcribing one of the DNA strands (the 'sense' strand) in the $3' \rightarrow 5'$ direction. Note that the RNA is formed in the opposite $(5' \rightarrow 3')$ direction. As transcription proceeds, σ dissociates from the complex and can be used to associate with another molecule of core enzyme to initiate transcription elsewhere.

Termination of transcription also occurs at specific sites. These contain a run of A residues in the sense strand of the DNA which are preceded by a pair of inverted repeat sequences which can form a stem and loop structure (Fig. 2.2). These stems are generally stabilised by a high proportion of G and C residues. Some transcripts require the factor ρ for termination, but this is not a universal requirement. The function of ρ in this process and the reason for its irregular requirement are unknown.

During transcription the newly formed RNA strand dissociates fairly rapidly from the complementary DNA strand as the polymerase passes along the DNA, and the double-stranded DNA reforms behind the enzyme. Initiation generally commences with a purine nucleoside triphosphate, using ATP more commonly than GTP. The rate of elongation of the RNA is about 60 nucleotides per second – an order of magnitude slower than the replication of DNA.

The majority of genes in a bacterium are transcribed into mRNA. mRNAs contain a ribosome-binding site a few bases upstream from the initiation codon (AUG or very much less commonly, UUG or GUG). This is known as the *Shine–Delgarno sequence*, and has the consensus AGGAGGU. It is complementary to a sequence of seven bases at the 3'-end of the 16s rRNA (Fig. 5.3). It plays an important role in positioning the mRNA correctly on the ribosome.

Translation of the mRNA starts while the mRNA is still growing by

Fig. 5.3. Interaction of Shine–Delgarno sequence on prokaryotic mRNA with the 3'-end of 16s rRNA. A similar interaction is believed to occur when eukaryotic mRNA is positioned on the ribosome.

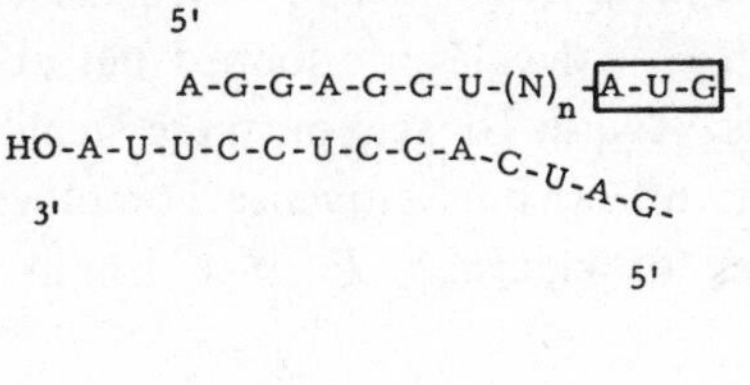

5'
A-G-G-A-G-G-U-(N)$_n$ -A-U-G-

HO-A-U-U-C-C-U-C-C-A-$_{C-U-A-G}$
3'

5'

Table 5.1. *Ribosomal RNA operons in* E. coli

Symbol	Map position (min)	Order of genes present					
rrnA	86	rrsA	alaT	ileT	rrlA	rrfA	
rrnB	89	rrsB	gltT	rrlB	rrfB		
rrnC	84	rrsC	gltU	rrlC	rrfC	aspT	trpT
rrnD	72	rrsD	alaU	ileU	rrlD	rrfD	thrV
rrnE	90	rrsE	gltV	rrlE	rrfE		
rrnG	56	rrsG	gltW	rrlG	rrfG		
rrnH	5	rrsH	alaV	ileV	rrlH	rrfH	aspU

rrs, rrl, and rrf stand for 16s, 23s and 5s rRNA respectively.
The other genes code for tRNAs for the amino acids indicated by their standard abbreviations.

transcription from the DNA. In prokaryotes mRNA is generally very un-stable, being rapidly hydrolysed by abundant ribonucleases that are present in the cell, so that the average half-life of prokaryotic mRNA is 2–4 minutes.

5.3   Some RNAs are processed after transcription

Some of the RNA that is synthesised by transcription is used directly, either as tRNA or as rRNA. In *E. coli* there are seven genes coding for long RNA molecules which contain sequences of all three of the different rRNA species (Table 2.1), as well as some kinds of tRNA (Table 5.1). The sequences of the rRNAs encoded by these genes are all identical. The transcripts of three of them contain the tRNAs for isoleucine and alanine in part of the spacer region between the 16s and 23s RNAs, while the other four transcripts contain a tRNA for glutamate in that position. In addition, one gene cluster of each type contains a gene for a tRNA for aspartate downstream from sequences encoding the rRNAs. This suggests that multiple rRNA gene clusters have arisen by duplica-tions from a common ancestor which has had different tRNA genes inserted at various times during the course of evolution.

The transcripts of the rRNA genes are processed by hydrolysis of phos-phodiester bonds at specific sites by RNase III in *E. coli*. The rRNA precursor exhibits interesting secondary structure in which the sequences that form the mature 16s and 23s molecules are looped out at the end of fairly long base-paired stems. RNase III acts on these stems making slightly staggered cuts. There is no obvious sequence homology imme-diately surrounding these sites of cleavage. Further trimming of a comparatively small number of nucleotides at both ends of the immediate precursors of these rRNAs follows (Fig. 5.4).

The formation of the rRNAs leaves some tRNAs embedded in sequences of nucleotides which need to be removed to generate the mature molecules. Other tRNAs are also transcribed as larger precursors – some individually; others in small clusters of up to four. All these precursors require processing, and this seems to be carried out largely by two enzymes – RNase P, an endoribonuclease which cleaves the precursors at the junction with the 5'-end of the tRNA, and RNase D, an exonuclease which hydrolyses phosphodiester bonds from the 3'-end of the precursors until it reaches the 3'-terminal trinucleotide CCA where its action is stopped. This CCA terminus is found in all tRNAs and is essential for their function since it is the site at which the amino acyl residue is attached to the molecule.

Further changes are required before the tRNAs are completely mature, since they all contain bases which are modified from the A, G, C and U in the primary transcript (Table 2.2). These changes involve a number of different enzymes, catalysing a wide range of individual reactions.

5.4   Transposable genetic elements

Bacterial genomes carry multiple copies of sequences known as insertion sequences (IS) and transposons (Tns). They are found at various sites in the genome and can move from one site to another, though often

Fig. 5.4. Upper line, RNA transcript of rRNA gene cluster with two tRNAs in the spacer between the 16s and 23s rRNAs.

Lower line, details of the secondary structure of the tRNAs and surrounding nucleotides before processing.

Arrows show some of the places where processing is believed to occur. III is ribonuclease III; P is ribonuclease P.

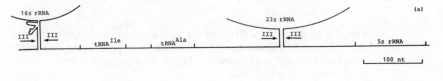

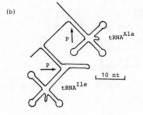

Table 5.2. *Antibiotic resistance genes carried by some transposons in* E. coli

Transposon	Size (bp)	Resistant to
Tn 3	4957	Ampicillin
Tn 5	5400	Kanamycin
Tn 9	2638	Chloramphenicol
Tn 10	9300	Tetracyclin

only at low frequency. There does not seem to be an absolute specificity of DNA sequence at the sites at which they are inserted but certain ones are favoured over others. IS are shorter than transposons and probably carry a sequence that codes for an enzyme called transposase that is required for insertion to take place. IS contain regions of inverted repeats at each end. Transposons also contain similar regions, but these tend to be much longer, and are often IS themselves. In addition Tns contain genes coding for enzymes conferring resistance to certain antibiotics (Table 5.2). Insertion of IS or Tns often leads to inactivation of genes of the cell's genome in the region of the insertion by interruption of transcription and/or translation. Less frequently they can activate transcription as they often contain promoter sequences required for transcription of their own DNA.

On insertion of IS and Tns there is a duplication of the target DNA for a short length (commonly 4–9 bp) on each side of the insertion, presumably because the insertion event involves a staggered break in the host DNA (Fig. 5.5).

The fact that these are movable elements implies that they can be excised from the genome, thus restoring any damaged functions. However, excision is not always exact, so reversion to the previous state does not always occur.

Fig. 5.5. Integration of IS1 into the *E. coli* chromosome generates a direct repeat of 9 bases at each end of the inserted element as a result of a staggered cut in the genomic DNA. This example shows insertion into the *lac*I gene.

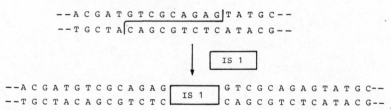

These elements obviously have properties in common with phages, but they have no capacity for autonomous replication. However, some small phages, such as Mu, have properties that are extremely similar to those of transposons so it is difficult to draw hard and fast lines between these two types of element.

6

The operon concept

6.1 Genes for sets of metabolically related enzymes are transcribed as one long message

In bacteria, genes specifying enzymes which are all part of a single metabolic pathway are commonly transcribed from adjacent lengths of DNA with only short non-coding stretches between them. Such 'super-genes' are known as *operons*, and are generally transcribed as single units giving rise to *polycistronic mRNAs*. Since translation immediately follows transcription this results in the rapid production of several functionally related enzymes in equivalent amounts – a process known as *co-ordinate control*. Some of these operons and the enzymes transcribed from them are synthesised more or less constantly and are known as *constitutive*; others are subject to precise control, signalled by the presence or absence of actual or potential metabolites in the bacterial cell and these are known as *inducible* when they are switched on by a particular metabolite, or *repressible* when they are switched off by a particular metabolite.

Much of our knowledge about operons has come from experiments on mutants in which they are not functioning normally. These have allowed the identification of two classes of genes. *Structural genes* encode information for either stable RNAs or mRNAs (see Chapter 2.3), while *regulatory genes* do not themselves give rise to any products, but are immediately involved in the regulation of transcription of structural genes.

These regulatory genes are situated immediately upstream from the operon whose activity they control. In addition to the promoter (Chapter 5.3), there is another region known as the *operator* which is either directly adjacent to the promoter, or even overlaps it, and there may also be other regions involved in regulatory control.

Inducible operons are said to be negatively controlled when they are normally not transcribed due to the binding of a specific *repressor* (always,

66

Table 6.1. *Type of control of operons*

Type of operon	State of regulatory molecules	Binding to operator	Trans- cription	Examples
Negative inducible	Repressor free	+	−	*lac*
	Repressor binds to inducer	−	+	*gal*
Positive inducible	Apo-activator free	−	−	*ara*
	Apo-activator binds to co-activator	+	+	*mal*
Negative repressible	Apo-repressor free	−	+	*trp*
	Apo-repressor binds to co-repressor	+	−	*arg*

as far as is known, a protein) to the operator. Induction occurs when an *inducer* (a small molecule) binds to the repressor, altering its conformation so that it now dissociates from the operator and allows transcription to proceed.

Positively controlled inducible operons are not normally transcribed. They become active when a *co-activator* (small molecule) binds to an *apo-activator* (protein) altering its conformation in such a way as to permit it to bind to a site near the operon so that transcription is initiated.

Negatively controlled repressible operons are normally transcribed, but when a co-repressor (small molecule) binds to the apo-repressor (protein) this complex then attaches to the operator resulting in inhibition of transcription.

It is theoretically possible to envisage positively controlled repressible operons in which an activator protein is normally bound to the operator, allowing transcription to occur. Interaction with an inhibitor (small molecule) would cause the activator to dissociate from the operator so that transcription would be inhibited. However, no example of this last situation is known. The various possibilities are shown in Table 6.1. The first two modes of regulation are not mutually exclusive, and several examples are known in which control is exerted in both these ways.

The operator is generally situated between the promoter and the site at which transcription starts. It is believed that binding of a repressor causes a physical block to the passage of RNA polymerase from the promoter to the site at which it would otherwise start transcription. Repressors are encoded by structural genes which may or may not be adjacent to the operon they control, and they are generally present in very few copies per cell.

Table 6.2 *Trans-acting mutations of the genes encoding regulatory proteins*

Regulatory protein	Function inhibited by mutation	Result
Repressor	DNA binding	Constitutive expression of operon
Repressor	Inducer binding	Operon becomes non-inducible
Apo-activator	DNA binding	Operon becomes non-inducible
Apo-activator	Co-activator binding	Operon becomes non-inducible
Apo-repressor	DNA binding	Operon becomes non-repressible
Apo-repressor	Co-repressor binding	Operon becomes non-repressible

A further method of regulation is seen in a number of operons coding for enzymes required for the synthesis of amino acids and other essential metabolites. This is called *attenuation*, and is discussed later (6.7).

As a general rule, repressible operons code for enzymes required for the synthesis of various small molecules, such as amino acids, purines and pyrimidines which are needed for other synthetic processes in the cell. These may be present in widely varying concentrations in the milieu in which the bacteria are living. When they are present in adequate amounts, considerable economy results from switching off the synthesis of the enzymes that would be required to make them. Conversely, inducible operons code for catabolic enzymes for metabolising certain sugars and amino acids which may sometimes be present in excess and can be used as energy sources. There is obviously no point in making these enzymes when such substrates are not available and the bacterium is utilising more generally available energy sources such as glucose.

Much of the information about the regulation of operons has been gained from studies of mutations mapping in the various regulatory regions. Mutations in the gene encoding repressor, apo-repressor or apo-activator can have various effects, depending on the site of the mutation. They can affect the interaction of the protein product with the operator. Thus, an altered repressor may no longer bind to the operator, in which case a negatively controlled inducible operon can become constitutively expressed. Alternatively, a mutation which causes defective binding of an inducer by the repressor will lead to an operon never being expressed. These and other possibilities are presented in Table 6.2. Mutations in the operator can obviously affect the binding of repressors and other regulatory proteins, as shown in Table 6.3.

Mutations in genes coding for regulatory proteins are said to act in *trans* because the products that they supply can act on chromosomes other than the one in which they are found. Operator and promoter

Table 6.3. *cis-acting mutations of the operator*

Type of operon	Likely effect of mutation	Result
Negative inducible	Inhibits binding of repressor	Constitutive expression of the operon
Positive inducible	Inhibits binding of apo-activator + co-activator	Operon is not expressed
Negative repressible	Inhibits binding of apo-repressor + co-repressor	Loss of ability to repress operon

mutations are said to act in *cis* because they will only affect the functions of the chromosomes on which they are sited. *trans*-acting mutations can be corrected (complemented) by the introduction of a plasmid bearing the un-mutated gene, while *cis*-acting mutations cannot be corrected in this way.

A common feature of many inducible operons which are positively controlled is that the induction is caused by the binding of a protein known alternatively as Catabolite Activator Protein (CAP) or Cyclic-AMP Receptor Protein (CRP). This protein, as its alternative name implies, also has a binding site cAMP, and it will interact only with regulatory sites on DNA after it has bound this nucleotide. These regulatory sites are generally situated just upstream from the RNA polymerase recognition site (the −35 box), and it is suggested that the binding of CAP makes it easier for the polymerase to bind here. A feature of some of these sites is that there is a region of symmetry in the DNA sequence extending over about 16 bp (Fig. 6.1). This may be important in the recognition of appropriate DNA sequences by CAP, since this is a dimer of two identical subunits.

Mutants which are defective in the gene coding for CAP (*crp*) or in that encoding adenylate cyclase (*cyc*) fail to show induction of operons that are normally induced by CAP.

Fig. 6.1. Region of inverted symmetry at sites where CAP is believed to bind in the *lac* (top) and *gal* (bottom) operons. Lines are drawn between the base sequences which exhibit the symmetry, and large dots mark the centres of symmetry. Note that stems giving rise to cruciform structures can be formed between the over- or under-lined bases in one strand.

```
G T G A G T T A G C T C A C
C A C T C A A•T C G A G T G

G T G T A A A C G A T T C C A C
C A C A T T T G•C T A A G G T G
```

Table 6.4. *Some examples of the control factors for operons*

Operon	Inducer	Co-activator	Co-repressor	Type of operon
lac	isopropyl-thio-galactoside allo-lactose	cAMP	—	{ Negative inducible { Positive inducible
gal	galactose	cAMP	—	{ Positive inducible { Negative inducible
mal	maltose	cAMP	—	Positive inducible
hut	urocanate	cAMP	—	{ Negative inducible { Positive inducible
trp	—	—	tryptophan	Negative repressible
arg	—	—	arginine	Negative repressible

Adenylate cyclase is stimulated when the concentration of glucose falls to a low level. This leads to the synthesis of cAMP which is then bound to CAP making it possible for this protein to bind to sites on inducible operons which provide enzymes for the catabolism of alternative energy sources (e.g. lactose, galactose, arabinose, histidine).

The foregoing description gives a general picture of some of the features that are frequently found in operons, but there are many individual variations. Several operons are discussed in more detail in the following sections so as to exemplify some of these differences (Table 6.4).

6.2   The *lac* operon

This was the first operon to be discovered, and it has been intensively studied over the past twenty years. It consists of three structural genes coding for proteins involved in the uptake and catabolism of lactose (Fig. 6.2). Upstream of the structural genes are an operator, a promoter and a CAP-binding site, and also the structural gene for the *lac* repressor protein. The operon is normally induced when lactose appears in the medium and when the concentration of glucose (a preferred energy source) drops to a low level. It is controlled in two ways – negatively by the *lac* repressor protein which dissociates from the operator when lactose becomes available, and also positively by CAP.

The operator site where the *lac* repressor binds is mostly in the region 5' to the initiation codon and extends just 5' to the site where transcription starts. It contains a 35 bp sequence with a high degree of inverted symmetry (only 7 bases do not match). This symmetrical arrangement may facilitate the binding of the *lac* repressor which is a tetramer of four identical sub-units. In fact, a completely symmetrical operator has been synthesised and shown to bind the *lac* repressor with an affinity ten-fold greater than the natural operator.

The gene for the *lac* repressor has its termination codon at or very near the 5'-end of the CAP-binding site. The gene is preceded by very weak promoter sequences and therefore is poorly transcribed. It has been estimated that there are only about ten molecules of *lac* repressor in a cell of *E. coli*. However, because of the high affinity it displays for the *lac* operator, this is a sufficient quantity to keep the *lac* operon repressed in the absence of lactose. In order to obtain enough of this interesting protein for study, the gene has been engineered into a plasmid immediately down-stream from a strong promoter. When the plasmid is grown in a suitable host large quantities of the repressor are produced.

The actual inducer of the *lac* operon is not lactose, but an isomer of this sugar called allo-lactose (Fig. 6.3). Its synthesis from lactose is cata-lysed by the first enzyme specified by the operon – β-galactosidase – and, like lactose, it can also be hydrolysed by this enzyme. Some other, non-metabolisable inducers have been synthesised, such as isopropyl-β-thio-galactoside (Fig. 6.3). Such inducers are called gratuitous inducers.

Fig. 6.2. The *lac* operon. Top: Map of the complete operon. I, Z, Y, A are the symbols for the structural genes with the names of the proteins that they encode given below the operon. c is the control region.

Bottom: Detail of the control region. The numbers refer to the conventional numbering of nucleotides, beginning at +1 at the 5'-most transcribed nucleotide. Numbering is negative from this point into the 5'-flanking region.

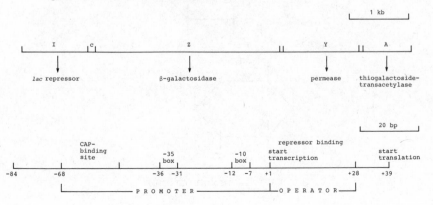

6.3   The *gal* operon

The *gal* operon encodes three structural genes for the catabolism of galactose (Fig. 6.4), and is derepressed by interaction of free galactose with the *gal* repressor protein. However, it has some features which distinguish its mode of control from that of the *lac* operon. An operator site was originally identified 5′ to the RNA polymerase-binding site, and therefore it did not seem possible that when the repressor was bound it could block transcription. However, a second operator has recently been discovered which is actually situated within the first structural gene. It is believed that the dimeric *gal* repressor protein binds to each operator

Fig. 6.3. Structure of lactose, the physiological inducer of the *lac* operon; allo-lactose, the actual inducer, produced by enzyme action on lactose; β-isopropyl-thiogalactoside, a synthetic gratuitous inducer.

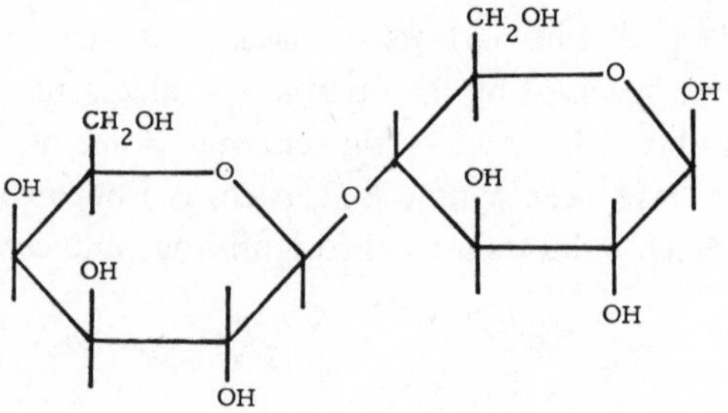

LACTOSE

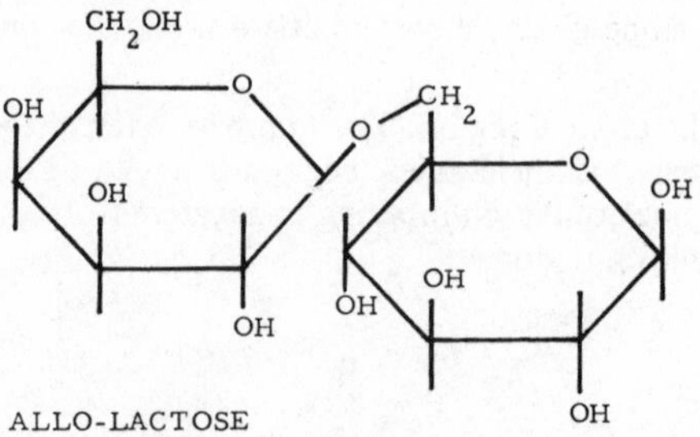

ALLO-LACTOSE

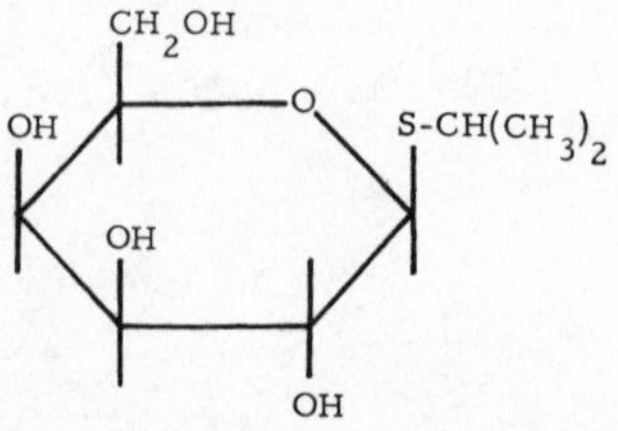

β-ISOPROPYL-THIOGALACTOSIDE

through one of its monomers, thus effectively preventing transcription of the operon (Fig. 6.5).

Another interesting feature of this operon is that there are two promoter sites, separated by 5 bp (Fig. 6.6). The Pribnow boxes overlap slightly, and are used to direct the RNA polymerase to start transcription at one or other of the alternative sites. There are also adjacent -35 boxes. The 3'-most promoter site is known as P_1 and transcription using this promoter is stimulated by CAP binding at a site overlapping the -35 box. There is a second binding site for CAP to the 5'-side of the first. Binding of the second molecule is facilitated by the prior binding of RNA polymerase. CAP binding inhibits transcription from the other promoter site (P_2), which is used only in the absence of CAP. CAP-independent transcription

Fig. 6.4. The *gal* operon. Top: Map of complete operon. Bottom: Details of the control region. For conventions, symbols and numbering see Fig. 6.2.

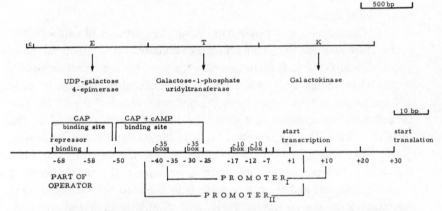

Fig. 6.5. The operator of the *gal* operon. The *gal* repressor binds at both the sequences shown, which are situated at approximately equal distances upstream and downstream from the initiation codon, which is shown on the right.

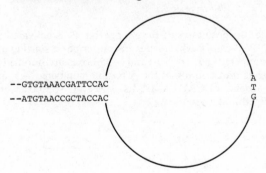

of the operon occurs at a low rate normally because the first two enzymes encoded by the operon are required for the production of galactose from glucose, as well as for the catabolism of galactose. The former process is essential, since small quantities of galactose are needed for the synthesis of some cell wall constituents of *E. coli*. The expression of the operon shows a marked degree of *polarity* – that is to say the distal enzyme (galactokinase, which is needed only for the catabolism of galactose – not for its production from glucose) is produced in considerably smaller amounts than the two proximal enzymes (the epimerase and transferase). On full induction of the operon, in the presence of CAP-cAMP, production of the kinase is stimulated to a greater extent than that of the other two enzymes.

A further difference between the *lac* and *gal* operons is the site of the genes coding for their repressors. That for the *gal* operon is situated on the opposite side of the *E. coli* chromosome.

6.4   The *ara* operon

Once again, an operon controlling the synthesis of catabolic enzymes (this time for the sugar arabinose) contains three structural genes (*ara*B, A and D) (Fig. 6.7). In this case the gene for the repressor (*ara*C) is very close to the structural genes. It is separated from them by only 147 bp, and it is transcribed from the opposite strand of the DNA, and therefore in the opposite direction. The 147 bp stretch contains all the regulatory sites. The *ara* repressor binds to the operator just 5′ to its own gene and represses its own transcription, so this is an example of autoregulation. However, in the presence of arabinose, the sugar combines with the repressor and the complex binds to a second operator (the BAD operator) 5′ to the structural genes, and stimulates their transcription. Transcription of both sets of genes is also stimulated by the binding of CAP-cAMP at a single site between the two operators. The BAD genes are transcribed when RNA polymerase binds between their promoter

Fig. 6.6. The two promoters of the *gal* operon. P_1 is activated in the presence of CAP-cAMP which leads to transcription starting at the A residue numbered $+1$. Its -10 and -35 boxes are outlined in full lines. P_2 directs transcription at the A residue numbered -5, and is used in the absence of CAP-cAMP. Its -10 and -35 boxes are shown outlined in dashes.

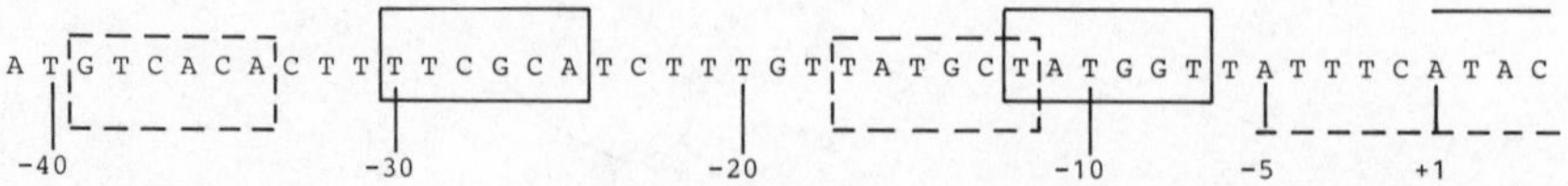

(P_{BAD}) and the CAP-binding site. The enzyme also binds in a second place between the other end of this site and the promoter site for the repressor gene. This second site overlaps the operator for the C gene. Each promoter contains a Pribnow box and a -35 box. From their sequences it is predicted that P_C is a fairly strong (though not ideal) promoter, while P_{BAD} is a poor one.

In the presence of arabinose there is probably a small amount of transcription of *ara*C, especially as the *ara* repressor binds to O_C site about ten times less strongly than the complex of repressor + arabinose.

In summary, there is positive control of P_{BAD} by CAP and by the complex between arabinose and its repressor. P_C is also positively controlled by CAP, but negatively controlled by the repressor especially when it is complexed with arabinose.

6.5   The *hut* operons

The catabolism of some amino acids is also inducible with negative control by a repressor protein. An example of this is found in the *hut* operons (histidine utilisation) which are switched on when there is a plentiful supply of histidine in the medium. Excess histidine can be degraded by bacteria to provide energy. This system has been most studied in *Salmonella typhii*.

Fig. 6.7. The *ara* operon. Top: Map of the complete operon. For conventions about symbols and numbering, see Fig. 6.2. The horizontal arrows show the directions of transcription. Bottom: Detail of the control region. Numbers above the line are with references to the transcription start site of the *ara* BAD gene cluster: numbers below the line are with reference to the transcription start site of the C gene.

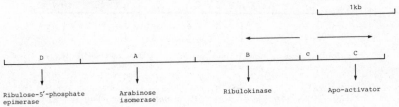

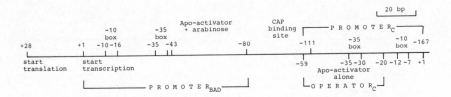

The four genes required for this catabolic pathway are found in two operons which are adjacent; the left-hand one also contains the gene for the repressor (Fig. 6.8). This binds to operators designated m and p which are situated immediately 5' to the left-hand and right-hand operons respectively. It binds less strongly to the m operator so that there is presumably some read-through of the left-hand operon in the presence of low concentrations of repressor, until a certain level of this protein is reached and further synthesis is inhibited.

The inducer is not histidine itself, but the first metabolite to be produced during its catabolism – urocanate – so there must be small amounts of histidine and the enzyme histidase available for the operon to be induced.

Catabolite repression is also involved in the control of these operons since CAP-cAMP stimulates transcription of the right-hand operon.

6.6   The *mal* regulon

The *mal* operons, coding for proteins involved in the uptake of maltose into the cell and its catabolism exhibit a further degree of complexity. There are two separate operons, located at different sites on the *E. coli* chromosome, which are under the common control of a single apo-activator (Fig. 6.9). Such a functionally linked series of operons is known as a regulon.

The *mal*A region contains the genes *mal*P and *mal*Q coding for enzymes involved in the catabolism of maltose, and also the *mal*T gene which codes for the apo-activator of this system. *mal*T is transcribed in the opposite

Fig. 6.8. Schematic diagram of the *hut* operon. The precise details are not yet known.

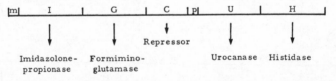

Fig. 6.9. Schematic diagram of the *mal* regulon. The A and B regions map to 74 min. and 90 min. respectively on the *E. coli* chromosome.

direction to *mal*P and *mal*Q, in a manner reminiscent of the repressor and structural genes of the *ara* operon. The *mal*B region encodes five structural genes. These are arranged in two transcription units which are transcribed in opposite directions with the interaction site(s) for the *mal*T gene product situated between them. The products of all these genes are involved in the uptake of maltose into the cell. The *lam*B gene product was originally shown to be required as a receptor for the uptake of the bacteriophage λ into the cell and only later found to stimulate the uptake of low concentrations of maltose.

The transcription of *mal*T is strongly stimulated by CAP-cAMP. After binding maltose, the *mal*T gene product activates transcription of *mal*P and *mal*Q, probably independently of CAP-cAMP. However, the transcription of the genes in the *mal*B region requires the binding of both CAP-cAMP and the apo-activator plus maltose. Thus a major physiological effect of the production of cAMP, which will occur when glucose concentration falls, is to allow entry of maltose into the cell in large amounts and this enhances the transcription of the genes encoding the enzymes required for its catabolism.

6.7   The *trp* operon: control by attenuation

There are a number of operons containing the structural genes coding for the enzymes required for the synthesis of amino acids. These are repressed, either by the amino acids themselves or by easily formed derivatives, so that unnecessary, energetically wasteful, synthesis of these compounds does not occur when there are adequate amounts available to the cell.

The *trp* operon encodes five structural genes required for the synthesis of tryptophan from chorismate (Fig. 6.10). An unlinked apo-repressor is encoded by the *trp*R gene, and when this combines with tryptophan the complex binds to an operator site just 5' to the site at which transcription

Fig. 6.10. The *trp* operon. In addition to the structural genes, the position of two terminators (t_1 and t_2) are shown. t_1 is a rho-independent terminator, while t_2 functions only in the presence of rho.

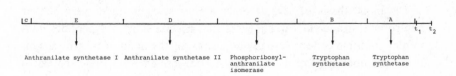

starts. This inhibits the passage of RNA polymerase from the promoter and so turns off transcription.

There is a rather long sequence of 162 bp between this start point of transcription and the codon at which translation of the structural genes begins. This is referred to as the leader sequence and is very important in a second method of control of this operon. The RNA transcribed from this leader sequence contains a short open reading frame of 42 bp, in which there are two consecutive in-phase codons for tryptophan. It ends with a typical termination signal – a region rich in G-C base pairs which can form a stem and loop structure – followed by a polyU tract (Chapter 5.3). This RNA transcript can form two alternative stable secondary structures by base pairing, which are mutually exclusive because part of one stem is common to both structures (Fig. 6.11). In the absence of other factors structure I is formed and permits the formation of the terminator stem and loop. However, in vivo, the newly transcribed RNA associates with a ribosome, and the short polypeptide encoded by the leader sequence will be synthesised. The presence of the ribosome on the RNA still allows the formation of the terminator stem and loop (Structure II). When there is a deficiency of tryptophan, and hence of charged $tRNA^{Trp}$, in the cell the ribosome will stall at one of the tryptophan codons and a new structure (III) arises so that the formation of the terminator stem and loop is pre-empted. In this situation the RNA polymerase will be about 150 nucleotides beyond the point where the RNA transcript emerges from the ribosome, and it will read through the potential terminator site, and continue transcribing the operon. The various sequences on the RNA that form the stem structures are known as the terminator, the pre-emptor (which includes part of one limb of the terminator), and a third one nearer the 5'-end of the RNA called the protector which can protect the terminator by forming a stem with one limb of the pre-emptor (Fig. 6.11).

This form of control, dependent on a mechanism which is responsive to the concentration of a particular charged amino acyl-tRNA in the cell is known as attenuation, and the region in which the RNA is terminated when the operon is repressed is called the *attenuator*. Both this system and the binding of the repressor to the operator seem to be operative in vivo for control of the tryptophan operon. Binding of the repressor causes a 70-fold reduction in transcription of the operon, while attenuation leads to an extra 10-fold reduction.

The biosynthesis of several other amino acids is under very similar control with leader DNA sequences that encode short polypeptides containing multiple codons (usually adjacent) for the amino acid whose synthesis the operon controls. The RNAs transcribed from these sequences

Fig. 6.11. The control region of the *trp* operon, showing the formation of alternative stem and loop structures on the mRNA transcribed from it.

Structure I can form in vitro in the absence of other factors.

Structure II forms in the presence of a ribosome which is actively translating the mRNA for the leader peptide. The formation of a terminator stem and loop proceeds and blocks further transcription of the operon by RNA polymerase.

Structure III forms in the presence of a ribosome which has paused at the *Trp* codons when there is an absence of charged tRNA[Trp]. The two *Trp* codons (UGG) and the termination codon (UGA) in the leader sequence, and the initiation codon (AUG) of the *Trp* E gene are shown. Some bases are numbered from the beginning of the leader sequence.

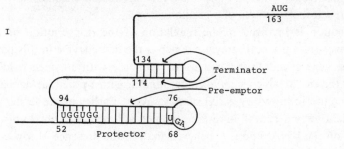

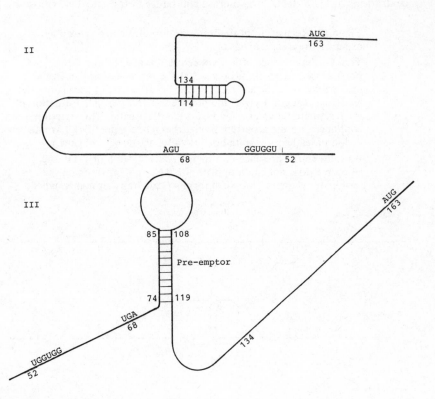

are all capable of base pairing to form protector, pre-emptor and terminator elements. Amino acids whose biosynthesis is regulated in this way include phenylalanine, threonine, isoleucine, leucine and valine (in *E. coli*), and histidine and leucine in *Salmonella typhii*.

Mutation of the genes coding for the amino acid-tRNA synthetases or the tRNAs generally leads to permanent induction of these biosynethetic operons, since there is likely to be a lack of the necessary charged tRNA so that the ribosomes will be permanently stalled on the leader sequence.

6.8   Pyrimidine biosynthesis

The whole pathway for the biosynthesis of pyrimidines requires six enzymes whose genes are dispersed over the *E. coli* chromosome and constitute another example of a regulon.

Attenuation is involved in the regulation of the transcription of the enzyme aspartate transcarbamylase which is the first enzyme in this pathway. The gene encoding it has a leader sequence with an open reading frame at the end of which is a typical terminator with a potential stem and loop structure followed by a run of A residues. Earlier in the leader sequence is a second run of A residues (which will be transcribed unto U residues on the RNA) and it is believed that, in the presence of low concentrations of UTP, RNA polymerase will pause at this site. The ribosome

Fig. 6.12. Regulation of the transcription of the gene encoding aspartate transcarbamylase.

Top: In the presence of low concentration of UTP the RNA polymerase (hexagon) has paused at the run of U residues in the leader sequence (vertical bars). The ribosome (circle) translating the leader has caught up with the polymerase and prevents formation of the terminator stem and loop (short vertical bars). Transcription continues into the aspartate transcarbamylase gene. (T = termination codon of leader; I = initiation codon of the structural gene.)

Bottom: In the presence of a higher concentration of UTP, the ribosome does not catch up with the RNA polymerase and the terminator stem and loop can form so that transcription is halted.

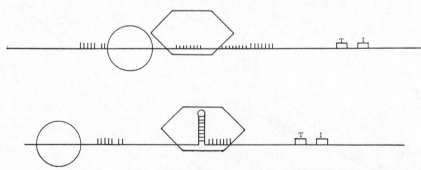

translating the mRNA catches up with the pausing RNA polymerase and when this enzyme moves forward to transcribe the region of the terminator, hairpin formation is precluded by the adjacent ribosome so that read through into the structural gene for aspartate transcarbamylase occurs. In the presence of large amounts of UTP the ribosome is further away and hairpin formation occurs at the attenuator so that transcription is terminated (Fig. 6.12).

6.9 Arginine biosynthesis

The biosynthesis of arginine requires at least nine different enzymes, two of which are oligomeric, so a total of eleven genes is needed. There is one cluster of four, though one of these is transcribed in the opposite direction to the other three, and a cluster of two, whilst the rest are scattered over a wide expanse of the *E. coli* chromosome (Fig. 6.13). There is also a separated apo-repressor gene (*arg*R). This set of genes constitutes a regulon under the control of the cellular level of arginine which acts as a co-repressor. However, the effectiveness with which the arginine–apo-repressor complex inhibits transcription of the different genes varies considerably (Table 6.5). Note that the transcription of *arg*R is also inhibited – another example of autoregulation. Carbamyl phosphate synthetase also provides carbamyl phosphate for the synthesis of pyrimidines, and expression of the two genes encoding this enzyme is additionally subject to inhibition by uracil nucleotides, presumably acting in concert with another apo-repressor.

Fig. 6.13. Map of the genes of the *Arg* regulon on the *E. coli* chromosome. The chromosome is conventionally numbered in a clockwise direction from 0 to 100, starting at the top centre of this diagram. Arrows show the direction of transcription of some of the genes.

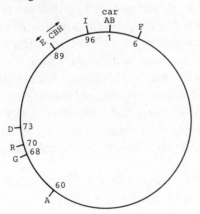

Table 6.5. *The repressibility of genes in the arginine regulon by combination of arginine with the apo-repressor*

Gene	Repressibility	
A	250	
CBH	60	
D	20	
E	17	
F	200–250	
I	400–500	
R	10	autoregulation
*car*AB	4	
	30	in presence of pyrimidines as well

Repressibility is defined as the ratio of transcription in the absence of arginine to that in the presence of 0.575 mM arginine.

There are reasonably well conserved sequences in the operator regions of five of these genes (including the *arg*ECBH and the *car*AB clusters) which are excellent candidates for binding sites for the arginine–apo-repressor complex.

There is no evidence for control by attenuation in this system.

6.10   Ribosomal proteins

There are 52 different proteins in the *E. coli* ribosome – 31 in the large subunit, 20 in the small subunit, and one common to both subunits. Many of these are encoded by genes linked together into several operons, some of which are very large (Table 6.6). A number of these operons also contain other genes encoding proteins involved in transcription or translation, including those for the subunits of RNA polymerase. A few of these genes seem to be unlinked.

There is good evidence that nearly all ribosomal proteins are produced in equal amounts, but it is not completely clear how this is achieved. Transcription of some of their operons is inhibited specifically by the binding of one of their protein products to the 5′-flanking region. These interactions occur with sequences which are homologous with part of the base sequence of the rRNA where the proteins are believed to be bound in the ribosome. This mechanism ensures that when there is an excess of a ribosomal protein which would be free and not actually incorporated into a ribosome, its synthesis and that of several other ribosomal proteins is shut off.

The operon coding ribosomal proteins L10 and L7/12 and the β- and β′-subunits of RNA polymerase has been much studied. Under normal

Table 6.6. *The ribosomal protein operons*

rplK	rplA	rplJ	rplL	rpoB	rpoC					
L11	[L1]	[L10]	L7/12	pol	pol					
rpsL	rpsG	fusA	tufA							
S12	[S7]	EF-G	EF-Tu							
rpsJ	rplC	rplD	rplW	rplB	rpsS	rplV	rpsC	rplP	rpmC	rpsQ
S10	L3	[L4]	L23	[L2]	S19	L22	S3	L16	L29	S17
rplN	rplX	rplE	rpsN	rpsH	rplF	rplR	rpsE	rpmD	rplO	
L14	L24	L5	S14	[S8]	L6	L18	S5	L30	L15	
rpsP	orf	trmD	rplS							
S16			L19							
rpsM	rpsK	rpsD	rpoA	rplQ						
S13	S11	[S4]	pol	L17						

Gene symbols are given above the lines: the products they encode are shown below the lines. L and S refer to the proteins of the large and small ribosomal subunits respectively. EF are elongation factors required in proteins synthesis (Chapter 2.5): pol are the RNA polymerase subunits. The proteins that are boxed can inhibit the transcription of all or most of the genes in the operon that encodes them (see text). The other ribosomal proteins are encoded either separately or in pairs. trmD encodes an enzyme which methylates G residues in some tRNAs. orf = open reading frame, coding for protein with no known function.

conditions of growth the ribosomal proteins are synthesised in much greater amounts than the polymerase subunits, although mRNA for all four proteins can be transcribed together, and the major promoter of this operon undoubtedly precedes the *rpl*J gene. There is a larger than usual non-translated spacer between the *rpl*L and the *rpo*B genes which contains several features of interest. First, it contains a potential rather weak terminator (attenuator) structure which could terminate translation transcription before the polymerase genes are reached. Secondly, there is a potential stem and loop structure, rather similar to that found in the intercistronic region between the 23s and 16s rRNA genes (Chapter 5.3). This might be a site where RNase III could cleave the primary transcript, splitting off mRNAs for the polymerase which could be translated separately from those for the ribosomal proteins. Finally, there are two Pribnow boxes in this region which could function as independent promoters for the transcription of the polymerase genes, especially in circumstances where there is an increased production of the enzyme (e.g. after treatment with rifampicin).

In the operon which encodes ribosomal proteins S16 and L19 their genes are separated by the gene coding for the transmethylase enzyme which methylates guanine residues in some tRNAs and by an open reading frame encoding a protein of unknown function. The two ribosomal proteins are expressed at a level of about 8000 molecules per cell, while there are only about 80 transmethylase molecules per cell. It is not certain how these different levels of expression arise from the same operon. The transmethylase is over twice as large as either of the ribosomal proteins, so that translation of its mRNA would take much longer, and in addition it uses a number of codons that are not commonly used in more abundantly expressed *E. coli* genes. However, these two factors alone are unlikely to reduce the rate of synthesis of this protein by a factor of 100 compared to that of the ribosomal proteins. It is probable that there are other controls operating at the level of translation.

The *str* operon encodes two elongation factors used in translation in addition to two ribosomal proteins. EF-Tu is produced in about sevenfold excess over both EF-G and the ribosomal proteins. The *tuf*A gene has a promoter site about 50 bp upstream from the 3'-end of the *fus*A gene, so the former gene could be transcribed on its own. There is also a *tuf*B gene at a different site on the chromosome which encodes a protein functionally similar to, and very nearly identical with, EF-Tu. This gene contributes about a quarter or a half of the EF-Tu to the cell. This second gene is at the 3'-end of an operon which also encodes four tRNAs, and is the only operon known so far which encodes a mixture of stable RNAs and mRNAs. There is a strong promoter upstream from the first tRNA gene in this operon, but also a weaker one between the last tRNA gene and the *tuf*B gene, so there could be some transcription of the *tuf*B gene alone from this operon.

The diverse features encountered in these operons suggest that their individual functioning can be finely controlled in appropriate ways to serve particular physiological requirements.

6.11   The stringent response

Under conditions of amino acid starvation a co-ordinated series of events, known as the stringent response, takes place. These include cessation of the production of rRNA and tRNA and ribosomal proteins and inhibition of synthesis of carbohydrates, lipids and peptidoglycans, while simultaneously there is increased transcription of some of the genes for the synthesis of amino acids (Table 6.7). This response is mediated by a pair of unusual nucleotides, formerly known as Magic Spots I and II (they were originally identified on chromatograms). They are guanosine

Table 6.7. *Some consequences of the stringent response*

Gene	Effect on transcription
rRNA	Inhibition
tRNA	Inhibition
Ribosomal proteins	Inhibition
EF-Tu and EF-G	Inhibition
lac operon	Stimulation
trp operon	Stimulation
galactokinase	Stimulation
ribulokinase	Stimulation
chloramphenicol-acetyl transferase	Stimulation

tetraphosphate – more strictly 3′-diphospho-, 5′-diphospho-guanosine (ppGpp), and guanosine pentaphosphate (pppGpp) – bearing an extra phosphate on the 5′-position (Fig. 6.14).

Mutant strains of bacteria which do not exhibit the stringent response are known as relaxed strains. They have mutations in the *rel*A gene which encodes an enzyme mediating this response. This enzyme is a pyrophosphoryl transferase which is associated with the ribosomes and becomes active in the presence of uncharged tRNAs. The concentration of ppGpp increases dramatically more than tenfold when the enzyme is stimulated by the binding of uncharged tRNAs to the A site on ribosomes (Chapter 2). It is not known how ppGpp exerts its effects on the transcription of certain parts of the genome. A number of genes whose expression is inhibited in the stringent response have the conserved sequence CGCCNCC encompassing the site of initiation of transcription. This could possibly be a site at which ppGpp binds, or more likely, a binding site for a hypothetical control protein. Alternatively, ppGpp could bind to RNA polymerase. It has been suggested that this enzyme can assume different conformations which might recognise different promoters: if this were so binding of ppGpp could stabilise certain of these conformations, thus altering the efficiency with which it transcribes certain genes or operons.

Fig. 6.14. The formation of pppGpp and ppGpp, the magic spot nucleotides which are the effectors of the stringent response.

7

Eukaryotic gene organisation and expression

7.1   DNA is in the nucleus in discrete linear chromosomes
Eukaryotic cells are distinguished from those of prokaryotes by the possession of a nucleus, separated from the rest of the cell by a membrane, and containing the cell's genetic material. (There is also a small amount of DNA in the mitochondria, discussed in Chapter 12.) The nuclear DNA is organised in linear chromosomes rather than in a single circular chromosome. The number of chromosomes varies considerably from one species to another (Table 1.1), and so does their size. There is little correlation between the total DNA content of a species and its evolutionary position. Many higher plants, amphibia and some fish have a much larger genome than humans and other mammals.

The chromosomes consist of two complementary anti-parallel strands of DNA and are generally associated with a considerable amount of protein. They exist in homologous pairs, though in metazoans (many-celled organisms) one pair which may not be homologous gives rise to the two sexes – in mammals females possess two X chromosomes, while males possess an X and a Y chromosome, differing from each other in both size and structure. In other orders of vertebrates and phyla an individual's sex may be determined in different ways.

When sexual reproduction occurs both ovum and sperm only contain a single copy of each of the parental chromosomes, so that the fertilised ovum comes to contain two copies of each chromosome, one derived from each parent. Germ cells are therefore said to be *haploid* (i.e. containing only half the genetic material of somatic cells) in contrast to the *diploid* state of somatic cells. The term *haplotype* is used to define the genetic constitution of the contribution from one parent to the offspring. Individual cells within a metazoan may divide repeatedly by binary fission, forming a *clone* in which all cells are genetically identical, unless any mutations have occurred spontaneously.

Table 7.1. *The major classes of histones*

Symbol	Number of amino acid residues	Percentage of	
		arginine residues	lysine residues
H1	215	1.4	28.4
H2A	132	9.1	9.1
H2B	124	6.5	16.1
H3	135	13.3	8.9
H4	102	13.7	9.8

7.2 Nuclear DNA is associated with proteins

The nuclear DNA of eukaryotes is associated with a number of proteins which are found only in the nucleus. The best studied of these are a group called the *histones* which are among the most abundant cellular proteins. There are five major classes (Table 7.1) and, for proteins, they are comparatively small molecules containing an excess of positively charged amino acids (lysine and arginine) over the negatively charged ones (glutamate and aspartate). They are thus ideally suited for binding to negatively charged nucleic acids by multiple ionic interactions. Spermatozoa are exceptional in containing protamines, a group of small highly positively charged proteins, in place of histones. They presumably aid in packaging the DNA into the small volume of the head of the sperm.

As a group, histones have been very highly conserved during evolution. This is particularly marked in the case of H4 in which there are only two conservative amino acid differences between the proteins of cow and lily although plants and animals diverged from each other just over 10^9 years ago. The rate of divergence of other proteins having similar functions in different species (e.g. cytochrome c or haemoglobin) is faster by two orders of magnitude. This suggests that there are very stringent constraints on the structure of histones which are essential for their particular functions.

Many other proteins are associated with nuclear DNA. As they are present in low concentrations they are difficult to isolate and study, and little is known about most of them. One group that has been investigated to some extent are those that migrate rapidly on electrophoresis. These are known as the high mobility group. They are given arbitrary numbers, and two of them – HMG 14 and HMG 17 – may play roles in the control of transcription.

7.3   Histones associate in a regular fashion with DNA to form nucleo-somes

If all the nuclear DNA of a single human cell were stretched out in its double helical form it would be about 230 mm long (on average about 5 mm per chromosome), but it is packed into a nucleus which is of the order of only 5×10^{-6} m in diameter. Thus the DNA must be packaged into a much more compact form. This is achieved in various ways which are not all clearly understood.

The fundamental unit of DNA structure in the nucleus is the nucleosome. Nucleosome preparations viewed in the electron microscope appear as a series of beads on a string. The beads are 10–11 nm in diameter, linked at regular intervals by a thin strand of DNA (Fig. 7.1). Mild digestion with micrococcal nuclease hydrolyses the DNA into units containing about 200 bp or multiples of this. Further digestion yields core particles which contain 146 bp. The DNA in these follows the course of one and three-quarter turns of a left-handed superhelix wrapped round the outside of an octamer consisting of two molecules each of histones H2A, H2B, H3 and H4 (Fig. 7.2). These protect the DNA from further digestion. The portions of DNA between these core particles are known as flexible linkers, and their length varies somewhat in different organisms and even between different tissues in the same organism. Histone H1 is bound to the outside of nucleosomes and probably plays a role in bringing individual nucleosomes together and stabilising them. This is facilitated because H1 shows co-operative binding to itself: that is to say, the binding of two or more molecules together enhances the ease of binding further molecules.

Further compaction of the DNA is achieved by packing the nucleosomes

Fig. 7.1. Chromatin fibres spilling out of rat thymus nuclei, showing nucleosomes between thin strands of DNA. (Reproduced from A. L. Olins & D. E. Olins, *Science* (1974), **183**, 330–2. © by the AAAS.)

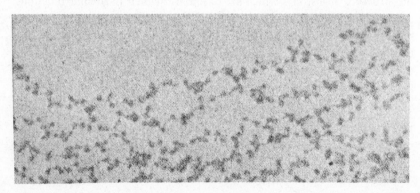

together in regular structures with a higher degree of order. Electron microscopy suggests that they form a fibre of 25–30 nm in diameter, in which
they are probably packed into a solenoid-like structure (Fig. 7.3). These
fibres, in turn, are formed into loops or domains of varying size containing
from 35 to 85 kbp of DNA. It is tempting to suggest that these loops may
be transcription units with some flanking sequences, since they are of the
correct order of size (e.g. the β-globins are transcribed from a region of
about 60 kbp in length – see Chapter 9.1), but there is no direct evidence
for this. It is believed that these loops are held in position by a scaffolding
of proteins.

Nucleosomes are probably in a dynamic state in which the higher orders
of structure can be temporarily disrupted by various processes, such as the
removal of H1. This may be necessary for transcription to occur, but the
evidence for this is largely circumstantial. However, in some situations the
DNA of genes that are being actively transcribed becomes sensitive to
DNases, while the same genes in quiescent tissues are resistant to such
digestion.

The charge properties of the histones can be affected by various means,
particularly by acetylation of lysyl residues and by phosphorylation of
seryl residues (Fig. 7.4). Both these processes reduce the positive charge

Fig. 7.2. The structure of a nucleosome. The circles and part circles
represent the monomers of H2a, H2b, H3 and H4, while the tape
winding round them represents the DNA double helix.

Fig. 7.3. Packing of nucleosomes. The larger circles represent the
nucleosomes with the DNA double helix coiling round them. The
smaller circles represent the H1 molecules bound to the flexible
linkers of DNA.

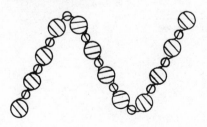

on the histones, so that they are likely to bind less tightly to DNA. Only certain residues are subject to modification in these ways, and they are all situated near the N-terminus of the proteins. The acetyl groups turn over rapidly. A small protein called ubiquitin (formerly HMG 20), on account of its widespread distribution, can be linked covalently to H2A, so that at any one time about a tenth of the H2A molecules are modified in this way. The significance of this modification is unknown.

7.4   Replication of DNA is a mystery

Three DNA polymerases, designated α, β, and γ, have been found in animal cells. The α and γ enzymes are large oligomeric molecules, while the β enzyme is much smaller and simpler. Their roles are not really known. Only the γ enzyme can catalyse the processive synthesis of long chains of DNA, so it is the most likely candidate for the enzyme involved in eukaryotic replication.

7.5   Transcription

Transcription in eukaryotes is a good deal more complicated than in prokaryotes. Eukaryotic cells produce three different RNA polymerases for transcribing nuclear genes. These are all very large proteins with molecular weights in the range of 500000 to 700000, and each is made up of 10–15 distinct subunits. Some of these with molecular weights of around 200000 are among the largest known single polypeptide chains. Each of the three different polymerases transcribes a particular set of nuclear genes. Polymerase I (Pol A) transcribes the precursors to ribosomal RNAs; polymerase III (Pol C) transcribes genes for tRNA precursors and some other small RNAs; Polymerase II (Pol B) transcribes genes into mRNAs coding for proteins. An RNA polymerase that transcribes mitochondrial genes is encoded by a nuclear gene. This enzyme seems to have a much simpler structure than those transcribing nuclear genes.

Fig. 7.4. Top: Acetylation of a lysyl residue in a protein.
Bottom: Phosphorylation of a seryl residue in a protein.

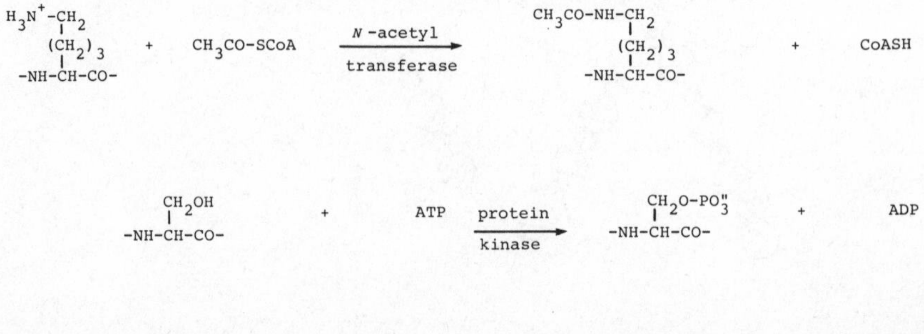

rRNA is encoded by a very large gene which contains the sequences that are transcribed into the 28s, 18s and 5.8s species of rRNA, as well as spacers and a 5'-flanking sequence. This gene is present in multiple copies which, in humans, are situated at the ends of the short arms of five different chromosomes. During interphase this DNA is looped out into the nucleolus for transcription by polymerase I, which is confined to this organelle. No definite promoter sequence recognised by the enzyme has been discovered, but about 150 nucleotides on both sides of the initiation site are required for optimal transcription. The primary transcript of mammalian rRNA contains about 13000 nucleotides and has a sedimentation coefficient of 45s. This molecule is first methylated and then processed in an ordered sequence of specific hydrolytic cleavages to give the mature 28s, 18s and 5.8s rRNAs (Fig. 2.3). Note that the 5s rRNA is transcribed from a different set of duplicated genes by polymerase III. This enzyme, which is not associated with the nucleolus, also transcribes the genes for tRNAs. These are often arranged in clusters in tandem, and again the primary transcripts require processing before yielding tRNA molecules. Some of the precursors contain extra nucleotides in internal regions which must be removed by endonuclease action with subsequent re-formation of the appropriate phosphodiester bonds (Fig. 7.5). Polymerase III does not recognise specific 5'-flanking sequences. In the transcription of genes for 5s rRNA it interacts with a protein (5s transcription factor) that binds directly to certain DNA sequences within the gene (Fig. 7.6). Polymerase III appears to recognise tRNA genes by binding to internal sequences of

Fig. 7.5. Portion of the primary transcript of the gene for yeast tRNA[Phe], showing the nucleotides which are excised during the formation of the mature tRNA.

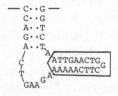

Fig. 7.6. Top: 5s RNA gene, showing the sequences recognised by RNA polymerase III.

Bottom: tRNA gene, showing consensus sequence of two stretches of DNA recognised by RNA polymerase III.

```
      50          60          70          80
————————AGCTAAGCTGGGTCGGGCCTGGTTAGTACTTGGA————
```

```
    8          18                  52          62
————TGGCNNAGTGG———————————————GGTTCGANNCC————
```

11 nucleotides which code for portions of the D and T stems and loops in the mature tRNA molecules (Fig. 7.6). Recognition is probably mediated through separate protein(s) to which polymerase III binds before initiating transcription.

Eukaryotic genes encoding proteins generally have a sequence homologous to the Pribnow box, called the Goldberg–Hogness box, which has the consensus sequence TATAAA^A_T. This is usually situated about 25–35 bp 5' to the site of initiation of transcription. It is believed to be required for correct positioning of the polymerase II at this site. There is some evidence that initiation generally starts with A, and neighbouring bases are more often pyrimidines than purines. It has not been easy to define a consensus sequence at this site, because it is difficult to map the actual site of initiation precisely (an error of up to ± 5 bases is commonly found), and also because initiation quite often starts at more than one site within a short sequence of bases. A second element which may be important in recognition by polymerase II is a sequence called the CAAT box. This generally occurs between 70 and 80 bases upstream from the putative initiation site. Limited sequence information suggests that the consensus sequence is CCAAT. However, this site is not always easily identifiable in a number of the genes that have been sequenced.

In the small number of yeast genes whose sequence is known initiation begins at an A residue in an A-rich region, and a Goldberg–Hogness box is frequently found up to 100, or even more, bp 5' to the initiation site. Thus there may be considerable variation in the detailed topography of promoters in different eukaryotic phyla.

It has recently been suggested that recognition of critical sites 5' to the initiation region may depend less on common sequences, but more on the collective properties of series of base pairs. The TATAA box is interpreted as a region of comparatively low stability, since A-T base pairs are only doubly hydrogen bonded, and would provide a 'bubble' surrounded by some more stable triply hydrogen bonded G-C base pairs. A G/C rich region which is often present about 20–40 bp 5' to the TATAA box could provide a region of high stability to give some kind of firm support for the attachment of the polymerase enzyme. This interpretation gives no role to the CAAT box, whose most invariant feature is the AA dinucleotide.

There is no information about signals for the termination of transcription in eukaryotes.

7.6   Enhancers

Some sequences situated near the promoters of a number of viral genes greatly increase the ability of these promoters to direct transcription.

Such sequences are known as enhancers and they have the remarkable properties of operating over a distance of up to 3 kb in either direction from the start point of transcription, and in either orientation (i.e. whether present in the 5'-3' direction or in the 3'-5' direction). Homologous structures with similar properties have been identified in the introns of immunoglobulin genes (Chapter 10.5). Although there is considerable divergence of sequence between these structures, a core sequence of GGTGTGG$\frac{AAA}{TTT}$G has been derived. A homologue of one of these viral enhancers in the human genome can act as an enhancer in an artificially constructed plasmid. It is not known whether it has any physiological function.

These enhancers show definite tissue and species specificity. Those associated with the immunoglobulin genes will only function in lymphoid tissue, and several viral enhancers will only function in the species in whose cells the parent viruses grow best. This may be because they need tissue-specific factors to function properly.

Enhancers may allow the entry of RNA polymerase onto the DNA. The enzyme would then move along the chromosome until it comes to a suitable promoter site where it could start transcription.

7.7 Many mRNA molecules have a cap and a tail added post-transcriptionally

A feature distinguishing eukaryotic from prokaryotic mRNAs is the presence of a cap at the 5'-end of the former. This cap is a 7-methylguanylyl radicle joined in a rather unusual way to the 5'-terminus of the transcribed mRNA (Fig. 7.7). Newly transcribed mRNA has a 5'-triphosphate (generally linked to adenosine) at the 5'-terminus. This is first enzymically hydrolysed to a 5'-diphosphate which acts as acceptor for a guanylyl residue to which it is linked by the activity of the capping enzyme (also known as guanylyl transferase) as shown in Fig. 7.7. The guanylyl residue is then methylated in position 7 by a specific methyl transferase. Some viruses produce a single enzyme in which all these three different catalytic functions reside in the same molecule. The 2'-positions of the ribose in the first two nucleotides at the 5'-end of the mRNA transcript are frequently, but not invariably, methylated. Capping occurs rapidly on nascent mRNA chains while they are still being formed by transcription.

Capping of mRNA may serve two functions. First, it promotes the binding of the mRNA to the ribosomes. This is mediated by a cap-binding protein which initially binds specifically to the cap, and is dependent on the positive charge bestowed on the guanylyl residue by its methylation. A second consequence of capping is to increase the stability of the mRNA, probably by protecting it against 5'-nucleotidase activity. It is well known that prokaryotic mRNA (lacking the cap) is very much more rapidly degraded than most species of eukaryotic mRNA.

A second type of modification found in most eukaryotic mRNAs is the addition of a polyA tail at the 3'-end. This is catalysed by enzymes called poly(A) polymerases. Typically, eukaryotic mRNAs have tails of about 100–200 adenylyl residues, though some mRNAs, particularly those coding for histones, are not generally polyadenylated. Polyadenylation takes place in the nucleus soon after the mRNAs are formed by transcription.

Most eukaryotic mRNAs contain the sequence AAUAAA located some way 3' to the termination codon, and polyadenylation starts about 13–20 nucleotides downstream of this. In some cases more than one AAUAAA sequence is present, and each may be used as a polyadenylation signal.

Fig. 7.7. The capping of eukaryotic mRNA.

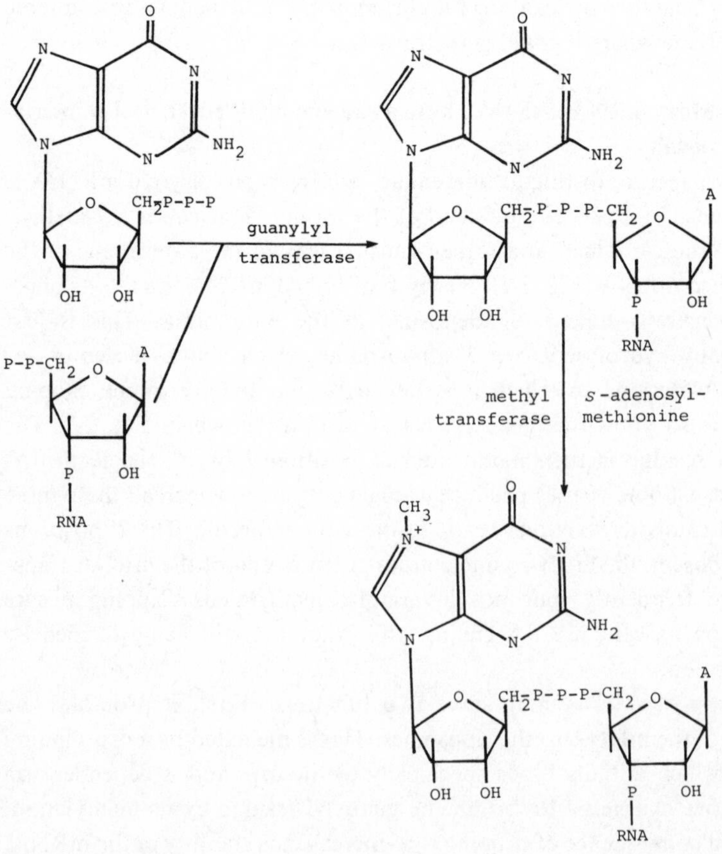

Thus, several species of mRNA can be produced which differ in the lengths of 3'-untranslated sequence. These AAUAAA signals are found at varying distances beyond the termination codon, and many mRNAs have several hundred untranslated nucleotides at their 3'-ends.

The importance of this AAUAAA sequence is emphasised by situations where it is mutated – either naturally or artificially. A mutation to AAGAAA in a viral system leads to extension of the RNA transcript into the succeeding gene. Mutation to AAUAAG has been observed in a case of α-thalassaemia (Chapter 9.2). This also results in the production of a longer mRNA than usual that is unstable so that insufficient α-globin mRNA is available.

There is some evidence that processing of the 3'-end of mRNAs requires the presence of a small nuclear RNA–protein complex (known as a snurp). It is possible that U4 (Table 2.1) may be the RNA component of this complex which recognises the AAUAAA sequence of the mRNA. The associated protein is presumably involved in the polyadenylation that follows.

It is probable that there are distinct termination sites to the 3'-side of the AAUAAA sequence where RNA polymerase II terminates transcription, but these have not so far been properly defined.

The function of polyadenylation is unknown, though it has been suggested that it may play a role in the transport of mRNA out of the nucleus. This is not likely to be correct because mRNAs which are not polyadenylated must leave the nucleus to be translated. Experiments using *Xenopus* oocytes for translation of micro-injected exogenous mRNAs have shown that polyadenylated mRNAs have a longer functional life time than deadenylated mRNAs. Interestingly, the life of histone mRNA (which is not naturally polyadenylated) in this system can be increased by artificial polyadenylation. Another function that has been suggested for the polyA tails is the alignment of splice sites at the beginning and end of introns (see next section) so that they can be processed correctly to produce mature mRNA.

7.8 The coding sequence of many genes is interrupted by non-coding sequences

In bacteria the genes, in the form of DNA, are exactly co-linear with the mRNA which is transcribed from them. It came as a great surprise when it was found that in eukaryotes the DNA is not always co-linear with the mRNA. Coding sequences of the DNA are interrupted by stretches of non-coding DNA, and the primary mRNA transcript must be processed by excision of the non-translated nucleotides.

These non-coding sequences in DNA were originally called intervening sequences (IVS), but are now more generally known as introns – the coding sequences are known as exons. Both introns and exons are very variable in length. Some introns may actually be longer than the whole of the exons in a particular gene. At the other extreme, some exons code only for two or three amino acids. Introns also occur in the flanking, non-translated sequences of genes.

Individual exons often code for well-defined domains of the polypeptide chain which may provide functional units within the protein that is encoded. This is particularly well exemplified in the case of the immunoglobulin genes (Chapter 10).

7.9   Introns are transcribed into RNA and then removed in the nucleus
Comparison of the base sequences at intron–exon junctions reveals a consensus sequence of ^C_AAG ↑ GU^A_GAGU at the 5'-end of the intron (arbitrarily known as the donor site), and Y_{11}NYAG ↑ G at the 3'-end (acceptor site). The GU at the 5'-end and the AG at the 3'-end are the only invariant residues. It is generally believed that the excision of introns (or splicing, as it is often called) begins with looping out of the introns so that the 5'- and 3'-ends are juxtaposed. This may be facilitated by limited base-pairing with part of the U1 RNA molecule which is one of a class of small RNAs which were discovered before any function could be assigned to them (Fig. 7.8). The final part of the process presumably involves endonucleolytic hydrolysis of the phosphodiester bonds at each end of the intron followed by formation of a new phosphodiester bond to join up the two exons. The enzymology of this process is not understood: indeed it is not even known whether one or two enzymes are involved.

It has also been suggested that polyA tails of mRNAs could play a role in the excision of introns. It is possible for a triple helix to be formed with polyA in a region overlapping an intron–exon junction in the ovalbumin gene (Fig. 7.9). This would require some unusual but quite feasible base-pairing between the A and U residues, and between two A residues. The recent isolation of an endoribonuclease that is activated in the presence of polyA lends some support to this idea.

Because the information content of the base sequences at exon–intron junctions is relatively small it is not surprising that splicing is sometimes incorrect. Some types of thalassaemia are known to be due to the use of potential (or cryptic) splice sites within an exon so that the mRNA produced cannot code for a normal globin chain (Chapter 9.2). Similarly, in a strain of rats lacking albumin in their plasma, the gene coding for this protein has a deletion of seven bp just 3' to the donor site of one of its

Fig. 7.8. Pairing between bases at intron–exon junctions and U1 RNA which may play a role in the excision of introns. Part of the U1 RNA is the inner sequence.

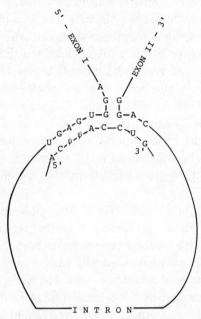

Fig. 7.9. Hypothetical formation of a triple helix at the intron–exon junction with the poly-A tail of the ovalbumin gene. The points of endonucleolytic scission and the hydrogen bonds which can form between the bases are shown. Introns run from the right-hand end of each top sequence to the right-hand end of each bottom sequence.

```
5'                        3'        5'                          3'
U A A A U A A G|G U G A G C C        A A G C U C A G|G U A C A G A
Å Å Å Å Å Å Å Å A A Å Å Å A A A        Å Å A A Å A A A A A Å A A A A A
Ů Ů G Ů Ů G|G A C A Ů Ů A A C        C G G Ů G Ů|G A C Ů Ů A Ů G Ů


U C C U G C C A|G U A A G U U        G A C A A A U G|G U A A G G U
Å A A Å A A A Å A Å Å Å A Å Å        A A A A Å A A A A A A Å A A A
C A Ů A A G|G A C A Ů Ů U C G        A Ů Ů A A G|G A A A Ů U C Ů Ů


U G A C U G A G|G U A U A U G        U U G A G C A G|G U A U G G C
Å A A Å Å Å A Å A A Å A Å Å A A        Å Å A Å A A A Å A A A A Å Å A A A
A A G A A C|G A C C Ů C G Ů Ů        G A G Ů Ů C|G A C G Ů Ů C Ů Ů
3'                        5'        3'                          5'
```

13 introns. This presumably prevents normal splicing which, in turn, leads to loss of the ability to synthesise plasma albumin.

Occasional situations have been found in which alternative splice sites can be used to produce similar, but distinct, proteins. For example, there are two variant forms of the γ-chain of fibrinogen which can arise depending on whether or not the seventh intron in its pre-mRNA is spliced out (Fig. 7.10). Two forms of human growth hormone (Chapter 11.2) also arise by the use of alternative splice sites, and this phenomenon also accounts for the production of two different forms of the immunoglobulins (membrane-bound and soluble) (Chapter 10.3).

It is surprising that there are apparently not more frequent errors in splicing. Perhaps there may be other features of the system, not yet apparent, which confer the high degree of specificity that is required for correct splicing.

Not all eukaryotic genes contain introns. The most notable exceptions are the histone genes. It also seems that introns are rather rare in yeast genes. No satisfactory explanation for these exceptions has been suggested. There has been considerable speculation about why introns have evolved and become such a general feature of eukaryotic genes. The most plausible suggestion is that their presence may lead to a higher rate of evolution by providing opportunities for new gene products to be produced by aberrant or alternative splicing.

Introns can be detected and visualised by electron microscopy (Chapter 3.7).

The rate of mutation in introns is higher than that in exons. This can be shown by comparison of sequences of homologous introns and exons in the corresponding protein in different species, or by hybridisation of germ-line DNA fragments containing the homologous gene from different species. In the latter case the exons will hybridise more strongly with each other than the introns. Introns provide a large pool of non-coding DNA available for mutation without the constraints imposed on coding sequences coding for specific proteins, so the production of new and potentially useful genes may occur.

Fig. 7.10. Alternative splicing in fibrinogen mRNA. The arrows show the 5′ and 3′ splice sites of the seventh intron. If it is spliced out the protein terminates with the methionine residue on the right. If it is not spliced out, it terminates with the lower sequence of amino acids, ending with the proline residue before the first termination codon.

- - CAG GUC - - - N_{30} - - - GGG UAA - - - N_{485} - - - UAG GUU GGA GAC AUG UAA - -

Gln   ↑   ↑Val Gly Asp Met Ter
     Val   Pro Ter

7.10   Post-translational modifications may be required to produce functional proteins

The primary translation products may need further processing to acquire their full functional properties. An mRNA specifies only the amino acid sequence of the polypeptide chain. It is believed that this determines the way in which the molecule will fold into its native secondary and tertiary structure. The folding probably occurs spontaneously so as to lead to a conformation that has the most thermodynamically favourable configuration.

Proteins that are destined for export from the cell are synthetised with an N-terminal extension of about 15–30 amino acid residues, generally known as the *leader sequence* or *signal peptide*. In the middle of this sequence is a stretch of amino acids which have hydrophobic side chains (leucine, isoleucine, valine, phenylalanine, tryptophan), which displays a high affinity for the membranes of the endoplasmic reticulum. The function of this leader sequence is to align the growing polypeptide chain so that it can be drawn through this membrane into a space from which it is ultimately secreted from the cell. Once the leader sequence has passed through the endoplasmic reticulum it is removed by hydrolysis with a proteolytic enzyme located there. Since all mature proteins do not possess the same N-terminal amino acid it is possible that there may be more than one enzyme which performs this task: alternatively a single enzyme may have a fairly low specificity for the amino acid residue at which it acts. The signal peptide is generally cleaved immediately before an amino acid residue with a small side chain.

Other kinds of proteolytic processing may occur, such as removal of an internal portion of the polypeptide to generate a protein with two (or more) polypeptide chains joined by disulphide links. This happens in the case of insulin (Fig. 11.1). A common signal for the action of proteolytic-processing enzymes is a sequence of two adjacent positively charged amino acids (arginine and lysine). Examples of this are given in Chapter 11.

A wide variety of chemical groups can be added to proteins, particularly carbohydrates, since many proteins are glycoproteins. Glycosylation occurs in the Golgi apparatus immediately after removal of the leader sequence. This is often a very complex process in which many glycosyl residues may be added to a protein, some of which are later trimmed off. Certain sequences of amino acids are recognised by the enzymes involved in these reactions (Fig. 7.11). Because glycosylation requires a defined amino acid sequence there are considerable evolutionary constraints if it is a necessary feature of the structure of a protein.

Phosphorylation of serine, threonine and (less frequently) tyrosine residues is another common modification of proteins. This is often used to control the activity of enzymes, and again takes place within specific amino acid sequences containing the residue which is phosphorylated (Fig. 7.11). Phosphorylation is generally a reversible process, with a protein kinase used to phosphorylate the protein and a protein phosphatase to remove the phosphate group.

A less widely studied modification is acylation (generally with an acetyl radicle) of amino groups – either the N-terminal ones, or those on lysine side chains (Fig. 7.4). This reaction changes the electric charge on a protein, and has already been discussed in connection with the histones (Chapter 7.3).

Fig. 7.11. Top: Amino acid sequence recognised by enzymes capable of glycosylating proteins on an asparagine residue. X can be any amino acid.

Bottom: Amino acid sequence recognised by protein phosphokinases. X, Y, Z can be any amino acid residue.

```
            Ser
    Asn X  Thr


Lys Arg X Y Ser Z
       or
Arg Arg X Ser Y
```

8

Repeated sequences and oncogenesis

8.1   Histone genes

Histone genes were among the earliest to be studied because histones are particularly abundant proteins and they are formed at a specific stage in the cell cycle just before cell division takes place. They are usually found in multiple copies in the genome (Table 8.1). In the invertebrate species that have been studied – several sea urchin species and also *Drosophila* – the genes are clustered into tandemly repeated units. However, in vertebrates and in yeasts there do not seem to be regularly repeating units, though the genes are still clustered to some extent.

As a general rule, histone genes contain no introns, nor are their mRNAs polyadenylated, though a few exceptions to both generalisations are known.

8.2   rRNA and tRNA genes

The genes for rRNA are present in multiple copies in eukaryotes (Table 8.2), presumably because of the great demand for ribosomes. In some amphibia and fish there is selective amplification of these rRNA genes during oogenesis so that there may be up to a million copies in a single egg. These amplified genes are found in extrachromosomal circular pieces of DNA.

Oocytes of some species of *Xenopus* may contain up to 20000 copies of the 5s rRNA gene which are encoded separately from the rest of the rRNA genes. These are mostly found in short repeating sequences which also contain a pseudogene separated from the true genes by spacers of varying lengths. This very large complement of 5s rRNA genes is needed to keep pace with the synthesis of the other kinds of rRNA that are transcribed from the genes which are amplified during the very early stages of growth immediately following fertilisation. Later in life different 5s rRNA genes

101

Table 8.1. *Histone gene clusters*

Species	Frequency of repetition	Repeat length (kb)	Organisation
Sea urchin	300–600	5–7	H1 H4 H2B H3 H2A
Drosophila	100	25	H3 H4 H2A H2B H1
Xenopus	20–50	5.8	H3 H4 H2A H2B
			H3 H4 H2A H2B H1 H3 H4
Chicken	10	> 10	variable
Mouse	10–20	5.2	H4
Human	10–20	> 10	variable

Table 8.2. *The numbers of repeated genes in various genomes*

Species	rRNA	5s RNA	tRNA
Yeast	140	150	360
Tetrahymena	200–300	300–800	800–1450
Drosophila	120–240	100–200	600–900
Xenopus	500	500[a]	7000
		20000[b]	
Rat	150	830	6500
Human	50–200	2000	1300

(*a*) in somatic cells; (*b*) in oocytes.

are used, and there are only about 900 copies of these spread throughout the genome.

tRNA genes are also present in multiple copies. In *Xenopus* many of these occur at about 150 sites each about 3200 bp long and containing 8 tRNA genes in tandem. There are also many other tRNA genes scattered throughout the genome, sometimes in clusters, sometimes singly. These unclustered tRNA genes are known as orphons. In mammals, where there are fewer copies of these genes, there is no evidence that they are clustered to any significant extent.

8.3   Repeated sequences

Perhaps as much as 20–30 % of the human genome consists of repetitive sequences of one kind or another. The most abundant of these are known as the *Alu* family because they all possess a site cleaved by the restriction endonuclease *Alu* 1. A few individual ones have been sequenced and, though not identical, they show a strong conservation of sequence. They are about 300 bp long, and there are 300000–500000 copies in each cell: thus they comprise about 3 % of the human genome. They are widely scattered and, on average, occur at intervals of about 5000 bp. They are composed of two repeated sequences of 130 bp, one of which has an insert of 31 bp. Similar structures are found in the rodent genomes, though these only possess a single sequence corresponding to the repeating unit in human *Alu* plus a non-homologous 32 bp insert (Fig. 8.1). Both human and rodent *Alu* sequences have dA runs of up to 40 or so nucleotides at the 3'-end, and both families are flanked by repeated sequences up to 19 bp long. It is not known whether there are related families in other mammals, but there is a little evidence that they may occur in birds.

RNA sequences that hybridise strongly to these *Alu* sequences are found as major constituents of the hnRNA. They are transcribed by RNA polymerase III which uses sequences internal to the DNA it transcribes as promoters. Thus the *Alu* family sequences can be transcribed from any position in the genome. Specific mRNAs have been detected which could have been transcribed from these sequences.

Human 7s RNA (about 300 nucleotides long) has a region which is about 80 % homologous to the *Alu* family. This RNA is used in processing the signal peptide which is present on all secreted proteins when they are first synthesised (Chapter 7.9). A 4.5s RNA found in rodents, but not apparently in humans, also shows considerable homology with rodent *Alu* sequences, but it has no known function.

It is believed that these *Alu* sequences may be moved about the genome in which they occur, but, apart from coding for the 4.5s and 7s RNAs, they do not seem to have any functions. It has been suggested that they might be regarded as 'parasitic' DNA which does not contribute to the phenotype of the organism.

Fig. 8.1. *Alu* sequence in the human (top) and rodent (bottom) genomes. D = Direct repeat at insertion site of the sequence; A = runs of dA nucleotides; H̄ = human specific insert; R = rodent specific insert.

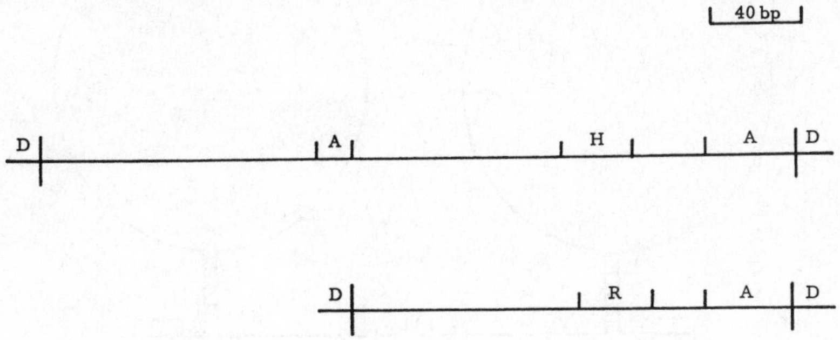

Other less highly repetitious DNA sequences have been characterised in insects (*Drosophila*), yeasts, maize and mammals.

In *Drosophila* three such families have been studied – *copia*, 412 and 297 – and there is evidence for at least 16 other families. Collectively they are known as transposons. Each occurs about 30 times in the genome. The positions at which they are found differ in different strains of fly, and even between individuals of the same strain. To some extent thay seem to be randomly inserted throughout the genome, though a definite preference is shown for certain sites. This widespread distribution has led to the suggestion that they are highly mobile elements which can sometimes take adjacent DNA sequences with them when they are moved around the genome. There has been genetic evidence for this behaviour for many years, and it is gratifying that this can be integrated with evidence obtained from studies at the molecular level.

In yeasts the Ty 1 family which has similar characteristics has been studied extensively.

These repeated sequences are flanked by direct repeats (called 'direct terminal repeats' or 'long terminal repeats') which are usually about 250–500 bp long. At the extreme ends of these direct terminal repeats are much shorter inverted repeats which can form the stem of a stem and loop structure by base pairing (Fig. 8.2). At the sites of integration of these transposons into the genome there is a short repetition (3–5 bp) of the genomic base sequence on each side of the insertion, presumably because a staggered cut is made in the target DNA prior to insertion (Fig. 8.2).

Fig. 8.2. *Copia* element in *Drosophila*. The open bars at each end are part of the regular genome. 5 = repeated pentanucleotide at insertion site: the stem and circle are the long direct repeats (276 bp) at each end of the element.

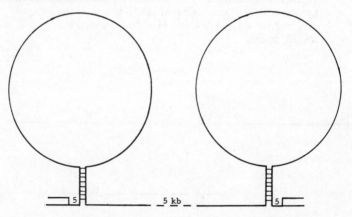

Table 8.3. *Some oncogenes found in retroviruses*

Oncogene symbol	Name of virus	Species where first found	Properties of gene product
v-*src*	Rous sarcoma virus	Chicken	Tyrosine protein kinase
v-*myc*	Avian myelocytomatosis	Chicken	Binds to DNA
v-*erb*	Avian erythroblastosis	Chicken	Homologies with EGF receptor
v-*abl*	Abelson leukaemia virus	Mouse	Tyrosine protein kinase
v-*Ha-ras*[a]	Harvey murine sarcoma	Mouse/rat	Serine protein kinase
v-*Ki-ras*[a]	Kirsten murine sarcoma	Mouse/rat	Serine protein kinase
v-*fes*	Snyder–Theilin feline sarcoma	Cat	Tyrosine protein kinase
v-*sis*	Simian sarcoma	Woolly monkey	Homologies with PDGF

(a) These are two closely related viruses with substantial homology in nucleotide sequence, and appear to be derived from two very similar, if not identical mammalian genes.

A striking characteristic of these transposon-like elements is their close analogy with bacterial transposons on the one hand (Chapter 5.4), and with retroviruses on the other (Table 8.3).

8.4 Retroviruses and oncogenic viruses

A great deal of interest has been aroused in the family of retroviruses because they can be oncogenic – that is to say they may cause cancer. Their genomes consist of a single strand of RNA which is generally fairly short (about 5–9 kb), and contains a very limited number of genes. They are encapsidated in a protein coat which is encoded by two of their genes. The *gag* gene directs the synthesis of its core protein, and the *env* gene codes for the glycoprotein occurring as a spike on the surface of the envelope.

Replication of these viruses is initiated by a reverse transcriptase which is encoded by their *pol* gene. They may also carry oncogenes (genes which cause cancer) in their genome (Fig. 8.3). When cells become infected with such a virus the *pol* gene is transcribed by the host's RNA polymerase II, producing an mRNA which directs the synthesis of reverse transcriptase on the host's own ribosomes. This enzyme then transcribes the viral RNA into a single-stranded DNA molecule. This, in turn, directs the synthesis of a complementary DNA strand making use of the host's DNA polymerase. This double-stranded DNA is then integrated into the host's genome as a provirus. Integration can occur at many sites within the chromosomes with duplication of 4–6 bp of the host DNA sequence at each end of the insertion. The provirus can now be replicated or transcribed in the usual way. This results either in the spread of the viral sequence among the host's cells or in production of new virus particles.

There are interesting structural features at the ends of retrovirus RNA. The 5'-end contains a cap structure (Chapter 7.7) followed by a short sequence of up to 80 bases (R), which is also found at the 3'-end of the genome. This is followed by a sequence of about 100 bases (U5), which precedes the coding portion of the genome. A longer sequence which may contain up to 1000 bases (U3) occurs at the 3'-end of the coding sequences, followed by the R sequence and a polyA tail. When the RNA is transcribed into DNA these end structures are duplicated and rearranged so that each end now has a long terminal repeat (LTR) consisting of U5-R-U3 (Fig. 8.4). This structure is very reminiscent of that found in transposons. The U3 unit in the integrated proviral DNA contains a sequence related to the TATA box, and also a polyadenylation signal (AATAAA) which are presumed to be important for transcription and for poly-A addition to the transcribed RNA respectively.

Oncogenic viruses (Table 8.3), in addition to causing cancer when injected into suitable hosts, often change the growth characteristics of certain types of cells in culture. This latter action is called transformation, and since it can be readily detected, it has been widely used in studies of these viruses.

Oncogenic viruses are believed to have arisen when a retrovirus has integrated very near to certain genes which have critical functions for cell

Fig. 8.3. Top: Retrovirus organisation. *gag*, *pol* and *env* are the viral genes (see text).

Bottom: Retrovirus after reverse transcription and integration into the host genome. Host genes can be incorporated anywhere into this structure.

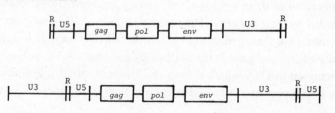

growth. By rare events such a gene can be incorporated into the proviral DNA and thence into the viral RNA genome. It is suggested that this gene is transcribed more efficiently when it is situated near a viral promoter than when it is under the control of the cell's own promoter. Thus, the gene's protein product will be synthesised in larger quantities than usual leading to unrestricted growth (cancer). A number of genes that are normal components of animal genomes and are very similar to the viral oncogenes have been detected. These cellular genes are sometimes known as proto-oncogenes, and are given the same symbol as the viral gene, except that they are preceded by the prefix c-. One very interesting feature of the viral oncogenes compared to their cellular counterparts is that the viral genes lack introns. This suggests that they have arisen by reverse transcription of cellular mRNA. They have also undergone a small number of point mutations. Because of the extensive homologies between the viral and cellular genes it is relatively easy to isolate the latter by hybridisation, and the sequences of several of them have been determined. They contain open reading frames that are capable of encoding proteins, but with one exception, the function of these putative proteins is not yet known. c-*sis* encodes a growth factor that has been found in platelets (Platelet Derived Growth Factor). If this were produced in excessive amounts uncontrolled cellular proliferation might result.

The protein products encoded by several viral oncogenes have been identified and studied. Some of them possess enzymic activity as protein kinases that catalyse the phosphorylation of tyrosine residues. This is a comparatively uncommon reaction, but it may be significant that both Epidermal Growth Factor and insulin (which acts as a growth factor in some circumstances) can both stimulate cellular protein kinases under some conditions. The viral oncogene v-*erbB* probably encodes a protein which is almost identical to the receptor for Epidermal Growth Factor, so the cellular homologue is presumably the true gene for that protein. It is possible that the production of a slightly altered receptor under the direction of the oncogene might generate signals which would cause increased cellular proliferation.

In cells cultured from a number of different carcinomas the normal human homologue of the rat Harvey sarcoma virus oncogene has suffered point mutations in a specific codon which has also mutated in related viral oncogenes (*Ha-ras* and *Ki-ras*) (Table 8.4). The normal human gene encodes glycine at this position, and mutations to codons for serine, cysteine, valine or arginine have taken place in three different carcinoma cell lines and in the viral oncogene. These seemingly small changes could have large effects on the secondary and tertiary structure of the protein that is

Table 8.4. *Mutations in c-Ha-ras found in v-Ha-ras and various human carcinomas*

Carcinoma or oncogene	Codon 12	Amino acid encoded
c-*Ha-ras*	GGT	Glycine
Bladder carcinoma	GTC	Valine
Colon carcinoma	GTT	Valine
Lung carcinoma	TGT	Cysteine
v-*Ha-ras*	AGA	Arginine
v-*Ki-ras*	AGT	Serine
Mammary carcinoma[a]	GAA	Glutamate

(*a*) Induced in the rat by treatment with nitroso-methyl urea.

encoded. It may also be significant that a second mutation in the viral genes leads to the incorporation of a threonine residue in place of an alanine residue, and this threonine is phosphorylated in the viral protein product. Again, this could have a very marked effect on the structure and properties of the protein. These fascinating discoveries suggest that cancer may arise not only by overproduction of some specific cellular proteins normally present in low concentrations, but also by alteration in the structure and function of some key proteins.

8.5   Chromosomal alterations in cancer
It is becoming increasingly clear that neoplasias (cancerous conditions) are very frequently associated with various chromosomal abnormalities, particularly deletions of parts of chromosomes, and translocations in which portions of non-homologous chromosomes are exchanged. Deletions are generally, though not exclusively, associated with solid tumours, while translocations are more usually found in leukemias and lymphomas where there are cancerous proliferations of various kinds of leukocytes. These translocations are sometimes near the known sites of proto-oncogenes (Table 8.5).

A very early discovery in this field was the so-called Philadelphia chromosome, found in many patients with certain types of leukemia. This results from a translocation between chromosomes 9 and 22 (symbolised t(9:22)). The break point on the former is near the proto-oncogene c-*abl*, and that on chromosome 22 is close to the gene for the immunoglobulin λ chain.

One of the most studied conditions is Burkitt's lymphoma in which a part of the end of chromosome 8 has been exchanged with a sequence near the end of chromosome 14 (Fig. 8.4). The segment from chromosome 8 contains the proto-oncogene c-*myc*, and it is translocated into the region

Table 8.5. *Chromosomal translocations which give rise to cancers*

Nature of disease	Chromosomal translocation	Proto-oncogene near to position of translocation
Leukaemia	t(9:22)	c-abl
Leukaemia	t(8:21)	c-mos
Burkitt's lymphoma	t(8:14)	c-myc
Burkitt's lymphoma	t(8:22)	c-myc

of chromosome 14 where the cluster of H-chain immunoglobulin genes is found (Chapter 10). These translocations are not all found at exactly the same point in different patients. In a few cases of Burkitt's lymphoma there is translocation of the same segment of chromosome 8 to sites on chromosomes 2 or 22 where the L-chains of the immunoglobulins are encoded. Analogous translocations of the chromosomal sites for c-*myc* (chromosome 15) and that for the H chain genes (chromosome 12) are found in some mouse plasmacytomas (see Chapter 10). Increased amounts of the RNA transcripts of the c-*myc* gene are frequently, though not invariably, found in cells where these translocations have occurred. The significance of this is not apparent, since we do not know the normal function of the protein encoded by the c-*myc* gene. If, as seems possible, the c-*myc* gene product is concerned with cell growth in some way, over-production could obviously cause increased cell proliferation and

Fig. 8.4. Structure of human c-*myc*, and part of the human Ig μ gene. The other structures are of various translocations between these two loci that have been isolated from cells of patients with Burkitt's lymphoma. The c-*myc* exons (E1–3) are shown below the line: the Ig exons above the line.

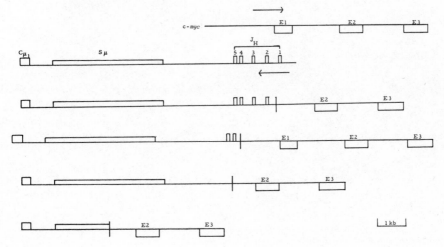

a cancerous condition. The detachment of the c-*myc* gene from its usual chromosomal position could result in loss of some control elements so that it is expressed in greater amounts than usual in its new chromosomal environment. The chromosomal deletions observed in other types of cancer could also lead to loss of control over the normal expression of genes concerned with normal growth.

Several other sites where translocation occurs in neoplasias are near oncogenes, so that similar mechanisms may operate in these cases. A few neoplasias are associated with the condition of trisomy, in which there are three copies of a particular chromosome instead of the usual two. This could result in over-expression of a gene giving rise to increased concentrations of a protein factor promoting uncontrolled cellular proliferation. In the mouse, trisomy of chromosome 15 which bears the c-*myc* gene, is frequently found in leukemia.

Cytogenetic studies have identified so-called fragile sites at which chromosomes are particularly likely to become broken, and some of these are the sites at which translocations frequently take place.

9
Haemoglobin

9.1   Genes for globins are found in two clusters

Haemoglobin consists of four polypeptide chains of two different though similar types, which are folded round each other in an orderly and compact fashion. The two types of chain, designated α and β, are present in equal amounts. They show a very substantial degree of homology – in humans 43 % of the residues are identical. During manufacture of haemoglobin these polypeptides (known as globins) are synthesised first and then each one binds a molecule of haem very firmly.

Several different β-like globins are synthesised at different stages of human life (Table 9.1). In the early embryo when haemoglobin is synthesised in the yolk sac, the ϵ chain is made. Later, synthesis switches to the foetal liver and two forms of the γ chain are made. In one of these an alanine residue is found at a position where the other contains glycine, and they are therefore known as $^A\gamma$ and $^G\gamma$ respectively. Finally, just before birth, β-chain synthesis commences, along with that of a very small amount of the nearly identical δ chain which differs from the β chain in only 10 residues.

The δ gene is transcribed at a much lower efficiency than the β gene so that δ chain mRNA is produced in much smaller amounts than β chain mRNA. There are sequence differences in the 5'-flanking regions of the δ and β genes which probably account for this (Fig. 9.1), but it is not yet possible to pinpoint the particularly crucial differences.

There is also a developmentally regulated family of α globins. In the yolk sac embryo ζ chain is synthesised, while both the foetus and the adult produce the α chain. Switching from ζ to α and from ϵ to γ is not precisely co-ordinated since small amounts of the tetramers $\alpha_2\epsilon_2$ and $\zeta_2\gamma_2$ are found during this transition.

These diverse forms of similar molecules have arisen during the course

Table 9.1. *Haemoglobins formed at different stages of human development*

β-like chain	α-like chain	Haemoglobin present	Stage at which synthesised
ε	ζ	$\zeta_2\ \varepsilon_2$	early embryo
$^A\gamma\ ^G\gamma$	α	$\left.\begin{array}{l}\alpha_2\ \varepsilon_2 \\ \zeta_2\ \gamma_2 \\ \alpha_2\ \gamma_2\end{array}\right\}$	foetus
β δ	α	$\left.\begin{array}{l}\alpha_2\ \beta_2 \\ \alpha_2\ \delta_2\end{array}\right\}$	end of foetal development onwards

of evolution because there must have been duplications of genes coding for primordial forms of these globins. Subsequent mutations have given rise to the present-day forms which have some survival value, so the mutations have become fixed.

The DNA containing the globin genes occurs in two clusters (Fig. 9.2). The β-like genes, on chromosome 11, are spaced out over a length of about 60 kbp, and are highly homologous with each other. They all possess a common pattern of three exons with two introns in exactly corresponding sites (Table 9.2). The α-like gene cluster, on chromosome 16, is shorter (rather less than 30 kbp), and again the individual genes have a common pattern of three exons and two introns, though the introns are

Fig. 9.1. Critical bases and spacers in the 5′-flanking region of some of the globin genes. The numbers over the lines designate the number of bases in these positions. del = deletion. AC is the cap site, where transcription starts: ATG is the initiation codon, which has mutated in ζ_2.

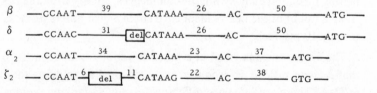

Fig. 9.2. Maps of the β- and α-globin gene clusters, on chromosomes 11 and 16 respectively, in the human genome.

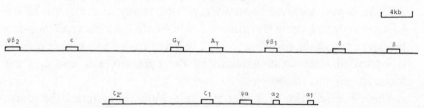

Table 9.2. *Organisation of the α- and β-globin genes in the human genome*

	Cap site to initiation codon	First exon	First intron	Second exon	Second intron	Third exon
α	38	96	127	204	133	126
β	54	92	138	223	889	123

shorter than those in the β-like genes. In each cluster the genes are arranged from 5' to 3' in the same order as their temporal expression.

Gene duplications occurring recently in evolutionary history are likely to have given rise to the $^A\gamma$ and $^G\gamma$ pair, and also the δ and β pair. There is no evidence for a duplication of the ε gene, though there is ample space for a duplicated gene 3' to the existing functional one. There has also been duplication of the α-like ζ gene, and triplication of the α gene itself. In the latter case, two of the genes present are identical, while the third α gene and one of the ζ genes have undergone mutations resulting in the production of non-functional pseudo-genes.

Pseudo-genes have DNA sequences which are highly homologous to those of transcribed genes, but which do not direct the synthesis of any functional polypeptides. Many pseudo-genes have been found, including two in the β-globin gene cluster. Pseudo-genes may contain insertions or deletions which cause premature chain termination of transcription products, or there may be mutations at the intron–exon junctions so that correct splicing cannot occur.

The pseudo α gene has mutations at the splice sites of both introns which make it unlikely that correct splicing could take place. Even if it did, a deletion of 23 bases in the second exon throws the reading frame out of phase so that a termination codon appears in phase. The pseudo ζ gene has a mutation giving rise to an in-phase termination codon in the first exon. It also has considerably longer introns than the true ζ gene. Introns in both ζ genes have numerous repeated sequences.

The globin genes of other species, such as the mouse and chicken, have also been studied and, in general, present a similar picture of developmentally regulated genes.

9.2 Thalassaemias

These are a collection of diseases in which the synthesis of one type of the globin chains is either reduced or absent. They are called α or β thalassaemias depending on the chain whose synthesis is deficient.

These defects result in the production of a relative excess of the chain whose synthesis is not affected, and unusual tetramers (i.e. β_4 or α_4) are found which tend to cause destruction of erythrocytes or red cell progenitors. Consequently anaemia develops and when the condition is homozygous, affected individuals are likely to die early unless treated with repeated blood transfusions. These diseases are widespread in the warm regions of the world. The heterozygotes have an advantage over the normal population because they are more resistant to the malarial parasite.

Examination of the globin genes of a number of patients has shown that many different mutational events can give rise to the thalassaemias. In some cases, point mutations in the β chain gene have converted amino acid codons to termination codons (e.g. AAG (Lys) $\rightarrow$ UAG at position 14, or CAG (Gln) $\rightarrow$ UAG at position 39). mRNAs produced from these mutated genes will obviously direct the synthesis of shortened proteins, which do not pair correctly with the α chain and are rapidly degraded.

In another type, common in SE Asia, a mutation has occurred in the termination codon of the α chain (UAA $\rightarrow$ CAA). Since another in-phase termination codon is reached 90 bp downstream the mRNA that is produced codes for a longer than usual α chain. In fact only very small amounts of this elongated chain are produced owing to the marked instability of the abnormal mRNA.

Other types result from mutations in or near the splice sites at which the introns are normally removed. A point mutation in codon 24 in the β chain leads to a pre-mRNA which is spliced incorrectly. This causes premature termination since the splicing generates a reading frame different

Fig. 9.3. Mutations (numbered) leading to various forms of β-thalassaemias. The sequence of the normal β gene is at the top with the amino acids which it encodes and their positions in the peptide chain shown above. Mutated nucleotides are shown below together with the amino acids specified by the mutated genes. d = deleted nucleotide.

```
      14              24   25  26   27   28   29  30                                                             31    32
      Lys             Gly  Gly Glu  Ala  Leu  Gly Ar. . . . . . . . . . . . . . . . . . . . . . . . . . . . . .  g Leu Leu
      AAG - - - - GGT GGT GAG GCC CTG GGC AG GTGG - - - - - - - GGTCTATTTTCCCACCCTTAG G CTG CTG - - -

      UAG             GGA G GTGAG - - - - - - - - - - - - - - - - - - - - - - - - - - - - - - - AG GC TGC TGG
      Ter             Gly                                                                           Gly Cys Trp

       1               2                                                       AG T CTA TTT TCC CAC CCT TAG
                                                                            3  Leu Phe Ser His Pro Ter

            39  40   41  42   43            57   58   59   60  61
            Gln Arg  Phe Phe  Glu           Asn  Pro  Lys  Val Lys
          - - - CAG AGG TTC TTT GAG - - - - - - AAC CCT AAC GTG AAG - - - - - - - - - GT - - - - - -

       2  - - - - - - - - - - - - - - - - - - - CAA CCC TAA                                        AT
                                                Gln Pro Ter                                         no
       4  - - - UAG                                                                                 splice
              Ter                                                                                   6

       5  - - - CAG AGG ddd d TTG AGT - - - - - - - - - - - -ACG TGA
              Gln Arg          Leu Ser                       Arg Ter
```

from that used in the normal mRNA. In addition, the pre-mRNA is processed more slowly. A mutation in the first intron has been discovered which generates a new acceptor splice site 5' to the usual one, and this also leads to premature termination because the reading frame is now out of phase. In other thalassaemias there are deletions of either the whole or 3'-portions of the β gene which lead to loss of ability to synthesise normal globin chains. Some of these mutations leading to β-thalassaemia are shown in Fig. 9.3.

Some very interesting cases have been studied in which the affected individuals produce chains containing the N-terminal portion of the δ chain coupled to the C-terminal portion of the β chain. Analysis of the DNA shows that crossing over has occurred with fusion of the corresponding parts of the two genes (Fig. 9.4). This is believed to have occurred at meiosis. These hybrid δ-β chains are synthesised at a slower rate than the α chains so that an excess of the latter appears and a mild thalassaemia results. This must be because of the relatively inefficient initiation of transcription of the δ-β chain gene, particularly in homozygotes. These hybrid chains are known as Lepore chains, named after the patient in whom they were originally discovered. There are several chemically distinct Lepore haemoglobins in which the crossover has occurred in different positions. Anti-Lepore chains are also known with the reverse crossover, i.e. the N-terminal part of the β chain coupled to the C-terminal part of the δ chain. These cause hardly any symptoms of thalassaemia because there should be more efficient initiation of these particular hybrid genes.

Fig. 9.4. Formation of Lepore and anti-Lepore globins by crossing over between the β and δ genes. The normal genes are shown in the middle, and may pair incorrectly because of their extensive homology.

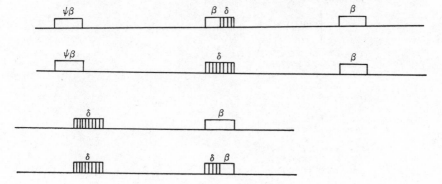

9.3   Other mutations

There are many forms of mutant haemoglobin in which single amino acid substitutions have occurred in either α or β chains. The vast majority of these are due to single base changes in one of the codons in the DNA. Some of these abnormal haemoglobins have undesirable properties, and may lead to various forms of anaemia. The best known occurs in the condition of sickle cell anaemia in which there is a change of glutamate to valine in position 6 of the β chain, caused by a point mutation of the codon GAG to GTG. This small change in amino acid sequence leads to a profound difference in the properties of the haemoglobin that is present in affected individuals. The dexoygenated form precipitates out very readily and may damage the red cell membrane leading to haemolysis and subsequent anaemia. This condition is quite widespread, especially among negroes, since the heterozygotes again have increased resistance to the malarial parasite. Anaemia is much more serious in the homozygotes.

Several hundred other single amino acid substitutions are known, and very many of them do not give rise to any observable pathological effects. However, if a substitution occurs in a position of the globin which is critical for its functions (e.g. involved in binding haem, diphosphoglycerate, or with another subunit) deleterious effects may be observed. In a few cases one or more amino acid residues are deleted. It is likely that these have occurred by unequal crossing over since they are found at sites

Fig. 9.5. Deletion of bases in β-globin Niteroi. This is believed to have occurred by mis-pairing of complementary bases following endonucleolytic chain scission and reformation.

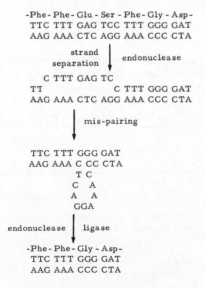

where there are repeats of several bases in the sequence of the DNA. At these sites mispairing of adjacent sequences after a chromosomal break, followed by degradation of an unmatched sequence, could occur (Fig. 9.5).

Finally, there are a number of haemoglobins with C-terminal extensions of either α or β chains. These have arisen either as a result of a mutation in the termination codon (see earlier), or because of a deletion or insertion near this site. Deletions or insertions occurring in the α gene (Wayne) or in the β gene (Cranston) have lead to frame-shift mutations which bring into use new termination codons several residues 3' to the usual ones (Fig. 9.6). These particular variants do not seem to have any undesirable consequences.

9.4   Prenatal diagnosis of anaemias

Prenatal diagnosis of genetically determined forms of anaemia is now technically feasible. Foetal cells can be obtained by amniocentesis and grown on to provide enough DNA for mapping by restriction endonucleases. This can often show up the presence of mutations leading to the expression of thalassaemic or other deleterious genetic changes (see Chapter 3.7). Parents can then be offered the chance of termination of a pregnancy which would result in the birth of a severely handicapped child. Screening of large populations in areas where thalassaemia or sickle cell anaemia is common may not at present be economically practicable, but this situation could change as simpler techniques are developed for this purpose.

Fig. 9.6. Deletion and insertion of bases near the termination codon leading to lengthened α chains (Wayne) and β chains (Cranston).

αCHAIN	Ser UCC	Lys AAA	Tyr UAC	Arg CGU	Ter UAA		
WAYNE	UCC Ser	AAU Asn	ACC Thr	GUU Val	AAG Lys	CUG Leu	GAG Glu
	CCU Pro	CGG Arg	UAG Ter				
βCHAIN	Lys AAG	Tyr UAU	His CAC	Ter UAA			
CRANSTON	AAG Lys	AGU Ser	AUC Ile	ACU Thr	AAG Lys	CUG Leu	GCU Ala
	UUC Phe	UUG Leu	CUU Leu	UCC Ser	AAU Asn	UUC Phe	UAU Tyr
	UAA Ter						

10

Proteins of the immune system

10.1   Immunoglobulins consist of H and L chains

Immunoglobulins (Ig), which are antibodies, are proteins consisting of four peptide chains – two identical light (L) chains of about 220 amino acid residues, and two identical heavy (H) chains of about 450–600 amino acid residues. Each L chain consists of two *domains* of approximately equal size. The N-terminal domain is variable in amino acid sequence, and is different in every individual L chain that has sequenced. The C-terminal domain is constant in sequence, though there are two types which exhibit a high degree of hololology. They are known as κ and λ, and either one (but not both) may be found in one Ig molecule.

The H chain contains four or five domains, each with about 110 residues. Again, the N-terminal domain is variable in sequence, and participates with the N-terminal variable domain of the L chain in forming the antigen-binding site, of which there are two per molecule of Ig (Fig. 10.1). A short region, generally about 20 amino acid residues long, between the second and third domains, is called the hinge region. It is devoid of secondary structure, and therefore very flexible. There are five types of H chain – γ, α, μ, δ, ε – whose constant portions are quite distinct though homologous (but μ and ε have an extra domain and no hinge region). There are subclasses of γ and α chains. Intact Ig molecules are identified by suffixes of Latin letters corresponding to the Greek letter of the H chain (G, A, M, D, E). Variable and constant domains of the L chains are designated Vκ or Vλ and Cκ or Cλ. Corresponding designations of the domains of the H chains, starting from the N-terminal end of the molecule are V_H, C_{H1}, C_{H2} etc. The chromosomal location of the L and H chain genes in mouse and humans is shown in Table 10.1.

The constant domains of all H chains and both classes of L chain are recognisably homologous, and it is also possible to discern weak homology

118

Table 10.1. *The chromosomal location of immunoglobulin genes*

Protein chain	Gene symbol	Mouse chromosome number	Human chromosome number
H	Igh	12	14
λ	Igl	16	22
κ	Igk	6	2

of these domains with the V_H and L_H domains. This suggests that the present-day molecules have evolved by repeated duplications and mutations from a primordial Ig-like domain.

When an animal starts producing antibodies after immunisation, the first ones to appear in the plasma are IgM, but as the immune response runs its course there is generally a switch to other classes (IgG, IgA etc.) which retain the original antigen-binding specificity, and therefore the same V domains.

Antibodies are highly specific molecules in that each one will only recognise and bind to one (or a very limited number of) chemically defined groupings in an antigen molecule. The total number of antigens which can be recognised is extremely large – estimates of the order of 10^5 to more than 10^6 different specificities seem quite reasonable. A central problem of immunology is to explain how so many antibody molecules, all of the same general pattern, can be synthesised with different specificities, which are presumably manifested through variations in the amino acid sequence.

Fig. 10.1. Structure of human immunoglobulin IgGl. The L chains are on the outside, joined to the H chains by disulphide bonds. The H chains are also joined together by a disulphide bond. The dark bands are the hypervariable regions. V_L and V_H are the variable domains: C_L, C_{H1}, C_{H2}, C_{H3} are the constant domains: h is the hinge region. There are two antigen-binding sites, each made up of the hypervariable region of one H and one L chain.

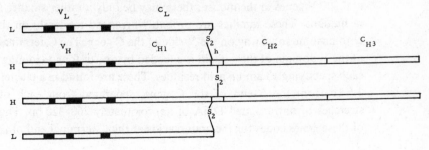

Detailed studies of both H and L chains have shown that the sequence variability is mainly confined to three sequences, each containing from 5 to 16 amino acid residues. These regions are known as *hypervariable* or complementarity-determining regions, since they are situated close together in the three-dimensional structure of Igs, where they form a pocket which is the combining site for the antigen. Intervening stretches of the peptide chain are referred to as framework regions, and hold the complementarity-determining regions in the appropriate configuration for antigen binding.

Igs are synthesised in plasma cells which are the end products of the differentiation of B-lymphocytes. In the adult animal there will be a very large number of clones of lymphocytes and the plasma cells descended from them. The cells of each such clone synthesise Igs possessing just one of the many possible variable domains of both H and L chains. Individual plasma cells can be caused to proliferate in mice by injection of certain mineral oils, which produce tumours called plasmacytomas. These consist of clones of cells derived from one single cell and therefore produce a single homogeneous specific type of antibody. In contrast, the normal Ig fraction of plasma contains a highly heterogeneous population of many individual Ig molecules produced by very many clones of plasma cells. Plasmacytomas can be cultured through many generations of mice by repeated inoculations of cells from affected animals. This condition is analogous to the disease of myeolomatosis in humans. Studies of the proteins produced in this condition have given us much of our knowledge of the detailed structure of Igs.

10.2   L chain genes

The V and C parts of the light chains are encoded by separate exons, and a third exon encodes most of the leader sequence (Fig. 10.2). Amino acid sequence analyses of a number of L chains suggests that there are likely to be families of V genes, and hybridisation studies of various cDNA probes to restriction fragments of genomic DNA have confirmed this. The best estimate is that there may be a total number of between 90 and 300 V genes in the mouse: there may be only a rather smaller number in humans. These families are probably arranged tandemly on the same chromosome some way on the 5'-side of the C gene. The C-terminal region of the V domain of the κ chain is encoded by one of four J (joining) genes, each specifying 13 amino acid residues. These are found in a cluster, about 2.5 kb from the 5'-end of the C gene, separated from each other by stretches of untranslated DNA of approximately 260–320 bp. The 5'-end of these genes codes for the amino acids at the C-terminal end of the third hypervariable region. Sequence analysis of the DNA encompassing the four J genes shows the presence of a putative fifth J gene, but the amino acid sequence corresponding to this has never been observed in any L chain that has been sequenced. There has been a mutation on the 5'-side of this gene such that it is unlikely (or impossible) for it to be combined with 3'-end of any V gene. The exact location of the V genes with respect to the J cluster is unknown. It is probable that any of the Vκ genes can be combined with any one of the four J genes, so the total number of κ-type L chains in the mouse is probably of the order of 400–1200.

The mouse produces very few λ chains, but there are three distinct types, and the arrangement of their genes is rather different from that of the κ chain genes. There are only two Vλ region genes; one of these is linked to either Jλ$_3$ and Cλ$_3$ or a Jλ$_1$ and Cλ$_1$ pair, while the other is linked to a Jλ$_2$ and Cλ$_2$ pair, and also to a non-functional Jλ$_4$ and Cλ$_4$ pair (Fig. 10.3). There are many more Vλ genes in humans, and at least six very similar Cλ genes, which are spaced out at intervals of about 5 kb

Fig. 10.2. The structure of a rearranged λ-gene. The boxes show the exons.

|_ 200 bp _|

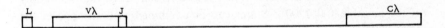

Fig. 10.3. Germ-line organisation of κ chain gene (top), and λ chain gene (bottom) in the mouse. There is a second similar cluster of λ chain genes. The distances between exons is not known where the lines are broken.

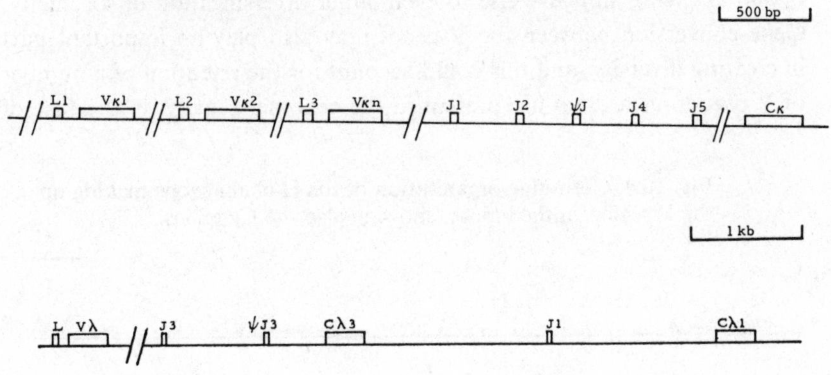

on a 50 kb stretch of DNA. Their mode of linkage to the Jλ and Vλ genes is not known at present.

10.3   H chain genes

The structures of the H chain genes are a good deal more complex, in keeping with the larger size of the H chain. Most of the signal peptide is again encoded by a separate exon. This is followed by bases coding for most of the V domain, which stops near the N-terminal end of the third hypervariable region. Part of this region is encoded by a small exon containing up to 50 bases, called the D (diversity) gene. There are a number of alternative D genes which encode sequences in length from 6 to 17 amino acid residues. Clusters of 11 D genes (mouse) and 4 D genes (humans)) have been detected, but there are probably additional D genes, particularly in humans. Finally, the V domain of the H chain is completed by one of four functional J genes (five in humans), analogous to those involved in the formation of the complete Vκ genes. One D gene is found about 700 bp from the 5′-end of the J region, while 11 others are clustered over a length of 60 kb further in the 5′ direction (Fig. 10.4). There are believed to be about 100–200 V_H genes in the mouse. Thus a total of $4 \times 11 \times (100–200) = 4400–8800$ is a reasonable estimate of the total number of possible H chains that can be formed in this species.

The amazing diversity of antibody specificities can be largely accounted for by the various combinatorial patterns that go to make the rearranged genes of the variable portions of the L and H chains. In the mouse any of the possible 400–1200 κ chains can be combined with any one of the 4400–8800 H chains. Since, as far as is known, any L chain can associate with any H chain to produce antibodies with differing specificities, this gives a potential total which could be as high as 10^7. This is achieved with a fairly modest utilisation of only a small fraction of the genome.

It is also probable that somatic mutations in the variable portions of H and L chains may give rise to even larger diversification of specificity. Gene conversion between the V genes may also play an important part in creating diversity, and this could account for the retention of a number of V pseudo-genes that are present in the genome. Finally, in the case of

Fig. 10.4. Germ-line organisation of the H chain exons making up the V region in the mouse, showing also the Cμ exons.

└ 1 kb┘

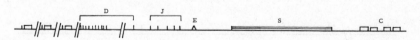

H chain genes, insertions and deletions may arise during the combination of D and J_H elements and, possibly, between V_H and D elements. Thus, both structures in the germ-line, and somatic mutations (non-heritable) can contribute to the amazing diversity of antibodies.

The coding region for the C_{H1} domain of the μ chain is located about 7.6 kb from the 3'-end of the last J gene. The other three C_H domains of the μ chain (which has no hinge region) are encoded by three separate exons which are tandemly arranged on the chromosome with fairly short introns between each other. Two forms of the μ chain exist which have different sequences at the C-terminus. $μ_m$ is membrane-bound with a predominantly hydrophobic C-terminal sequence. This part of the molecule anchors it to the cell membrane where it acts as an antigen receptor. When a cell encounters an antigen which can bind to it this somehow triggers the synthesis of $μ_S$, the other kind of μ-chain which is part of a soluble IgM molecule which is secreted from the cell. It has a short C-terminal extension to C_{H4} consisting predominantly of hydrophilic amino acid residues. In the genome the DNA sequence coding for the $μ_S$-C-terminal amino acid sequence is found directly adjacent to the 3'-end of the DNA coding for C_{H4} (Fig. 10.5). The C-terminal portion of $μ_m$ is encoded in two exons, the first of which is situated about 1800 bp on the

Fig. 10.5. Arrangement of the exons of the several classes of C region genes in the mouse. The α gene probably has exons coding for a membrane-binding peptide. The high boxes show the exons; the low ones the untranslated regions.

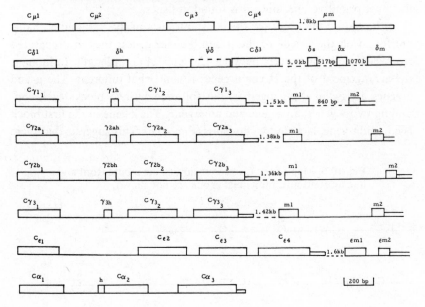

3'-side of that coding for μ_s. The second of these exons contains only two codons and a termination codon. Both the μ_s exons and the second μ_m exon are followed by untranslated bases including the AATAAA sequence just to the 5'-side of the poly-adenylation site. It has been suggested that the choice between producing μ_m and μ_s depends on which poly-A site is used, since it is believed that poly-A addition precedes mRNA processing and splicing.

The δ gene is located about 2.5 kb on the 3'-side of the μ gene, and contains three exons for the major part of the constant region. One of these codes for a longer hinge region than is found in other classes of H chains. A sequence of bases with homology to the second domain of other Ig classes ($\psi C\delta 2$) is spliced out of the primary transcript, so there is no such domain in the protein. 3' to the gene for the third domain are two short exons encoding alternative C-terminal ends of secreted forms of the protein (δs and δx). Further downstream still, there are two exons encoding the amino acids of the membrane-bound form of the δ chain.

The genes for the constant regions of the γ subclasses are fairly similar to each other, containing three exons for the major domains, an exon for the hinge region, and downstream exons for the membrane-bound forms of these proteins. The ε gene lacks an exon for the hinge region, but has exons for four domains, while the α chain gene has only three exons since the coding sequence for the hinge is already joined to the 5'-end of the exon for the second domain. Thus there has been considerable diversification during the evolution of the various classes of Igs. IgM is probably the most primitive one, since it is found in fishes.

In the mouse the genes for the constant parts of the γ, ε and α chains are found on the 3'-side of the μ and δ chain genes: they are separated from each other by spacers between 12.5 and 24 kb in length. In humans the arrangement of the H chain genes is somewhat different. The μ and δ genes are again close together, but followed by two blocks each containing two γ genes, an ε gene and an α gene. The ε gene in the first block is a pseudo-gene, lacking two domains (Fig. 10.6). This suggests that there

Fig. 10.6. Map of the C_H gene region in the mouse (top) and human (bottom). Individual exons are not shown.

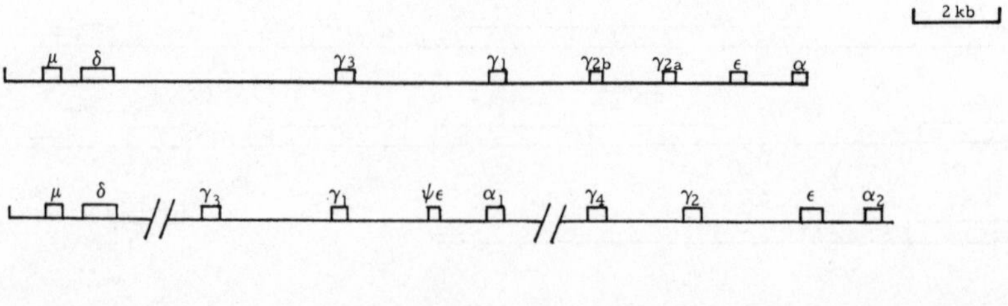

has been a large duplication of that part of the genome originally containing one (or possibly two) γ genes, an ε gene and an α gene during the descent of the human race, while in the lineage leading to the modern mouse it is only the γ genes that have been duplicated. In humans there is also an ε gene on another chromosome which lacks introns. It has presumably arisen by the action of reverse transcriptase on mRNA followed by integration of the cDNA into the genome. In addition there is a pseudo-gene which is closely related to the γ₁ and γ₃ genes, and is probably located near them.

10.4 DNA processing is employed during the course of the immune response

At least three types of nucleic acid processing are involved in the expression of Ig genes. First there is joining of V and J genes (and D genes in the case of the H chains). These genes are widely separated, and become juxtaposed in Ig-producing cells as the result of rearrangements of DNA, brought about by the pairing of complementary sequences located at the 3′-end of the V exons, at the 5′-end of the J exons, and flanking both ends of the D exons. A highly conserved heptamer (consensus sequence CACTGTG) is found to the 5′-end of all known D and J genes, and either 12 or 23 bases further upstream is a nonamer (GGTTTTTGT). Complementary sequences are present in the corresponding positions 3′ to several V genes that have been sequenced and to the D genes (Fig. 10.7). These sequences could readily pair and provide sites for an endonuclease to remove a looped-out intervening sequence. The significance of the gap of 12 or 23 bases between the heptamer and nonamer is that this corresponds to just over either one or two turns of the DNA helix so that these two sites would be on the same face of the molecule and reasonably close to each other. Transcription of this sort of structure from L chain genes yields

Fig. 10.7. VDJ joining. 9 and 7 represent the highly conserved nonamers and heptamers which can pair and align the DNA for the process of splicing.

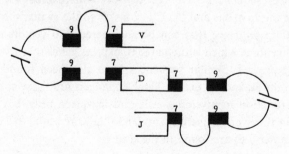

a pre-mRNA in which there is a small intron near the end of the sequence coding for the leader peptide, and a much larger one between the 3'-end of the J gene and 5'-end of the C gene. These introns are excised in the usual way. Thus the final production of the mature mRNA requires processing at the level of RNA rather than DNA.

In the case of the H chains there may also be rearrangements of the DNA in switching from the synthesis of one class of H chain to another. There are several switch sites about 4–5 kb on the 5'-side of the Cμ gene in the untranslated stretch between it and the J_H cluster (labelled S in Fig. 10.4). Different sites appear to be used for joining to the Cγ, Cε, and Cα genes, and there do not seem to be any obvious homologies in these base sequences. The sites so far identified are clustered just to the 5'-end of a stretch of 2.6 kb which consists almost entirely of two highly conserved pentamers (GGGGT and GAGCT).

Parts of the switch sites 5' to the Cγ, Cε and Cα chain genes have also been sequenced. There are highly conserved tandem repeats of 49–80 bases in their vicinity, containing the two pentameric sequences that also occur in the μ switch sites. These repeated base sequences are most probably ¡nvolved in the recognition by processing enzymes which will be activated when switching from the synthesis of one class of Ig to another occurs.

Tandem repeated sequences are known to increase the chance of recombination of DNA in bacteria, so they may also be used in switching from the production of one type of H chain to another. This could explain why deletions of part or all of the repetitive sequences 5' to the μ gene often occur during cloning of this stretch of DNA.

There are no switch sites on the 5'-side of the Cδ gene which is situated only 2.5 kb to the 3'-end of the Cμ gene. It is believed that simultaneous expression of μ and δ chains occurs before B-lymphocytes differentiate into plasma cells. Alternative modes of processing a pre-mRNA could lead to the production of either μ or δ chains.

10.5   Enhancers

Enhancer sequences have recently been discovered within the intron between the J_H exons and the Cμ exons, 5' to the switch sites. They reside in a 140 bp sequence that can be engineered into vectors derived from the SV40 genome which is then able to express genes that would not otherwise be expressed. Similar sequences have also been found in the intron between the Jκ and Cκ exons. These enhancers are tissue specific in that they will function in myeloma cells and in spleen cells, but not in fibroblasts or in the liver. Sequences similar to the core sequences of viral enhancers (Chapter 7.6) are found in these sites.

It is suggested that these enhancers are required for the expression of rearranged H and L chain genes. There are promoter sequences 5' to the V genes but they may only be able to function if they have been brought near enough to the enhancers situated downstream from them in the rearranged gene. Thus any un-rearranged genes are not able to be transcribed.

10.6 Allelic exclusion

Since all somatic cells possess pairs of homologous chromosomes, it might be expected that a fully differentiated lymphocyte would produce two types of Ig – one programmed by the genes on each chromosome. However, this never seems to happen, and only one of the pair of chromosomes directs the synthesis of Ig. This phenomenon is known as allelic exclusion, and has been investigated extensively. In some cases (more common with L chain genes) only one copy of the chromosomal DNA has been rearranged to contain the complete L chain gene ready for transcription. In other cases (more frequent with H chain genes) both copies are rearranged, but one undergoes an abortive rearrangement leading to inability to produce a functional polypeptide. In some cases proteins are secreted that are presumably products of mis-arranged genes, in the sense that they lack those amino acid sequences that are normally coded for by a separate exon.

10.7 The major histocompatibility complex

When a tissue (other than erythrocytes) is transplanted from one individual to another the graft will generally not survive for more than a few days. This is because of the presence of certain integral membrane proteins found on the surface of almost all nucleated cells. They are encoded by highly polymorphic genes which allow 'self' to be distinguished from 'non-self'. These transplantation antigens control, in various ways, co-operation between different types of lymphocytes in the immune system when it has been stimulated by molecules foreign to the organism. Their discovery, through work on tissue transplantation, was a serendipitous accident. Because of their major clinical importance they have been investigated widely in mice (as a convenient model system), and in humans. The antigens are part of the Major Histocompatibility Complex (MHC) which is encoded on chromosome 6 in humans. In the mouse this is known as the H-2 complex, and is encoded on chromosome 17 (Fig. 10.8). These complexes extend over two to four million bp, and they contain genes encoding two distinct groups of cell-surface antigens as well as some of the proteins of the complement system.

Class I molecules are all similar in structure and are non-covalently but firmly bound to a small protein called β_2-microglobulin (Fig. 10.9). They consist of two groups. One comprises the classical transplantation antigens (K, L, D in the mouse; HLA-A, HLA-B, HLA-C in humans), which are all highly homologous in sequence. All three of these are expressed as integral membrane proteins on all nucleated cells, but are found in very much larger amounts on the surface of all kinds of lymphocytes. They play an important role in events which lead to the killing by cytolytic T-cells of cells bearing non-self antigenic determinants.

Fig. 10.8. Schematic map of the Major Histocompatibility Complexes on chromosome 17 of the mouse (top), and on human chromosome 6 (bottom). The genes that have mapped relative to each other are represented by short vertical lines. Regions where other genes are found are delineated by longer vertical lines. B = factor B of the complement pathway: C2 = second component of complement: C4x, y, z = non-allelic genes for the fourth component of complement.

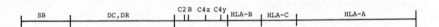

Fig. 10.9. Structure of the Class I (right) and Class II (left) transplantation antigens showing the domain structure. S_2 = disulphide bonds. $\beta_2 m$ = β_2-microglobulin. The cell membrane is in the middle between the dotted lines with some of the polypeptide chains passing through it into the cytoplasm at the very bottom.

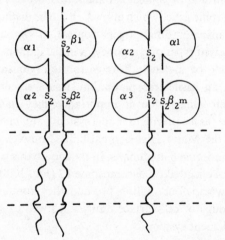

The second group, designated Qa and Tla in the mouse, are found mostly on lymphocytes, particularly those of the T-lineage (i.e. those that have matured in the thymus gland). Human homologues have been detected serologically but have not so far been fully characterised; nor have their genes been mapped. It has been suggested that these antigens may be involved in cellular differentiation. They are found on certain embryonic tissues. Their mRNAs are present in a wide range of tumour cells, including those transformed by oncogenic viruses and by chemical carcinogens. Malignant cells often display many characteristics of undifferentiated or partially differentiated cells, so it is not unreasonable that they should express the genes encoding these antigens.

Class II antigens show some structural similarities to the class I molecules. They contain two non-covalently associated polypeptide chains (α and β), but are not associated with β_2-microglobulin. In the mouse there are two sets – designated I-A and I-E – while in the human their counterparts are called DC and DR, with a third set called SB for which no murine homologue has so far been found. These molecules are again integral membrane proteins, found predominantly on cells of the B-lymphocyte and monocyte lineages. The genes encoding them were formerly known as the Immune Response (Ir) genes, since their products control the magnitude of this response to certain antigens. This effect requires the co-operation of helper and suppressor T-lymphocytes with B-lymphocytes.

10.8   Class I genes

These genes are highly polymorphic – about 50 have been described in the mouse, though a rather smaller number in the human. Alleles in the mouse are designated by lower case superscripts (e.g. K^h or K^d; L^b; D^b etc.), while the human alleles are given numerical designations (e.g. HLA-A2; HLA-B7; HLA-C5 etc.). The proteins consist of three extracellular domains of about 90 amino acid residues each, a trans-membrane domain with a hydrophobic core, and a cytoplasmic domain. Each domain is encoded by one or more separate exons. In the mouse the cytoplasmic domain is encoded by three short exons, while in humans there are only two exons in this part of the gene (Fig. 10.10). Most of the sequence variation between alleles occurs in the first and second extracellular domains. The third extracellular domain and also the β_2-microglobulin have distinct homology with the constant region domains of the immunoglobulins. β_2-microglobulin is invariant and is encoded on a different chromosome from the MHC antigens.

Limited examination of the polymorphism of these antigens has shown that the differences between allelic products are found as multiple substitutions which are clustered, rather than as individual substitutions which could have arisen from point mutations. This and other evidence has led to the suggestion that the multiple alleles in this system have arisen by gene-conversion events.

The Qa and Tla genes are much less polymorphic, and occur as clusters of distinct genes over a length of about 2000 kbp (Fig. 10.8). So far 10 genes have been mapped to the Qa region and 21 to the Tla region. It is not known how many of these are expressed (some could be pseudo-genes), and it has been suggested that they might provide a reservoir of donor sequences that can be used in gene-conversion events to give rise to the multiple alleles found in the K, L and D genes.

10.9   Class II genes

The overall structure of the α and β chains of the Class II proteins is somewhat reminiscent of that of the transplantation antigens with transmembrane and cytoplasmic domains, but they only contain two extracellular domains (Fig. 10.8). However, in each case one of these domains (α_1 and β_2) is homologous to a constant region domain of the immunoglobulins, and the other extracellular domain is of a similar size. These polypeptides are encoded by genes with exons corresponding to each of the domains, though in some cases the trans-membrane and cyto-plasmic exons are already joined, and in others there is more than one exon coding for the cytoplasmic domain. The mouse I-A and I-E genes have been located on a 100 kb length of chromosome 17. This contains genes coding for both α and β chains, and also two short stretches of DNA that hybridise with I-A and I-E probes which are believed to be single isolated exons and may therefore be pseudo-genes. The corresponding portion of the human genome has not yet been mapped.

Fig. 10.10. Structures of the H-2K^b gene of the mouse (top), and of a human HLA-B gene (bottom). L = leader peptide sequence: α_1, α_2, α_3 = extracellular domains: TM = transmembrane domain: CY = parts of the cytoplasmic domain: UT = untranslated sequences.

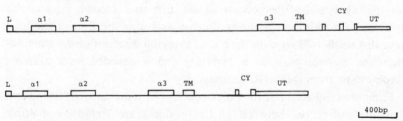

In all these antigens, the β-chains are polymorphic, as are the I-Aα and DCα-chains. In contrast, the I-Eα and DRα-chains seem to be conserved. Most of the observed polymorphisms in these molecules occur in the β_1, $A\alpha_1$ and $DC\alpha_1$ domains and are responsible for the differences in immunological responsiveness. Polymorphism probably arises mainly from allelic modifications at the particular locus in question, though there is some evidence that there may be as many as six different Class II genes in the mouse. There are likely to be multiple β genes and also DCα genes in the human.

Although the external domains of Class I and Class II genes are very similar in size, there is little homology in amino acid sequences except in the case of the immunoglobulin-like domains (α_3 in the transplantation antigens, and α_2 and β_2 in the Class II antigens). However, it is quite likely that all these proteins may have evolved by repeated duplication and mutation from a common primordial immunoglobulin-like domain. If this has occurred, the divergence must have started about 500 million years ago, considerably before the first mammals arose.

10.10   Complement genes

The genes for some of the components of the complement systems, which brings about the lysis of cells recognised as foreign to the organism, are found in the middle of the MHC (Fig. 10.8). In humans the genes for the second component of complement (C2) and a protein known as Factor B are extremely close together. These proteins are both atypical serine proteases which have analogous functions in two pathways by which complement may be activated. There are additionally two genes for the fourth component of complement (C4), though these are not quite so closely linked (Fig. 10.8). The two C4 genes are polymorphic (13 variants having been described at one locus and 22 at the other), but little is known of the precise differences between the alleles. The C2 and Factor B genes are also polymorphic, but rather less so. It has been suggested that these polymorphisms may be connected with the extreme polymorphism of the Class I and Class II antigens of the MHC in the recognition of self and non-self cells in the immune response.

Two C4-like genes in the mouse and also the Factor B gene have been mapped (Fig. 10.8). Only one of these C4 genes produces a product with haemolytic activity; the second encodes a protein known as Slp which is expressed only in males.

11

Hormone genes

The genes for a number of hormones have been widely studied for several reasons. First, there are valuable commercial possibilities in engineering these genes into cultured cells in the hope of producing large quantities of their products for therapeutic use. Secondly, it may be possible to link mutations in these genes with specific pathological states. Thirdly, since there is evidence that some of these genes are under transcriptional control, study of their expression may enlarge our understanding of the selective control of expression of the genome. Fourthly, several peptide hormones contain only a small number of amino acid residues, so it is of interest to discover how peptides so much smaller than the average protein are synthesised. In most cases it is relatively easy to isolate mRNAs for peptide hormones since they are the predominant mRNAs in the endocrine glands producing them. Some tumour cells which can be grown easily in culture also produce large quantities of certain hormones.

11.1   Insulin

Insulin is a small protein consisting of two polypeptide chains (A and B chains), joined together by two disulphide bridges. It is synthesised as a larger molecule (called proinsulin) containing an additional connecting peptide (C peptide) between the A and B chains (Fig. 11.1). It is believed that the C peptide allows the proinsulin to fold into the conformation with correct pairing of cysteine residues to form the disulphide bonds. Before secretion the C peptide is cleaved off by a processing peptidase.

In humans insulin is encoded by a single gene which has two introns – one in the 5'-flanking region, and a long one in the coding region (Fig. 11.2) This intron is, in fact, considerably longer than the coding part of the gene,

132

and is not located between two obviously functional domains of the protein.

Because of the presence of introns, the gene cannot be engineered directly into plasmids for growth in bacteria to produce insulin, since the bacteria would be unable to process the primary transcript. Production of insulin in bacteria has been achieved by using a cDNA prepared from mRNA to programme the organisms.

The 5'-flanking region sequence of the gene is highly polymorphic, with DNA from different individuals yielding a variety of restriction fragments. Preliminary analysis suggests that certain polymorphisms may be associated with either increased or decreased risks of developing some forms of diabetes. There are a number of repeated sequences of up to 14 bp in this region.

In the rat there are two allelic forms of the insulin gene, giving rise to two very similar but distinct insulin molecules. One lacks the second intron, and in the other one it is much shorter than in humans. Both these genes are transcribed, but the proportions of the two insulin molecules that are synthesised differ in various physiological and pathological states.

It has recently been shown that control sequences other than the TATAA and CAAT boxes are essential for transcription of insulin genes.

Fig. 11.1 The structure of proinsulin. The A and B chains are the upper and lower straight lines respectively, while the C peptide is the curved line joining them. Arrows point to the positions of cleavage.

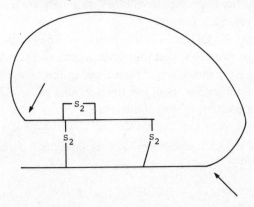

Fig. 11.2. The human insulin gene. The open boxes are untranslated sequences in the exons: hatched ones are the translated portions.

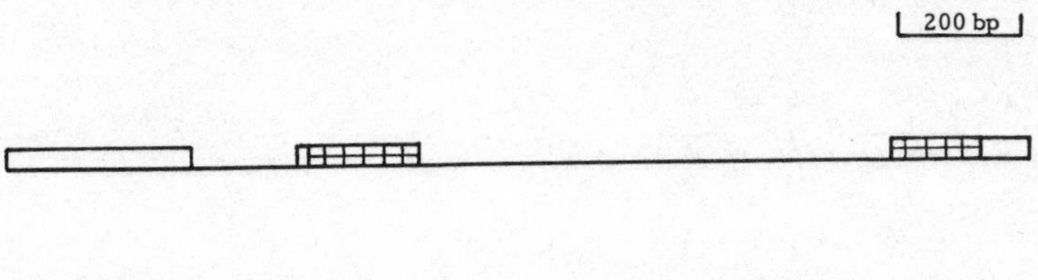

Deletion of sequences in the region situated between 120 and 260 bp 5' to the transcription start site inhibits transcription by about 90 %. Furthermore, there are tissue-specific factors which control transcription since both human and rat insulin genes are transcribed in cells derived from a tumour of the hamster pancreas, but not in cells derived from a rat exocrine pancreatic tumour, or from hamster ovary.

A few other peptide hormones show considerable homology with insulin and have presumably arisen by duplication and separate evolution from a primordial insulin gene. The only one that has been studied at the gene level is relaxin – a hormone that is involved in the relaxation of the muscles of the birth canal during labour. In the guinea pig this gene is very similar in overall structure to the insulin genes that have been examined in other mammalian species.

11.2   Growth hormone family

The growth hormone family encompasses three hormones – Growth Hormone (GH) itself and prolactin, both secreted from the anterior pituitary, and somatomammotrophin (SMT) (also known as Placental Lactogen) which is secreted by the placenta, and may function as a growth hormone for the developing foetus. Prolactin and GH are believed to have arisen by gene duplication about 400 million years ago, while SMT probably diverged from GH much more recently. During the course of evolution the structure of human GH has diverged from that of non-primates to such an extent that the hormones from other mammals are inactive in humans, though the reverse is not true.

The rat has one gene for GH and one for prolactin. They both show a similar organisation of five exons and four introns, but all the introns in the prolactin gene are very much longer than those in the GH gene (Fig. 11.3). Thus, the prolactin gene is about five times as long as the GH gene, although they encode proteins of very similar size.

Fig. 11.3. The growth hormone (top) and the prolactin (bottom) genes of the rat.

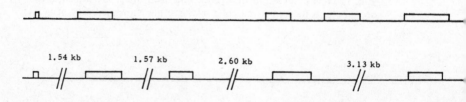

The human genome contains a single prolactin gene, which has not yet been sequenced, but there is a cluster of three GH and four SMT genes on chromosome 17. The precise linkage of all these genes has not yet been determined, but linkage between one GH gene and two SMT genes, and between a second GH gene and a third SMT gene has been established by restriction mapping of the surrounding DNA. Two of the GH genes have been sequenced. They encode proteins that differ in 15 positions in their amino acid sequence, one of which corresponds exactly to GH. It is not known whether the second gene is expressed. If it is, it is not functional since it has been detected in the genome of an individual exhibiting GH deficiency, in whom the normal GH gene has been deleted. It is not known whether the third GH gene is expressed or whether it is a pseudogene.

Two forms of GH are found in human pituitaries which differ in that one has a deletion of amino acid residues 32–46. This sequence is encoded at the beginning of the third exon, and the variant form arises by an alternative form of splicing which results in excision of the bases coding for these amino acids (Fig. 11.4).

Two of the SMT genes are expressed and encode proteins which differ by only a single amino acid residue in the leader sequence. They contain four other base changes, but these are all silent in their coding properties. It is likely that the two other genes are not expressed, though the reasons for this are unknown.

11.3 Polyproteins

Some hormonally active polypeptides are synthesised as part of much larger proteins which are then processed by proteolytic enzymes yielding the smaller active hormones. Such precursors are sometimes

Fig. 11.4. Alternative splicing in human growth hormone mRNA. The upper line shows the sequence round the splice sites at the ends of the second intron which are used to produce the full-length hormone. The lower line shows the alternative 3'-splice site which is used to produce the shorter growth hormone molecule. Note that both 3'-splice sites are very similar, and show a high degree of homology with the consensus sequence (Chapter 7.9). The numbers are the numbers of the codons counted from the beginning of the mature hormone.

```
31                                      32
UUU GUAAGCUCU . . . . . . . . . UCCUUCUCCUAG GAA


31                                      47
UUU GUAAGCUCU . . . . . . . . . UCAUUCCUGCAG AAC
```

known as polyproteins, though this is not a very good name for them
since the products of processing are generally peptides rather than
proteins.

The anterior pituitary synthesises a protein known as pro-opiomelanocortin (POMC), which is a precursor of several physiologically active
peptides, including adrenocorticotrophin (ACTH), three melanotrophins
(α, β and γ MSH), and β-endorphin (Fig. 11.5). The sequences of genes
encoding this polyprotein in several mammals all have the same general
structure with one long intron in the 5'-flanking region, and a second one
in the protein-coding sequence. Scattered throughout the protein are
adjacent pairs of the two basic amino acid residues – lysine and arginine
located at the beginning and end of the mature polypeptides (Fig. 11.5).
This sequence of two basic amino acids acts as a processing signal for a
peptidase (as yet uncharacterised) which is found in many tissues where
processing of precursor proteins takes place. The processing of POMC
may occur in a number of ways in different parts of the pituitary or at
different stages of development so that the peptides derived from the
molecule arise in different proportions (Table 11.1).

In the mouse a pseudo-gene has been found which contains only part
of the protein-coding sequence. Mutations that have accumulated within
this gene give rise to a premature termination codon and the replacement
of one of the basic amino acids in a processing signal sequence by cysteine,

Fig. 11.5. The structure of the primary translation product of the
POMC gene (upper line). The short vertical lines show the positions
of pairs of basic amino acid residues. The three lower lines show the
products that can be formed by cleavage at some of these points.
MSH = melatonin; ACTH = adrenocorticotrophin; LPH = lipo-
trophin; END = endorphin; CLIP = corticotrophin-like intermediate
lobe peptide.

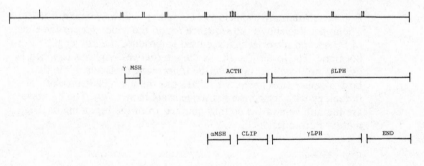

Table 11.1. *Ratio of products formed by alternative processing of POMC at different stages of development in the rhesus monkey pituitary*

Peptide	Foetus	Adult
ACTH	1.00	1.00
CLIP	1.21	0
α MSH	0.34	0
β MSH	1.01	0.17
β LPH	0.59	0.83
β END	1.71	0.98

Abbreviations as in Fig. 11.5.

so that even if the gene were transcribed it would lead to the production of a non-functional protein.

β-endorphin, which has potent analgesic properties, contains a penta-peptide sequence at its amino terminus which is identical to met-enkephalin. This is one of a pair of pentapeptides, the other being known as leu-enkephalin, which are distinguished by the C-terminal amino acid in their otherwise identical sequence (Fig. 11.6). They also have potent analgesic properties and may have other specific effects in the central nervous system. Several other peptides of varying lengths all containing C-terminal extensions of one of the enkephalins have similar analgesic properties. The related peptides in this family are all synthesised as larger polyproteins. Two precursors have recently been discovered, and are called preproenkephalins A and B.

The gene for preproenkephalin A from the adrenal medulla has been sequenced. It contains two introns – one in the 5'-flanking region, and one interrupting the protein-coding sequence. The protein it encodes (Fig. 11.7) contains no fewer than seven enkephalin sequences. One of these is leu-enkephalin, four are met-enkephalin, and the remaining two contain met-enkephalin with short C-terminal extensions. All these pep-tides are situated in the carboxyl two-thirds of the protein.

Preproenkephalin B is synthesised in the hypothalamus, and the cDNA to its mRNA from humans and pigs has been sequenced. The protein it specifies contains three of the enkephalin family of peptides – leu-enke-

Fig. 11.6. The structure of leu-enkephalin (top) and met-enkephalin (bottom).

Tyr – Gly – Gly – Phe – Leu

Tyr – Gly – Gly – Phe – Met

phalin, dynorphin and β-neoendorphin, which both contain C-terminal extensions to leu-enkephalin (Fig. 11.7). Again these peptides are situated in the C-terminal part of the protein.

In both these precursor proteins the individual enkephalin-like peptides are bounded by the usual pairs of the basic amino acids. The N-terminal parts of the proteins, which do not contain any enkephalin-like peptides, exhibit considerable homology especially in the positioning of six cysteine residues which may be involved in folding the proteins into analogous conformations.

There are two octapeptide hormones – vasopressin and oxytocin – which are chemically very similar, but possess very different physiological actions. They are synthesised in the hypothalamus and transported along axons in the pituitary stalk to the posterior pituitary, where they are stored before secretion. During transport they are non-covalently bound to two homologous proteins – neurophysins I and II. The synthesis of each octapeptide is directed by a gene coding additionally for the associated neurophysin, and the small peptides are split off the primary translation products at the usual signal dipeptide lysyl-arginyl sequence (Fig. 11.8). At the same time the C-terminal residues of these peptides are amidated – a reaction which involves a glycine residue which precedes the lysyl-arginyl pair (Fig. 11.9).

The two genes have similar structures with three exons, the first of which encodes a leader peptide, the hormone and the N-terminal part of the associated neurophysis. The second exon encodes nearly all the rest of the neurophysin and the third exon in the oxytocin gene codes for about 20 amino acids at the C-terminal end of neurophysin I. In the vasopressin gene the third exon encodes both the C-terminal end of neurophysin II as well as a glycoprotein of unknown function which is cleaved from the vasopressin-neurophysin II precursor post-translationally. Nearly all the second exon as well as 135 nucleotides immediately preceding it in the first intron are identical in the two genes (Fig. 11.8). This completely

Fig. 11.7. The structure of the primary translation products of preproenkephalin A (top), and preproenkephalin B (bottom). The tall bars at the left show the sites of cleavage of the leader peptides. The double short bars represent pairs of adjacent lysine and/or arginine residues at which further cleavage occurs. M = met-enkephalin; L = leu-enkephalin; N = neo-endorphin; D = dynorphin.

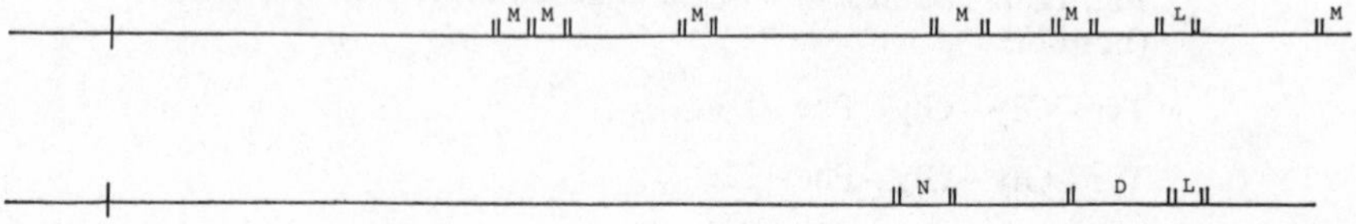

conserved sequence of 332 nucleotides spanning an intron–exon junction is extremely unusual, and is probably the result of a gene-conversion event which has occurred much more recently than the original divergence of the genes for the two hormones.

The Brattleboro strain of rats is a well studied strain which lacks vasopressin and its associated neurophysin: consequently the animals suffer from diabetes insipidus – an inability to control the excretion of water, so that they excrete very large volumes of dilute urine and have an extremely high water intake to balance this. The gene for vasopressin in this strain has a single base pair deletion in the second exon. This throws the rest of the reading frame out of phase and there is now no termination codon in the mRNA that is presumably transcribed from the mutated gene. It is probable that this somehow inhibits translation of the mRNA, since

Fig. 11.8. Bovine genes for the precursors of vasopressin and oxytocin and their protein products. Lines 2 and 3 show the structure of these genes, aligned to show the region of identity encompassing the 3′-end of the first intron and most of the second exon (indicated by vertical lines below the structures). Open rectangles are exons, with untranslated sequences lined vertically. Lines 1 and 4 show the structure of the protein products. The region of identity is again indicated by lines below the structures. The short vertical lines to the right of vasopressin and oxytocin indicate glycyl, lysyl and arginyl residues which are involved in the processing of the primary translation products. VP = vasopressin; OT = oxytocin; NP = neurophysin; GP = glycoprotein.

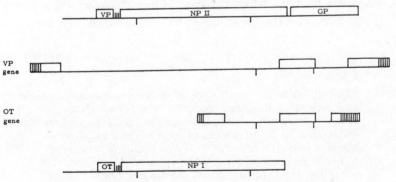

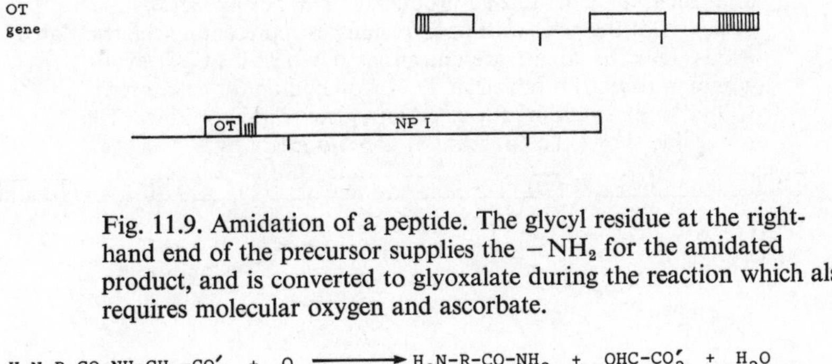

Fig. 11.9. Amidation of a peptide. The glycyl residue at the right-hand end of the precursor supplies the $-NH_2$ for the amidated product, and is converted to glyoxalate during the reaction which also requires molecular oxygen and ascorbate.

$$H_2N-R-CO-NH-CH_2-CO_2' \; + \; O_2 \longrightarrow H_2N-R-CO-NH_2 \; + \; OHC-CO_2' \; + \; H_2O$$

while the mRNA has been shown to be present in the hypothalamus of Brattleboro rats, only minute amounts of its translation product are found there.

The hormone calcitonin, contains 32 amino acids and is involved in calcium homeostasis. It is made as a considerably longer precursor molecule. Again, the mature polypeptide is split from this precursor by peptidase action at pairs of basic amino acid residues. Calcitonin is produced in the thyroid gland, but RNA that hybridises with its cDNA is also found in the hypothalamus, though the hormone itself is not made there. The cDNA to this hypothalamic RNA has a sequence identical with that of the cDNA for calcitonin at its 5'-end, but the sequences of the two cDNAs diverge sharply about half way along the molecules. The calcitonin gene contains three exons before that which encodes the mature calcitonin molecule, and in addition there are two further exons on the 3'-side of the calcitonin-specific exon (Fig. 11.10). The transcription products of these last two exons appear in the hypothalamic mRNA, while mRNA containing the calcitonin-specific sequence is not found in that tissue. The production of two distinct mRNAs in different tissues must occur by a mechanism involving alternative exon splicing (Fig. 11.10). Note that a similar phenomenon occurs in generating alternative types of immunlglobulin molecules (Chapter 10.4). The processed hypothalamic gene product also contains pairs of basic amino acid residues which presumably direct the formation of a smaller peptide of 37 amino acids, known as calcitonin gene-related product (CGRP). This has been detected immunochemically in a number of sites within the brain which also contain its mRNA, so that it is very likely that it has important functions in the central nervous system.

Fig. 11.10. Alternative processing of the gene coding for calcitonin and calcitonin gene-related product. The upper line shows the structure of the gene with open rectangles representing the translated exons: lined rectangles are untranslated exons. E_1, E_2 = exons common to both products; CT = exon coding for calcitonin; CGRP = exon coding for calcitonin gene-related product. The lower line shows the structure of the two mRNAs.

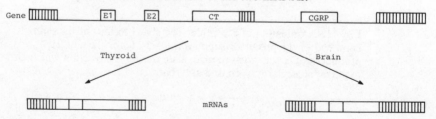

Table 11.2. *The glycoprotein family of hormones*

Hormone	Abbreviation	Site of production	Major sites of action
Thyroid-stimulating hormone, Thyrotrophin }	TSH	Anterior pituitary	Thyroid
Follicle-stimulating hormone, Follitrophin }	FSH	Anterior pituitary	Ovary and testis
Luteinising hormone, Lutrophin }	LH	Anterior pituitary	Ovary and testis
Chorionic gonadotrophin	HCG	Placenta	Ovary

Note: The first three have alternative names, as shown.

11.4 Glycoprotein hormones

A family of hormones, known as the glycoprotein hormone family, are synthesised in the anterior pituitary. They regulate the functions of the thyroid gland and the gonads (Table 11.2). A similar hormone, synthesised in the placenta of a few mammals (chorionic gonadotrophin), is homologous in structure and function with luteinising hormone and is also a member of this group.

All these hormones consist of two polypeptide subunits which are bound together non-covalently. The α-subunit is common to all forms of the hormones in a single species, while the β-subunits, though obviously related, differ from each other and account for the biological specificity of the molecules. There is a low degree of homology between the mRNAs for the α-subunits and the β-subunits, suggesting that all the individual polypeptides have probably evolved from a common ancestor. The β-subunit of chorionic gonadotrophin is 24 amino acids longer than the other β-subunits, and its mRNA has a very short 3'-flanking sequence in which the third, fourth and fifth nucleotides of the signal for poly-adenylation

Fig. 11.11. The structure of the 3'-terminal parts of the human genes for the β-subunits of LH (top) and HCG (bottom). The amino acids specified by the DNA sequences are shown above and below them. d = deletion; m = mutation.

```
     Gln Leu Ser Gly Leu Leu Phe Leu Ter
--CAA CTC TCA GGC CTC CTC TTC CTC TAA --- N₇₁ --- AATAAA
     d   mm                      m
--CG C TTC CAG GCC TCC TCT TCC TCA AAG --- N₆₉ --- CAA TAA A
     Arg Phe Gln Asp Ser Ser Ser Ser Lys         Gln Ter
```

(AAUAAA) are also used as the termination codon. The read-through of the codon used for termination in the β-subunits of the other hormones results from deletion of a single base pair a little way upstream from the site where it occurs in the β-subunit of LH (Fig. 11.11). Seven β-HCG like genes are present in the human genome, and some are closely linked to each other and to that for LH.

12
The mitochondrial genome

Mitochondria contain a single circular chromosome which directs the synthesis of a small number of proteins, and also the various RNAs that are required for this process.

12.1   Yeast mitochondrial genome

In yeasts and other fungi the mitochondrial genome is about 78 kbp long. Fairly large portions of the yeast genome have been sequenced, although the complete sequence is not yet known. It encodes two rRNA molecules, a complete set of tRNA molecules, and mRNAs directing the synthesis of at least nine proteins (Fig. 12.1). The polymerases for the synthesis of mitochondrial DNA and RNA, nearly all the mitochondrial ribosomal proteins, and all the tRNA synthetases are encoded by nuclear genes and synthesised on cytoplasmic ribosomes. In addition, the majority of the functional mitochondrial proteins (e.g. enzymes of the citrate cycle and electron-transport chain) are encoded by nuclear genes. It is believed that all these proteins are made in the cytosol and contain special amino acid sequences that are involved in their translocation into the mitochondria.

The mitochondrial rRNAs (21s (3200 nt) and 15s (1660 nt)) are even somewhat smaller than prokaryotic rRNAs, and there is no rRNA corresponding to the 5.5s rRNA that is found in cytoplasmic ribosomes. The two rRNAs are encoded on widely separated portions of the genome, and are not transcribed together. There are genes for 24 tRNAs, some of which are clustered together, but others are found singly. There are separate tRNAs for *N*-formyl-methionine (which is used as an initiator amino acid as in prokaryotes), and methionine when it is incorporated into the middle of a polypeptide. Otherwise only threonine, serine and arginine have two tRNAs. The 5'-position in each anticodon is always either U or G, so

143

that U-G pairing must occur in the reading of the codons. When four codons are used for one amino acid, the 5′-base in the anticodon is always U. Presumably only the second and third bases of the anticodon are used when the third base in the codon is U or C (so-called 'two out of three' base pairing). The genetic code is slightly different from the one that is universally used by nuclear genes, with UGA coding for tryptophan, rather than termination, and the codons beginning CU code for threonine (Table 12.1). There is a very strong bias against using codons with G or C in third position. This is in keeping with the unusual base composition of yeast mitochondrial DNA which has only 18 % G+C.

Fig. 12.1. The yeast mitochondrial genome. For clarity it has been drawn twice. The inner circle shows the sites of the genes for the tRNAs, labelled according to the one-letter code for amino acids:
A = Ala; C = Cys; D = Asp; E = Glu; F = Phe; G = Gly;
H = His; I = Ile; K = Lys; M = Met; N = Asn; P = Pro;
Q = Gln; R = Arg; S = Ser; T = Thr; V = Val; W = Trp;
Y = Tyr. The outer circle shows the exons as outwardly placed boxes and introns as inwardly placed boxes. 21s and 15s = rRNA; Cox I, II and III = subunits I, II and III of cytochrome oxidase; Cyt b = cytochrome b; ATPase 6 and 9 = subunits 6 and 9 of ATPase; RAP = ribosome-associated protein; U = open reading frames for which no corresponding proteins are yet known; m = maturase. There are also open reading frames in some of the Cox I introns. (After L. A. Grivell, 1983.)

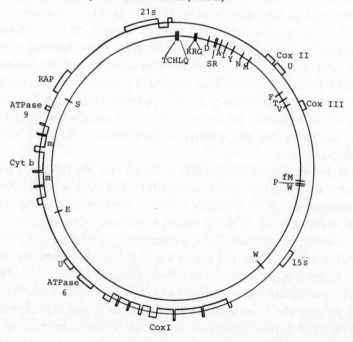

Table 12.1. *Differences in codon usage between mitochondrial and nuclear genes*

Codon	Nuclear genes	Amino acid encoded in	
		Yeast mitochondrial genes	Mammalian mitochondrial genes
CUU	Leu	Thr	Leu
CUC	Leu	Thr	Leu
CUA	Leu	Thr	Leu
CUG	Leu	Thr	Leu
UGA	Ter	Trp	Trp
AUU	Ile	Ile	Ile initiation?
AUA	Ile	Met	Met
AUG	Met	f Met	Met
AGA	Arg	Arg	Ter?
AGG	Arg	Arg	Ter?

Other codon usages are common to nuclear and mitochondrial genes.

Table 12.2. *Proteins encoded by yeast and mammalian mitochondrial genomes*

Subunits I, II and III of cytochrome oxidase
Apocytochrome *b*
Subunits 6, 8 and 9[a] of ATPase complex
Maturases from introns 2 and 4 of apocytochrome *b* gene[a]

(*a*) Only in yeast mitochondria.

Transcription of mitochondrial genes is performed by a single nuclear-encoded RNA polymerase which is much smaller and simpler than those used for transcription in the nucleus. It transcribes all the mitochondrial genes, irrespective of whether they encode rRNAs, tRNAs or mRNAs. The highly conserved sequence ATATAAGTA is used as promoter. This is found at the initiation sites of all the transcripts that have so far been identified. The last A in this sequence is the actual base at which transcription starts. Many of the transcripts contain more than one gene product, so there must be secondary processing to produce the mature final products.

The genome also encodes various proteins found in the inner mitochondrial membrane (Table 12.2). The genes for the cytochrome oxidase subunit I and cytochrome b are both very much longer than required to encode these proteins, and usually contain five and seven introns respectively.

(Remember that introns seem to be rather rare in nuclear-coded proteins of yeast – Chapter 7.6.)

The cytochrome b gene is found in two forms in different strains of yeast. Some have a short form of the gene in which the first four exons are already adjacent, with loss of introns I–III, while the long form, found in other strains, has the full complement of six exons and five introns.

The processing of this gene has a number of interesting features. The excision of the first three introns is controlled by at least three genes encoded by the nucleus. Excision of the first intron leaves an RNA which is transcribed to give a chimaeric protein consisting of 143 amino acids encoded by the first two exons of the gene plus another 280 amino acids encoded by a continuous open reading frame which terminates near the 3'-end of intron II. This is known as a maturase, since it catalyses the excision of intron III – a stage in the maturation of cytochrome b messenger. It thus acts to destroy its own mRNA – a phenomenon known as splicing homeostasis. Another open reading frame in intron IV is also transcribed into a second maturase, which plays a part in the excision of its intron. The second maturase, as well as the products of the nuclear genes mentioned above, are also involved in the excision of some of the introns of the *oxi I* gene. It is noteworthy that some of these introns contain open reading frames that are more than 50 % homologous with the second maturase of the cytochrome b gene. These may also be involved in the processing of primary RNA transcripts.

The introns in the cytochrome b, cytochrome oxidase I genes and also a single intron in the 15s rRNA gene, can be divided into two groups, each of which contains a unique set of highly conserved sequences which can assume common secondary structures (Fig. 12.2). Highly homologous structures are found in the corresponding genes from other fungal species and also in nuclear genes coding for tRNA molecules in some Protista. None of these introns obeys the GU–AG rule, so their removal by splicing is believed to follow a path different from that which occurs in other eukaryotic pre-mRNAs. The mitochondrially encoded maturases and the products of the nuclear genes mentioned above are involved in the splicing of these introns.

The small size of the mitochondrial ribosomal subunits and the small number of tRNAs required for protein synthesis in mitochondria suggest that there have been evolutionary constraints tending to keep the mitochondrial genome small, yet it is odd that there appears to be a comparatively large proportion of the fungal mitochondrial genome with no coding functions. It is also not clear why there has been selective pressure to maintain the separate synthesis of just a few mitochondrial proteins within the organelles. It has been suggested that those that are synthesised are more hydrophobic than most known proteins: it is possible that if these were synthesised in the cytosol their transport into the mitochondria might pose problems. However, this may not be the true or only explanation, since in *Neurospora crassa* the most hydrophobic of these proteins, the ATPase proteolipid subunit 9 is synthesised in the cytosol.

12.2 Mammalian mitochondrial genome

Human, bovine, mouse and rat mitochondrial genomes have been completely sequenced and they are all highly homologous and extremely similar in size (about 16.5 kbp). This small genome again only codes for a very limited set of proteins. Most of the mitochondrial proteins, including all those involved in mitochondrial transcription and translation, are encoded by the nuclear genome and synthesised on cytoplasmic ribosomes.

These mitochondria display an even more striking concentration of

Fig. 12.2. Secondary structure of the two groups of introns found in fungal mitochondrial genes, and also in some protistan nuclear genes. The thickened lines represent very highly conserved sequences. The short lines across the main sequences denote the actual sites of splicing. The size of the large loops and some of the stems is somewhat variable. (Modified from F. Michel & B. Dujon. *EMBO J.* (1983) **2**, 33. figs. 2 & 3, reprinted by permission.)

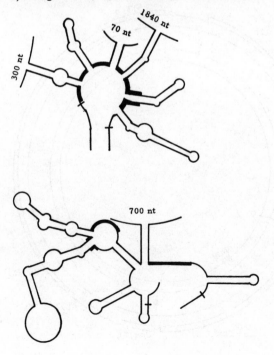

function into an extremely compact, highly utilised genome. The mito-chondrial DNA can be separated into its two strands under denaturing conditions by taking advantage of their different densities due to their overall differences in base composition. These are referred to as the heavy (H) and light (L) strands. The map (Fig. 12.3) compiled from sequence data shows that most of the genetic information is encoded on the H strand: the L strand seems to code only for a few tRNAs and a single mRNA.

The mammalian mitochondrial genome has several features which set it apart from both nuclear and also fungal mitochondrial genomes. It uses a slightly different genetic code, in which AUA codes for methionine, and can be used as an initiation codon (it does not use N-formyl-methionine for initiation, unlike fungal mitochondria): AUU, coding for isoleucine,

Fig. 12.3. The human mitochondrial genome. For clarity it has been drawn twice. The two inner circles represent the H strand (outer) and the L strand (inner), and show the sites of the genes for tRNAs (labelled as in Fig. 12.1). The two outer circles show the sites of genes coding for proteins and rRNA. 12s and 16s = rRNA; cyt b = cytochrome b; Cox I, II and III = subunits I, II and III of cytochrome oxidase; ATPase 6 = subunit 6 of ATPase; URF 1–6 = open reading frames whose protein products have not yet been identified. (After L. A. Grivell, 1983.)

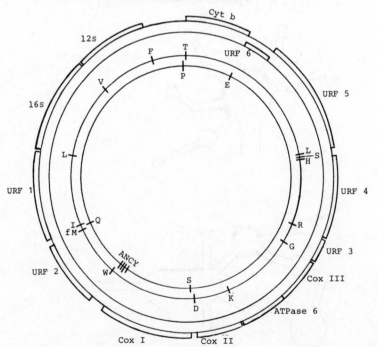

can probably also be used as an initiation codon: UGA is not used for termination, but codes for tryptophan, while AGA and AGG are used as termination codons, rather than codons for arginine (Table 12.1). In other respects the code is the same as that used in nuclear genes, including the use of codons beginning in CU for leucine (not threonine as in yeast). Only 22 tRNAs are used, and in those that have to read four different codons U is always found at the 5'-end of the anti-codon. Leucine and serine are the only amino acids which require two tRNAs. There is a strong bias in favour of codons ending in A or C (77 %), rather than in G or U.

Mammalian mitochondrial DNA is believed to be transcribed into one long precursor RNA molecule corresponding to the full length of the genome. There are no normal flanking sequences for the mRNAs. They are generally found immediately adjacent to the genes for some of the tRNAs, and it is believed that the tRNAs themselves may serve as punctuation signals. It seems plausible that the secondary structure adopted by the tRNAs should be recognised by an endonuclease that will hydrolyse the newly transcribed RNA at appropriate places. A number of the mRNAs so generated lack a complete termination codon at the 3'-end. However, they are all polyadenylated immediately after they have been transcribed, and this polyadenylation can generate a suitable termination codon (UAA). They all lack the characteristic cap structure of most other eukaryotic mRNAs (Chapter 7.7).

The H chain can code for twelve proteins containing between 68 and 603 amino acids. The two smallest ones (68 and 98 amino acids) overlap out of phase slightly with two of the larger ones. A thirteenth polypeptide is encoded by the L chain, and there might possibly be others. None of these putative reading frames appears to contain any introns. Only five of the proteins have been identified – they are three of the subunits of cytochrome oxidase, subunit 6 of the ATPase, and cytochrome b. After transcription and translation of highly purified mitochondrial DNA no fewer than 26 polypeptides can be consistently identified as definite translation products. The surplus 13 polypeptides could be primary products that have been post-translationally modified.

The amino acid composition of the mitochondrially encoded proteins, as predicted from the DNA sequence of their genes, is strongly hydrophobic (more than 50 % of the amino acid residues have hydrophobic side chains). This suggests that these proteins are firmly anchored in the inner membrane of the mitochondria.

The rRNAs in these mitochondria are much smaller than those made in yeast or those found in the cytosol, having only 954 and 1559 nucleotides (12s and 16s). Again, there is no molecule corresponding to the 5.5s RNA.

Most of the tRNAs have unusual structures with less base pairing than the corresponding cytosolic ones. They also vary in the length of the constant TΨC loop (Chapter 2.3). It is surprising that there seems to be considerably less homology between the corresponding pairs of human and bovine mitochondrial tRNAs than between the corresponding pairs of cytosolic ones. However, the overall structure of the mitochondrial tRNAs is recognisably similar to their cytosolic counterparts, even though the former tend to be smaller molecules than the latter.

12.3   Mitochondrial genome of higher plants

What little is known about the mitochondrial genomes of higher plants suggests that they are different from those of both fungi and animals. In the very few species that have been investigated the size seems to vary considerably, but they are much larger than those of fungal mitochondria, containing more than 200 kbp. They are composed of a heterogeneous set of circular molecules which are all derived from a single circular molecule containing the whole genome. They encode a larger number of mRNAs than animal or fungal mitochondrial DNA. They also contain a gene for a 5s rRNA so that their ribosomes are more like cytosolic ribosomes.

13

The control and plasticity of the genome

13.1 Sequences 5' to genes control their expression

An early event in the expression of genetic information is undoubtedly the binding of RNA polymerase which is generally controlled by DNA sequences 5' to the actual coding sequences. The promoter sequences which are commonly found 5' to many genes in both prokaryotes (the -10 and -35 boxes) and in eukaryotes (the TATA and CAAT boxes) are important sites in directing the binding of prokaryotic RNA polymerase and eukaryotic RNA polymerase II. However, the sequence requirements in eukaryotes seem to be much less stringent than in prokaryotes, since it is not uncommon for there to be no discernible CAAT box 5' to the site where transcription is initiated, and there is appreciable variation in the distances between this site and the TATA and CAAT boxes. Even in the sequences that seem to be essential, appreciable variation in base sequence can be tolerated. Out of 137 TATA sequences collected in eukaryotic genes no position is invariant, with the second base in the sequence (A) being most highly conserved in 134 of these cases (Table 13.1). Lower conformity with a consensus sequence is tolerated in the case of the CAAT box, with the most highly conserved bases being the two As (62 and 56 in a collection of 67).

13.2 Control in prokaryotes

The expression of a number of prokaryotic genes seems to be controlled by the strength of their promoter elements, i.e. the degree of congruence with the consensus sequence of the promoter. Even so, some genes that are abundantly transcribed in some circumstances do not appear to have particularly strong promoters. In these cases the binding of activators upstream from the promoter sites may be able to overcome the liability of a weak promoter element.

151

Table 13.1 *Sequences of TATA and CAAT boxes in some eukaryotic genes.*

TATA box									
A	25	10	134	5	116	95	106	67	45
G	52	5	1	1	1	0	7	26	48
T	19	105	2	131	17	41	19	36	18
C	40	16	0	0	3	1	5	8	25
consensus		T	A	T	A	$\frac{A}{T}$	A	$\frac{A}{T}$	

CAAT box								
A	19	11	8	2	62	56	12	14
G	21	27	6	7	2	2	11	15
T	17	11	15	6	2	5	38	18
C	10	17	38	52	1	4	6	20
consensus			C	C	A	A	T	

A second mode of regulation of the expression of prokaryotic genes lies in the choice of codons for some of the amino acids that they specify. Not all tRNAs are expressed equally abundantly, and some proteins that are found only in low concentrations in bacterial cells contain amino acids that are specified by codons for which tRNAs are present in only small amounts. This is an inflexible way of controlling the amounts of these proteins present since the proportions of the different tRNAs seem to be fairly constant, but it does minimise the wasteful synthesis of proteins that are required in only very small amounts (e.g. repressors). There is some evidence that a similar form of control may operate in the reverse sense in yeast where the codons used for directing the incorporation of some amino acids into abundant glycolytic enzymes are predominantly of only a few molecular species. However, there is not yet enough information to decide whether these particular forms of control of gene expression are used at all widely in eukaryotes.

A fairly frequent feature of the 5'-flanking sequences of transcription units is the presence of inverted repeats which could lead to the formation of stem and loop types of structures. This is by no means universal, but it may facilitate the binding of RNA polymerase and other proteins.

Proteins other than RNA polymerase are frequently involved in the control of transcription. These have been well characterised in the case of prokaryotes, where CAP and various repressors and activators bind to operator and promoter sites. A very important feature of these interactions is that the proteins will only bind to DNA when they are in certain configurations as the result of either binding or being free of small effector molecules.

Table 13.2. *Some situations in which hormones affect the transcription of DNA into mRNA*

Hormone	Species and tissue	Gene
Ecdysterone	*Drosophila* cultured cells	Actin
Oestrogens	Hen oviduct	Ovalbumin
		Ovomocuid
		Lysozyme
	Xenopus laevis male liver	Vitellogenin
Progesterone	Hen oviduct	Avidin
	Rabbit uterus	Uteroglobulin
Glucocorticoids	Rat liver	Tyrosine transaminase
		Tryptophan oxygenase
	Mouse mammary gland	Mouse mammary tumour virus
	Rat anterior pituitary	Pro-opiomelanocortin[a]
Tri-iodothyronine	Rat anterior pituitary	Growth hormone[b]
Prolactin	Rat mammary gland	β and γ casein[c]
Insulin	Rat liver	Pyruvate kinase
		Phospho-enol pyruvate carboxykinase[a]
Glucagon	Rat liver	Phospho-enol pyruvate carboxykinase[d]

All these effects are stimulation of transcription, except for (a) which is inhibition.
(b) Potentiated by glucocorticoids.
(c) This is a complex situation, in which insulin and glucocorticoids are also required for maximal stimulation.
(d) Cyclic AMP is used as a second messenger in this situation.

13.3   Control in eukaryotes

There is no direct evidence for the existence of operons in eukaryotic genomes, though there may be some co-ordinate control of the expression of metabolically related enzymes. However there is strong evidence that binding of proteins can affect the efficiency of transcription. The most noteworthy cases are numerous examples where the cellular concentration of a particular mRNA is controlled by the local concentration of a hormone (Table 13.2). In most of these cases the hormones themselves enter the cell and bind to proteins which are found in the nucleus. These probably interact directly with chromatin or DNA. Steroid hormones and

Table 13.3. *Effect of deletions in the 5′-flanking sequence on the transcription of the Mouse Mammary Tumour Virus genes induced by dexamethasone*

Length of 5′-flanking sequence from initiation site	Fold induction after exposure to 1 μM dexamethasone[a]
451	3.0
236	3.3
202	1.6
137	1.6
50	1.0
37	1.0

(*a*) Dexamethasone is a synthetic glucocorticoid.
The 5′-flanking sequence of the genes transcribed from the virus was artificially placed in front of a thymidine kinase gene for ease of assay and incorporated into a plasmid which was grown in a suitable bacterial host.

tri-iodothyronine enter the cell and act in this way. Polypeptide hormones such as insulin are believed to exert their actions through release of a second messenger (probably a small molecule) after binding to specific receptor on the cell membrane. The second messenger then transmits some kind of message that stimulates or inhibits transcription of certain genes. In none of these cases has the detailed mechanism been worked out at the molecular level. It has been possible to show that certain regions of the DNA are involved in the response to steroid hormones since their deletion leads to unresponsiveness (Table 13.3). They contain some well conserved sequences lying upstream from both the TATA and CAAT boxes, usually situated between about 100 and 200 bp 5′ to the site where initiation is initiated. Parallel studies have shown that steroid hormone–receptor complexes also bind to these regions. It is likely that these sites contain the sequences that are important for the control of expression by these hormones. The binding of hormone–receptor complexes may open up the rather tight structure of the nucleosomes to make the DNA available for interaction with RNA polymerase.

13.4   DNA methylation

In the DNA of higher animals about 70 % of the cytosine residues that are found in the dinucleotide sequence CpG are methylated in the 5-position, whereas it is very unusual to find other cytosine residues methylated. The methylation of cytosine does not affect its ability to base pair

Table 13.4. *Genes whose transcription is inhibited by methylation*

Gene	Species and tissue	Effect
Thymidine kinase	Herpes simplex (virus)	Gene is hypomethylated when expressed. Expression of gene is inhibited by prior methylation
Adenosine-phosphoribosyl transferase	Mouse many tissues	Sequence 5′ to the coding portion of the gene is unmethylated in many mouse tissues where it is expressed
ε-globin	Human	Sites 5′ to the gene are unmethylated during expression in yolk sac and foetal liver. They become methylated at later times in foetal liver and in adult bone marrow when it is not expressed
γ-globins	Human	Sites 5′ to the genes are methylated in early foetal liver and in adult bone marrow when not expressed. They are unmethylated in later foetal liver when they are expressed
Vitellogenin	Cock liver	Methylation of the gene is decreased after oestrogen treatment which leads to its expression
β- and γ-casein	Rat liver and mammary gland	Genes are methylated in liver where they are not expressed, but hypomethylated during expression in the mammary gland

with guanine, so CpG base pairs are formed during replication whether or not the original cytosine residue is methylated. Many of these cytosine residues are maintained stably in a methylated state through many rounds of cell division. This is accomplished by the methylation of cytosine residues that have been newly incorporated into DNA by the activity of a methylating enzyme that specifically catalyses the methylation of cytosine residues in a CpG pair that is base paired to a GpC pair that is already methylated (Fig. 13.1).

It has been discovered that in a number of cases where genes are not

Fig. 13.1. Maintenance methylation of newly incorporated cytosine residues after DNA replication.

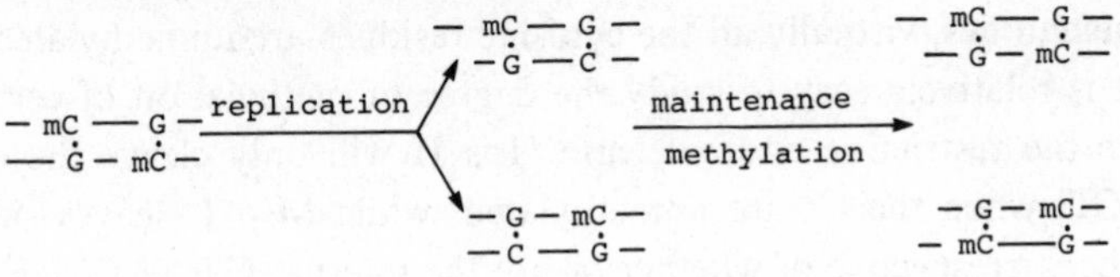

Table 13.5. *Genes where there is poor correlation between transcription and methylation*

Gene	Species and tissue	Effect
Immunoglobulin	Mouse plasmacytomas	Undermethylated in all expressed and in some unexpressed genes
Immunoglobulin	Mouse liver	Genes are methylated and unexpressed
δ-crystallin	Chick embryo	Demethylation follows induction of gene transcription in both lens and other embryonic tissues
rRNA	*Xenopus laevis* sperm	Gene expressed when non-transcribed spacer is fully methylated
rRNA	*Xenopus laevis* other tissues	Gene expressed when non-transcribed spacer is hypomethylated
Vitellogenin	*Xenopus laevis* liver	Gene expressed after stimulation by oestrogen with no change in level of methylation
Insulin	Rat pancreas, insulinoma, and liver	No correlation between expression of the two insulin genes and their level of methylation

being transcribed (e.g. in certain tissues or at certain stages in development), there are methylated cytosine residues in their 5'-flanking regions (Table 13.4). It is even possible to methylate cytosine residues 5' to certain genes that are normally transcribed, introduce them into cells and show that they are no longer expressed. Thus, it has been suggested that methylation of these residues provides a means whereby gene expression is controlled. While this may be true in certain cases, there are also a number of situations in which there is no apparent correlation between the degree of methylation and the level of transcription (Table 13.5). In no cases are there any clues to the factors in the cell which determine the degree of methylation of cytosine residues that might be critical for gene expression. In fact, it is even possible that cytosine methylation or demethylation could be the effect of changes in the level of transcription rather than its cause.

Cytosine methylation is not used universally as a means of regulating gene expression, since in some invertebrates, notably in *Drosophila* and in nematodes, virtually all the cytosine residues are unmethylated.

It is relatively easy to study the degree of methylation of certain sites since the restriction endonuclease Hpa II will only cleave the sequence CCGG when the Cs are unmethylated, while Msp I cleaves similar sequences irrespective of whether or not the internal C is methylated. Thus, comparison of restriction maps produced with these two enzymes allows the detection of methylated cytosine residues. Unfortunately, some cytosine residues, which are not in the sequence CCGG recognised by this pair of enzymes, may become methylated, so that some changes in methylation will go undetected by this simple test. Such changes might be occurring in situations like those listed in Table 13.5 where there appears to be no correlation between the degree of methylation and gene expression.

A valuable tool in the study of methylation of DNA is the cytidine analogue 5-aza-cytidine (Fig. 13.2). It can be incorporated into DNA in place of cytidine, but it is not capable of being methylated. It also inhibits the methylation of cytosine residues in DNA. Transcription of the γ-genes of haemoglobin (normally expressed only in the foetus) can be switched on in reticulocytes from adults (where they would not normally be transcribed) by incubation with 5-aza-cytidine. This leads to demethylation in the 5'-flanking sequence of the γ-genes. It has been suggested that this compound might be used to treat certain forms of thalassaemia so that the affected individuals would synthesise the foetal forms of haemoglobin and so overcome their lack of the normal adult form. Although foetal haemoglobin is better adapted to the physiological state of the foetus, its presence in the adult would provide a significant advantage over the defective thalassaemic forms of the protein. Unfortunately, this kind of therapy is probably not very specific, and other changes in the expression of the genome might occur following treatment with 5-aza-cytidine.

13.5   DNase sensitivity

Regions of eukaryotic genomes that are being actively transcribed often show increased susceptibility to digestion with various DNases. This would be expected if the nucleosome structure in these regions is opened up so that the proteins and enzymes necessary for transcription are able to gain access to the DNA. This increased susceptibility can be

Fig. 13.2. Cytidine (left) and 5-aza-cytidine (right), showing the 5-site of methylation in the former, and the impossibility of methylation at the 5-site in the latter.

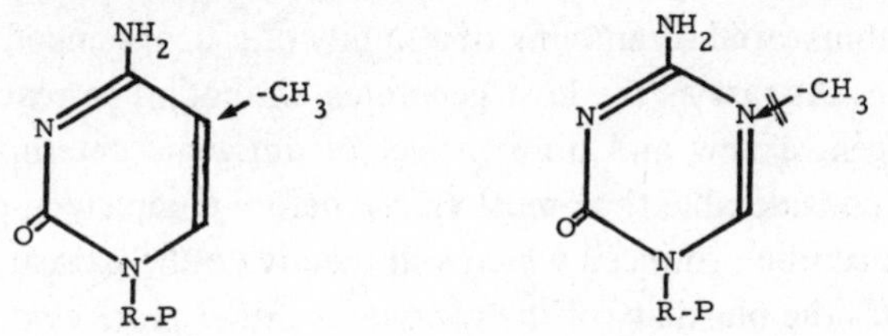

used as a valuable tool to demonstrate transcriptionally active portions of the genome, but as it is probably a consequence of their activation it tells us nothing of the factors which actually promoted the transcription of these regions.

13.6   The plasticity of the genome

One of the most striking findings that have emerged from recent studies of the genome is that it is continually in a dynamic state and not a fixed stable body of information that is always handed on intact from one cell to its descendants. This is largely because there are a number of ways in which recombination between the various portions of genomic DNA can take place. Insertion sequences and transposons move around the genome, often taking neighbouring stretches of DNA with themselves, and they can be inserted at sites where they may disrupt and inactivate various functional DNA sequences. Perhaps more occasionally they can induce the transcription of certain DNA sequences in circumstances in which they are not normally transcribed. Gene conversions and crossovers can generate new genes and link together genes that were previously un-linked. Translocations between chromosomes can cause drastic effects in activating genes that are not normally transcribed, and are probably one of the mechanisms by which cancers arise. Finally, viruses can be inte-grated into the host chromosomes and subvert the metabolism of the infected cell through the production of virally encoded products. Many of these changes, particularly those occurring in somatic cells, may never be observed since they may be so deleterious to the cell in which they occur that these affected cells do not survive, or they may reproduce so slowly that the effects are swamped out by more rapidly growing 'normal' cells. Effects on the genome of gametes are more likely to be observed, though even here, where many million spermatozoa die for each one that actually fertilises an ovum, the chances of observing changes are extremely low.

Occasionally, advantage has been taken of this plasticity of the genome as in the production of the vast number of antibodies of differing specificity that can arise by random reshuffling of the cistrons that go to make up the complete H and L chain genes (Chapter 10.6).

When the protozoal trypanosomes parasitise a host, there is rapid and repeated recombinatorial scrambling of a family of cistrons encoding their surface antigens. As fast as the host generates antibodies to react with a particular antigen, a new and different set of antigenic determinants is synthesised. A considerable time must elapse before complementary anti-bodies to these can be produced which will inactivate the parasite.

In prokaryotes the plasticity of the genome is rather more evident since they usually divide more rapidly than eukaryotic cells so that favourable mutations can spread rapidly through a population. It is comparatively easy, for example, to select bacteria to grow on an unusual nutrient by cultivating them continuously in the presence of steadily increasing concentrations of it. Equally, resistance to toxic substances such as antibiotics is fairly rapidly acquired, especially through the transmission of plasmids, and this has become a major hazard especially in hospitals throughout the more developed countries.

So it seems that the genome is subject to considerable changes and modifications and it is probably valuable to have considerable portions of it (such as introns and non-coding sequences) which provide a pool of material which is of potential value for the production of new materials which can be selected by natural selection if they are found to have survival value.

13.7 Evolution

The plasticity of the genome provides an explanation for the evolution of species, regardless of the means by which this is driven. Comparison of the sequences of homologous proteins (or, better still, the genes encoding them) from different species can suggest the pathways by which present-day proteins have evolved. Where there is good palaeontological evidence of the time of divergence of species it is possible to compare this with the rate of accumulation of mutations in the DNA to provide a molecular clock for evolution. Introns and non-coding portions of the genome which are not involved in the controlling functions provide the best parts for the study of mutation rates. A recent estimate of this rate is about 1 mutation in 2×10^8 sites per year. Even in the DNA actually coding for proteins mutations may be silent because of the redundancy of the genetic code. In certain portions of most proteins constraints on the precise structure may not be very great. Mutations which conserve the properties of the amino acids that are encoded (e.g. leucine for isoleucine or valine; glutamate for aspartate) are commonly found. However, in certain critical regions intimately concerned with the protein's function very little change can be tolerated. These regions tend to be highly conserved, and will evolve at a much slower rate (if at all) than other regions.

In situations where the structures of a large number of homologous proteins or genes are available it is possible to construct trees showing the routes by which they have probably evolved. A specially favourable case is that of the antibody family of proteins (Fig. 13.3).

The overall rates of evolution observed in different proteins are strikingly different (Table 13.6). At one extreme are the histones H3 and

H4, while at the other are the immunoglobulins and snake venoms. There are probably two major factors accounting for these wide differences. Conservation of function is obviously important and presumably limits the changes that can be tolerated in proteins like enzymes with highly specific functions. In this context, it is interesting that there has been wide diversification of some classes of enzymes which have identical catalytic centres with very strong conservation of amino acids directly involved in the catalytic act. Neighbouring parts of the molecules which function as binding sites for the substrates have evolved much faster and led to the ability to catalyse reactions of different substrates. This is seen, for example, in the large family of serine proteases (e.g. trypsin, chymotrypsin, elastase), and also in the phosphokinases where the binding site for ATP has been conserved while that for the second substrate varies quite widely from enzyme to enzyme.

Dispensibility is also an important factor. Serum albumin is an excellent example of this. It has evolved relatively rapidly, and to a large extent, its functions can be taken over by other plasma proteins, since in the admittedly rare condition of analbuminaemia the total lack of the protein

Fig. 13.3. Tree showing the probable evolutionary path of some κ and λ chains of immunoglobulins. The circles represent points at which lineages diverge. The linear distances between them are proportional to the number of changes in amino acids that have occurred. (Modified from L Hood. *Fed. Proc.* (1976) **35**, 2162. Fig. 6, reprinted with permission.)

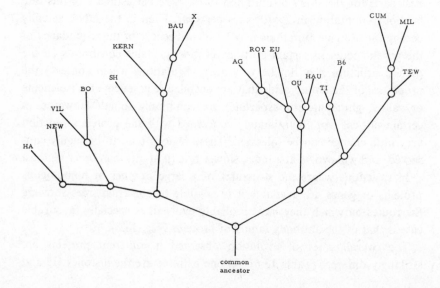

Table 13.6. *Rates of evolutionary change in the amino acid sequence of some proteins*

Protein	(*a*)
Histones	
H4	400
H3	330
Structural proteins	
Collagen	36
Crystallin	22
Intracellular enzymes	
Glutamate dehydrogenase	55
Lactate dehydrogenase	19
Cytochrome c	15
Carbonic anhydrase	4
Extracellular enzymes	
Trypsin	6
Ribonuclease	2.3
Hormones	
Glucagon	43
ACTH	24
Insulin	14
Insulin C peptide	1.9
Growth hormone	4
Serum proteins	
Albumin	3
Immunoglobulin V regions	0.7–1.0
Immunoglobulin C regions	0.9–1.7
Snake venoms	0.8–0.9

(*a*) The numbers refer to the unit evolutionary period, the numbers of millions of years for 1 % change in sequence.

does not seem to lead to impaired viability. Snake venoms and immuno-globulin are also free to evolve rapidly provided that toxicity and antigen-binding properties respectively are maintained. In fact, diversification of the latter is likely to be advantageous in conferring protection against a wider range of noxious molecules.

Gene duplications have been a common feature during evolution. This is especially evident in duplications of a number of globin genes (Chapter 9.1). Similar events must have given rise to the various forms of the H chains of antibodies. Here it is of interest to note how the H chain clusters in the mouse and human genomes have undergone duplications in different ways (Fig. 10.6).

Banks of DNA sequences have been assembled. By comparing them with the aid of specially written computer programs it is possible to infer the existence of super-families of genes that seem to have evolved from a common ancestor. One such example is the super-family of immunoglobulin genes which also includes the genes for the proteins of the Major Histocompatability Complex, as well as the gene for a protein which transports some immunoglobulins across cells, and, surprisingly, the enzyme superoxide dismutase.

13.8   Future developments

With the advent of rapid and reliable DNA sequencing the sequence of many long stretches of DNA is now known, and it is likely that the complete sequence of the *E. coli* genome will be available in the not too far distant future.

Although elucidation of the complete structure of the human genome is a much more distant prospect, it is likely that as knowledge of this accumulates benefits will become available in the treatment of some heritable diseases, and, perhaps, of cancer. For these purposes it is equally important to understand how the expression of genetic information is regulated and controlled.

Many unexpected and fascinating concepts and mechanisms have come to light as a result of experimental work on the genome, and it is almost certain that Nature still has a number of surprises in store for us as we gain further information.

Bibliography

Reading list for Chapter 1

Zimmerman, S. B. The three-dimensional structure of DNA. *Ann. Rev. Biochem.* 1982, **51**, 395–427.

Bauer, W. R., Crick, F. H. C. & White, J. H. Supercoiled DNA. *Sci. Amer.* 1980, **243**(1), 100–13.

Cozzarelli, N. R. DNA gyrases and the supercoiling of DNA. *Science*, 1980, **207**, 953–60.

Gellert, M. DNA topoisomerases. *Ann. Rev. Biochem.* 1981, **50**, 879–910.

Cold Spring Harbor Symposium on Quantitative Biology, 1983, vol. 47, entitled 'Structures of DNA' is a valuable source of material, particularly the Introduction and the Summary, and the sections on 'The handedness of DNA' (pp. 13–285) and 'Gyrases and topoisomerases' (pp. 763–93).

Reading list for Chapter 2

rRNAs
Wittmann, H. G. Components of bacterial ribosomes. *Ann. Rev. Biochem.* 1982, **51**, 155–83.

tRNAs
Kim, S.-H. Three dimensional structure of transfer RNA and its functional implications. *Adv. Enzymol.* 1978, **46**, 279–315.

Schimmel, P. R., Soll, D. & Abelson, J. (eds.). *Transfer RNA*. Cold Spring Harbor, N.Y.: Cold Spring Harbor Laboratory, 1980.

Small RNAs
Busch, H., Reddy, R., Rothblum, L. & Choi, Y. C. Sn RNAs, Sn RNPs, and RNA processing. *Ann. Rev. Biochem.* 1982, **51**, 617–654.

Translation
Lake, J. A. The ribosome. *Sci. Amer.* 1981, **245**(2), 56–69.

Wool, I. G. The structure and function of eukaryotic ribosomes. *Ann. Rev. Biochem.* 1979, **48**, 719–54.

Kozak, M. Comparison of initiation of protein synthesis in prokaryotes, eukaryotes and organelles. *Microbiol. Rev.* 1983, **47**, 1–45.

163

Reading list for Chapter 3

Watson, J. D., Tooze, J. & Kurtz, D. T. *Recombinant DNA – a short course.* Scientific American Books. New York: W. H. Freeman and Co., 1983.

Reverse transcriptase
Verma, I. M. Reverse transcriptase. In *The Enzymes*, vol. XIV, A (ed. P. D. Boyer). New York: Academic Press, 1981.

Nucleases
Linn, S. Deoxyribonucleases: survey and perspectives. In *The Enzymes*, vol. XIV, A (ed. P. D. Boyer). New York: Academic Press, 1981.

Restriction endonucleases
Smith, H. O. Nucleotide sequence specificity of restriction endonucleases. *Science,* 1979, **205**, 455–62.
Wells, R. D., Klein, R. D., & Singleton, C. K. Type II restriction enzymes. In *The Enzymes*, vol. XIV, A (ed. P. D. Boyer). New York: Academic Press, 1981.

Base sequence determination
Gilbert, W. DNA base sequencing and gene structure. *Bioscience Rep.* 1981, **1**, 353–75.
Sanger, F. Determination of nucleotide sequences in DNA. *Bioscience Rep.* 1981, **1**, 3–18.

Reading list for Chapter 4

Plasmids
Bolivar, F. & Backman, K. Plasmids of *E. coli* as cloning vectors. *Methods Enzymol.* (ed. R. Wu), 1979, **68**, 245–67.

Cosmids
Collins, J. *Escherichia coli* plasmids packageable *in vitro* in λ bacteriophage particles. *Methods Enzymol.* (ed. R. Wu), 1979, **68**, 309–25.
Hohn, B. *In vitro* packaging of λ and cosmid DNA. *Methods Enzymol.* (ed. R. Wu), 1979, **68**, 299–308.

Phages
Hohn, T. & Katsura, I. Structure and assembly of bacteriophage lambda. *Curr. Topics Microbiol. Immunol.* 1977, **78**, 69–110.
Nash, H. A. Integration and excision of bacteriophage λ. *Curr. Topics Microbiol. Immunol.* 1977, **78**, 174–99.
Blattner, F. R. et al. Charon phages: safer derivatives of bacteriophage Lambda for DNA cloning. *Science,* 1977, **196**, 161–9.

Viruses
Anderson, W. F. & Diacumakos, E. G. Genetic engineering in mammalian cells. *Sci. Amer.* 1981, **245**(1), 60–93.
Berg, P. Dissections and reconstructions of genes and chromosomes. *Bioscience Rep.* 1981, **1**, 269–87.
Howard, B. H. Vectors for introducing genes into cells of higher eukaryotes. *Trends Biochem. Sci.* 1983, **8**, 209–12.

Reading list for Chapter 5

Glass, R. E. Gene function. *E. coli* and its heritable elements. London: Croom Helm, 1982.
Bachman, B. J., has published a map of the *E. coli* chromosome in *Microbiol. Rev.* 1983, **47**, 180–230.

Replication
Kornberg, A. DNA replication. San Francisco: W. H. Freeman and Co., 1980.
McHenry, C. & Kornberg, A. DNA polymerase III holoenzyme. In *The Enzymes*, vol. XIV, A (ed. P. D. Boyer). New York: Academic Press, 1981.
Ogawa, T. & Okazaki, T. Discontinuous DNA replication. *Ann. Rev. Biochem.* 1980, **49**, 421–57.

Promoters
Hawley, D. K. & McClure, W. R. Compilation and analysis of *Escherichia coli* promoter DNA sequences. *Nucleic Acid Res.* 1983, **11**, 2237–56.
Rosenberg, M. & Court, D. Regulatory sequences involved in the promotion and termination of RNA transcription. *Ann. Rev. Genet.* 1979, **13**, 319–53.

RNA processing
Abelson, J. RNA processing and the intervening sequence problem. *Ann. Rev. Biochem.* 1979, **48**, 1035–69.

Transposons
Papers in *Cold Spring Harbor Symposium on Quantitative Biology*, 1980, **45**, 1–297.

Reading list for Chapter 6

Miller, J. H. & Reznikoff, W. S. (eds.) *The Operon.* Cold Spring Harbor, N.Y.: Cold Spring Harbor Laboratory, 1980.

Attenuators
Crawford, I. P. & Stauffer, G. V. Regulation of tryptophan biosynthesis. *Ann. Rev. Biochem.* 1980, **49**, 163–95.
Yanofsky, C. Attenuation in the control of bacterial operons. *Nature*, 1981, **289**, 751–8.

Ribosomal proteins
Nomura, M. The control of ribosome synthesis. *Sci. Amer.* 1984, **250**(1), 72–83.

Stringent response
Gallant, J. A. Stringent control in *Escherichia coli*. *Ann. Rev. Genet.* 1979, **13**, 393–415.
Travers, A. Conserved features of coordinately regulated *E. coli*. promoters. *Nucleic Acid Res.* 1984, **12**, 2605–17.

Reading list for Chapter 7

Histones
Isenberg, I. Histones. *Ann. Rev. Biochem.* 1979, **48**, 159–91.

Nucleosomes
Kornberg, R. D. & Klug, A. The nucleosome. *Sci. Amer* 1981, **244**(2), 48–60.
McGhee, J. D. & Felsenfeld, G. Nucleosome structure. *Ann. Rev. Biochem.* 1980, **49**, 1115–56.

Transcription

Bensimhon, M., Gabarro-Arpa, J., Ehrlich, R. & Reiss, C. Physical characteristics in eucaryotic promoters. *Nucleic Acid. Res.* 1983, **11**, 4521–40.

Breathnach, R. & Chambon, P. Organisation and expression of eucaryotic split genes coding for proteins. *Ann. Rev. Biochem.* 1981, **50**, 349–83.

Darnell, J. E. Jr. Variety in the level of gene control in eukaryotic cells. *Nature*, 1982, **297**, 365–72.

Lewis, M. K. & Burgess, R. R. Eukaryotic RNA polymerases. In *The Enzymes*, vol. XIV, B (ed. P. D. Boyer). New York: Academic Press, 1982.

Paule, M. R. Subunit composition of the eukaryotic nuclear RNA polymerases. *Trends Biochem. Sci.* 1981, **6**, 128–31.

Enhancers

Gillies, S. D., Morrison, S. L., Oi, V. T. & Tonegawa, S. A tissue specific transcription enhancer element is located in the major intron of a rearranged immunoglobulin heavy chain gene. *Cell*, 1983, **33**, 717–28.

RNA processing

Darnell, J. E. Jr. The processing of RNA. *Sci. Amer.* 1983, **249**(4), 72–82.

Edmonds, M. Poly(A) adding enzymes. In *The Enzymes*, vol. XIV, B (ed. P. D. Boyer). New York: Academic Press, 1982.

Filipowicz, W. & Gross, H. J. RNA ligation in eukaryotes. *Trends Biochem. Sci.* 1984, **9**, 68–71.

Proudfoot, N. The end of the message and beyond. *Nature*, 1984, **307**, 412–13.

Nevins, J. R. The pathway of eukaryotic mRNA formation. *Ann. Rev. Biochem.* 1983, **52**, 441–66.

Shuman, S. & Hurwitz, J. Capping enzyme. In *The Enzymes*, vol. XIV, B (ed. P. D. Boyer). New York: Academic Press, 1982.

Reading list for Chapter 8

Histone genes

Kedes, L. H. Histone genes and histone messengers. *Ann. Rev. Biochem.* 1979, **48**, 837–70.

Repeated genes

Long, E. O. & Dawid, I. B. Repeated genes in eucaryotes. *Ann. Rev. Biochem.* 1980, **49**, 727–64.

Jelinek, W. R. & Schmid, C. W. Repetitive sequences in eukaryotic DNA and their expression. *Ann. Rev. Biochem.* 1982, **51**, 813–44.

Schmid, C. W. & Jelinek, W. R. The *Alu* family of dispersed repetitive sequences. *Science*, 1982, **216**, 1065–70.

Papers in *Cold Spring Harbor Symposium on Quantitative Biology*, 1980, **45**, 519–673.

Retroviruses and oncogenes

Bishop, J. M. Oncogenes. *Sci. Amer.* 1982, **246**(3), 68–78.

Bishop, J. M. Cellular oncogenes and retroviruses. *Ann. Rev. Biochem.* 1983, **52**, 301–54.

Land, H., Parada, L. F. & Weinberg, R. A. Cellular oncogenes and multistep carcinogenesis. *Science*, 1983, **222**, 771–8.

Leder, P. et al. Translocations among antibody genes in human cancer. *Science*, 1983, **222**, 765–71.

Robertson, M. Paradox and paradigm: the message and meaning of *myc. Nature*, 1983, **306**, 733–6.

Varmus, H. E. Form and function in retroviral proviruses. *Science*, 1982, **216**, 812–20.

Weinberg, R. A. A molecular basis for cancer. *Sci. Amer.* 1983, **249**(5), 102–16.

Reading list for Chapter 9

Globin genes

Efstratiadis, A. et al. The structure and evolution of the human β-globin gene family. *Cell*, 1980, **21**, 653–68.

Flavell, R. A. The globin genes of rabbit and man. *Biochem. Soc. Trans.* 1983, **11**, 111–18.

Proudfoot, N. J., Shander, M. H. M., Manley, J. L., Gefter, M. L. & Maniatis, T. Structure and *in vitro* transcription of human globin genes. *Science*, 1980, **209**, 1329–36.

Thalassaemias

Forget, B. G. Molecular studies of genetic disorders affecting the expression of the human β-globin gene. *Rec. Progr. Hormone Res.* 1982, **38**, 257–74.

Weatherall, D. J. & Clegg, J. B. The thalassaemia syndromes, 3rd edn. Oxford: Blackwell Scientific Publications, 1981.

Mutations

Vella, F. Human haemoglobins and molecular disease. *Biochem. Educ.* 1980, **8**, 41–53.

Reading list for Chapter 10

Immunoglobulins

Davis, M. M., Kim, S. K. & Hood, L. E. DNA sequences mediating class switches in α-immunoglobulins. *Science*, 1980, **209**, 1360–5.

Honjo, T. Immunoglobulin genes. *Ann. Rev. Immunol.* 1983, **1**, 499–528.

Rabbits, T. H. The human immunoglobulin genes. *Biochem. Soc. Trans.* 1983, **11**, 119–26.

Tonegawa, S. Somatic generation of antibody diversity. *Nature*, 1983, **302**, 575–81.

Wall, R. & Kuehl, M. Biosynthesis and regulation of immunoglobulins. *Ann. Rev. Immunol.* 1983, **1**, 393–422.

Major histocompatibility complex

Hood, L., Steinmetz, M. & Malissen, B. Genes of the major histocompatibility complex of the mouse. *Ann. Rev. Immunol.* 1983, **1**, 529–68.

Steinmetz, M. Structure, function and evolution of the major histocompatibility complex of the mouse. *Trends Biochem. Sci.* 1984, **9**, 224–6.

Steinmetz, M. & Hood, L. Genes of the major histocompatibility complex in mouse and man. *Science*, 1983, **222**, 727–33.

Reading list for Chapter 11

Insulin

Bell, G. I., Pictet, R. L., Rutter, W. J., Cordell, B., Tischer, E. & Goodman, H. M. Sequence of the human insulin gene. *Nature*, 1980, **284**. 26–32.

Growth hormone

Moore, D. D., *et al.* Structure, expression and evolution of growth hormone genes. *Rec. Progr. Hormone Res.* 1982, **38**, 197–222.

Polyproteins
Richter, D. Vasopressin and oxytocin are expressed as polyproteins. *Trends Biochem. Sci.* 1983, **8**, 278–81.

Calcitonin
Rosenfeld, M. G., *et al.* Calcitonin, prolactin and growth hormone gene expression as model systems for characterisation of neuroendocrine regulation. *Rec. Progr. Hormone Res.* 1983, **39**, 305–48.

Glycoprotein hormones
Talmadge, K., Vamvakopoulos, N. C. & Fiddes, J. C. Evolution of the genes for the β subunits of human chorionic gonadotrophin and luteinising hormone. *Nature*, 1984, **307**, 37–40.

Reading list for Chapter 12

Anderson, S. et al. Sequence and organisation of the human mitochondrial genome. *Nature*, 1981, **290**, 457–65.
Attardi, G., Borst, P. & Slominski, P. P. *Mitochondrial Genes.* Cold Spring Harbor, N.Y.: Cold Spring Harbor Laboratory, 1982.
Fox, T. D. Multiple forms of mitochondrial DNA in plants. *Nature*, 1984, **307**, 415.
Grivell, L. A. Mitochondrial DNA. *Sci. Amer.* 1983, **248**(3), 60–73.
Michel, F. & Dujon, B. Conservation of RNA secondary structures in two intron families including mitochondrial-, chloroplast- and nuclear-encoded members. *EMBO J.* 1983, **2**, 33–8.
Wallace, D. C. Structure and evolution of organelle genomes. *Microbiol. Rev.* 1982, **46**, 208–40.

Reading list for Chapter 13

Bird, A. P. DNA methylation – how important in gene control? *Nature*, 1984, **307**, 503–4.
Doerfler, W. DNA methylation and gene activity. *Ann. Rev. Biochem.* 1983, **52**, 93–124.
Hood, L. Antibody genes and other multigene families. *Fed. Proc.* 1976, **35**, 2158–67.
Ringold, G. M. et al. Glucocorticoid regulation of gene expression; mouse mammary tumour virus as a model system. *Rec. Progr. Hormone Res.* 1983, **39**, 387–421.

Index

F refers to Figures, T refers to Tables

acetylation 89
ACTH (adrenocorticotrophin) 136, 137T
acylation 100
albumin 96, 160, 161T
allele 5
allelic exclusion 127
alternative splicing 98, 135, 140
Alu family 103
amber codon 21
amidation 138
AMP (adenosine monophosphate) 14, 15F
amphibia 17, 101
annealing 38, 50
antibiotic 47, 48T, 51, 54, 64, 158
antibody – *see* immunoglobulin
anticodon 18, 19F, 21, 22, 23, 144, 149
apo-activator 67, 68
apo-repressor 67, 68, 77, 81
ara operon 74–5
arginine biosynthesis 81–2
ATP (adenosine triphosphate) 8, 14, 42
attenuation 68, 77, 78, 80
autoregulation 74, 81
Avery 1
aza-cytidine 157

B-lymphocyte 120, 126, 129
β-endorphin 136, 137T
β$_2$-microglobulin 128, 129
bacteria 4, 29, 47, 51, 52, 66
bacteriophages 23, 30, 52–5, 56, 65
bacteriophage M13 44, 45, 52
bacteriophage λ 52, 53
base-pair 3
blunt ends 34, 49

Brattleboro rats 139
Burkitt's lymphoma 108, 109F
Bacillus subtilis 5T, 56, 60

c-*myc* 108, 109, 110
CAAT box 92, 133, 151, 152T
calcitonin 140
calcitonin gene-related product 140
cAMP (cyclic AMP) 69, 70, 74
cancer 106, 108, 158, 162
cap 93–5, 106, 149
catabolite activator protein (CAP) 69, 70, 73, 74, 75, 76, 77, 152
chloramphenicol 47, 48T
chorionic gonadotrophin 141, 142
chromatin 153
chromosome 5, 6, 12, 47, 52, 88, 108–110, 127, 158
chromosome, circular 5, 6, 56, 143
chromosome, *E. coli* 52, 53F, 56, 58, 64F, 80, 81
chromosome, homologous 5, 12, 86, 127
chromosome, linear, 6, 86
chromosome, sex 5, 86
chromosome walking 37
cistron 10
clone 29, 86, 120
co-activator 67, 68T, 69T
CMP (cytidine monophosphate) 14, 15T
codon 21, 22, 23, 26, 144, 145T, 149, 152
codon, initiation 22, 23, 24, 148
codon, termination 21, 22, 27, 113, 114, 117, 124, 136, 139, 142, 144, 145T, 149
cohesive ends 53
colicin 48
complement 131

complement genes 127, 131
complementation (genetic) 5, 69
concatamer 54
constant region 118, 161T
constant region genes 120, 123, 130
co-ordinate control 66
copia 104
co-repressor 67, 68T, 69T, 81
cosmid 48T, 54
CRP – *see* catabolite activator protein
cross-over 12, 115, 116, 158
CTP (cytidine triphosphate) 14
cytochromes 12
cytochrome b 144F, 145, 146, 149
cytochrome oxidase 144F, 145, 146, 149
cytosine 11, 154–7

D-genes 122, 125
dAMP (deoxyadenosine monophosphate)
    1, 2F
dCMP (deoxycytidine monophosphate)
    1, 2F
deletion 12, 22, 36, 37, 39, 108, 113,
    115, 117, 123, 154
denaturation 30, 38, 148
dGMP (deoxyguanosine monophosphate)
    1, 2F
dideoxynucleoside triphosphate 44
diploid 6, 86
direct terminal repeat 104
DNA 1, 11, 12, 13, 14, 17, 21, 28, 29,
    30, 33, 36, 37, 38, 40, 42, 43, 47, 50,
    54, 58F, 64, 69, 86, 87, 88, 95, 106,
    125, 152, 158, 159
DNA cloning 29
DNA, complementary (cDNA) 28, 31,
    46, 47, 125, 133, 140
DNA, double-stranded 56, 61
DNA, extra-chromosomal 17, 101
DNA gyrase 6, 7F, 56
DNA, mitochondrial 143, 147, 149
DNA, passenger 49, 50F, 51, 53, 54, 55
DNA polymerase 8, 31, 32, 45, 57, 59,
    90, 106, 143
DNA polymerase, proof reading 33, 59
DNA primase 57, 59
DNA, recombinant 29, 48
DNA repair 57
DNA, single-stranded, 8, 30, 56
DNase (deoxyribonuclease) 31T, 32, 89
DNase sensitivity 157–8
domain 96, 118, 124, 129, 130, 131
Drosophila 5T, 101, 104, 156

elongation 25, 61
elongation factors 25, 84
electrophoresis 6, 7F, 29, 30, 34, 36, 37,
    40, 42, 51
endonuclease 16, 31T, 96, 125, 149
endonuclease, restriction 28, 30, 32, 33–7,
    43, 48, 49, 50F, 51, 52, 54, 156
enhancers 92–3, 126–7
enkephalins 137
epidermal growth factor (EGF) 107
Escherichia coli (*E. coli*) 5T, 10, 15F, 16,
    17, 23, 29, 31, 48, 56, 80, 162
ethidium bromide 30
eukaryotes 5, 11, 16, 20, 23, 24, 32,
    151, 153
evolution 11, 98, 125, 159–62
exon 96, 112, 113T, 120, 124, 126, 127,
    129, 130, 134, 138, 140, 146
exonuclease 31T, 32, 59, 63

fibrinogen 98
flanking sequences 10, 20, 111, 133, 136,
    149, 154, 156
FSH (follicle cell stimulating hormone)
    141T
fragile sites 110
free radicals 11
fungal genes 146, 147F

gal operon 69F, 70T, 72–4
gene 5, 10, 48, 53, 66
gene conversion 12, 122, 130, 139, 158
gene duplication 113, 131, 134, 161
gene, eukaryotic 40
gene, mutant 5
gene, plasmid 47
gene, regulatory 66
gene, structural 66, 67
gene superfamilies 162
genetic code 21–3, 144, 145T, 148, 149
genetic code, degeneracy 21, 159
genome 10, 30, 36, 39, 158, 162
genome, bacteriophage 52
genome, plasmid 47
globin 27, 111
globin genes 12, 22T, 37, 89, 111–17,
    155T, 157, 161
glycoprotein 99
glycoprotein hormones 141
GMP (guanosine monophosphate) 14,
    15F
Goldberg–Hogness box 92
growth hormone 98, 134–5

GTP (guanosine triphosphate) 14, 24, 25, 26, 85F
guanosine pentaphosphate 85
guanosine tetraphosphate 85

H chain 29, 118, 120, 122, 126
H chain genes 109, 122–5, 127, 158, 161
HLA 128, 129
haemoglobin 12, 22F, 29, 87, 111–17, 157
haemoglobins, abnormal 116–17
haemoglobins, Lepore 115
hairpin 14
haploid 6, 86
haplotype 6, 86
helix-destabilising proteins 8, 56
helix, double 2, 3, 5, 6, 8, 125
heteroduplex 39, 40
high mobility group proteins (HMG) 87
hinge region 118, 123
histones 87, 88, 159, 161T
histone genes 89, 101
homoduplex 39
hormones 29, 132–42, 153, 161T
Huntington's chorea 38
hut operon 75–6
hybridisation 38, 39, 51, 52, 98, 107, 120
hybridisation, stringency of 39

immune response genes 129
immunoglobulins 29, 98, 118–27, 130, 131, 160F
immunogloblin genes 12, 93, 96, 109, 120–7, 156T, 161T, 162
immunoprecipitation 29
inducer 67, 68T
inducer, gratuitous 71
initiation 16, 22, 23, 24, 61, 92, 115, 145
initiation factors 24, 25
inosine 18F
insects 4
insertion 12, 22, 36, 37, 39, 113, 117, 123
insertion sequences 63–5, 158
insulin 99, 132–4, 154, 161T
intervening sequences 96
introns 40, 93, 95–8, 101, 107, 112, 113T, 114, 115, 125, 132, 134, 136, 137, 145, 146, 147F, 149, 159
inverted repeat 57, 61, 64, 152
isoschizomer 33

J genes 120, 121, 122, 125

L chain 29, 118, 120, 122, 160F
L chain genes 109, 120–2, 125, 127, 158
lagging strand 59
leader sequence 78, 80, 99, 120
leading strand 59
Lepore haemoglobin 115
leukaemia 108, 109T, 110
ligase 9F, 10, 49, 53, 59
linker 49
long terminal repeat 104, 106
luteinising hormone (LH) 141, 142

magic spot nucleotides 84
mal regulon 76–7
major histocompatibility complex (MHC) 12, 127–31, 162
maturase 145T, 146
melanotrophin (MSH) 136, 137T
melting temperature 38
metazoans 11, 86
methionine 22, 23, 26, 143, 145T, 148
methionine, *N*-formyl 23, 143, 145T, 148
methylation 33, 41, 91, 93, 154–7
mitochondria 143–50
mutagen 11
mutation 11–13, 22, 37, 58, 68, 80, 108, 113, 115, 116, 131, 132, 136
mutation, *cis*-acting 69
mutation, point 11, 27, 36, 37, 38F, 60, 107, 108, 114
mutation rate 13, 98, 159
mutation, silent 13, 22
mutation, somatic 122, 123
mutation, *trans*-acting 68, 69
myelomatosis 120

neoplasia 108, 110
neurophysins I and II 138
nitrous acid 11
Northern blotting 39
nucleolus 91
nucleosomes 88–90, 154, 157

ochre codon 21
Okazaki fragments 9F, 10, 59
oncogenes 105–9
opal codon 21
operator 66, 67, 68, 69T, 70, 71, 72, 74, 82, 152
operon 62T, 66–85, 153
operon, constitutive 66
operon, inducible 66, 67, 68
operon, repressible 66, 67, 68

oriC 56, 57, 59
orphon 102
ovalbumin gene 97F
oxytocin 138

palindrome 34
phage – *see* bacteriophage
phenotype 10
phenylketonuria 38
Philadelphia chromosome 108
phosphodiester links 1, 14, 31, 57, 59, 62, 96
phosphokinase 100F, 160
phosphorylation 89, 99, 107, 108
placental lactogen 134
plasma cells 120
plasmacytoma 29, 109, 120
plasmid 6, 30, 36, 47–51, 52, 56, 57, 69, 71, 159
plasmid, yeast 51
platelet-derived growth factor (PDGF) 107
poly-A 29, 94, 95, 101, 103, 106, 124, 142, 149
polarity 1, 3F, 4, 10, 74
polycistronic mRMA 66
poly-proteins 135–40
polysome 26F, 27
post-translational modifications 23, 99–100
pre-emptor 78, 79F
prenatal diagnosis 37–8, 117
preproenkephalin 137–8
Pribnow box 60, 73, 75, 83, 92
priming 57
prolactin 134–5
promoter 48, 57, 60, 61, 66, 67, 70, 73, 83, 84, 85, 91, 107, 127, 145, 151
proof reading 59
pro-opiomelanocortin (POMC) 136–7
protector 78, 79F
proteins 10, 23
protein kinase 100, 107
Protista 146, 147F
pseudogenes 101, 113, 122, 125, 130, 136
pyrimidine biosynthesis 80

R factors 47
reading frame 22, 113, 115
regulon 76, 80, 81
relaxin 134
replication 8–10, 14, 56–9, 90, 155F
replication fork 9F, 10, 57, 58, 59

replication, initiation of 57
replication origin 47, 48, 56
replication, plasmid 47
repressor 66, 67, 68, 69T, 152
repressor, *ara* 74
repressor, *gal* 72
repressor, *hut* 76
repressor, *lac* 70, 71
repressor, *trp* 78
restriction fragment length polymorphism (RFLP) 37–8
restriction maps 36, 37F, 48, 135, 157
retrovirus 105–8
reverse transcriptase 28, 30, 105, 107, 125
rho (ρ) 16, 59, 61
ribosomal proteins 17, 82–4, 143
ribosomes 16T, 17, 24, 26, 27, 30, 61, 78, 80, 143, 147
rifampicin 57, 83
RNA 10, 14–27, 28, 30, 32, 38, 39, 57, 95, 105
RNA, heterogeneous nuclear (hnRNA) 20
RNA, messenger (mRNA) 20, 24, 29, 30, 40, 46, 47, 61, 66, 84, 90, 93–5, 105, 114, 125, 129, 132, 133, 139, 140, 143, 145, 148, 149, 153
RNA polymerase 8, 14, 16, 31, 32, 57, 59, 61, 67, 69, 72, 73, 78, 80, 82, 85, 90, 91, 93, 103, 105, 143, 145, 151, 152, 154
RNA polymerase genes 82
RNA, ribosomal (rRNA) 16, 17F, 25, 29, 62, 82, 84, 90, 91, 101, 102T, 143, 145, 149
RNA, transfer (tRNA) 17, 18, 19F, 23, 25, 26, 27, 29, 31, 32, 62, 63, 84, 85, 90, 91, 92, 102, 143, 145, 146, 148, 149, 150, 152
RNA, transfer, iso-accepting 23
RNA, small 20T, 95
RNA, stable 21, 66
RNA, U1 96, 97F
RNase (ribonuclease) 17, 30, 31, 32, 59, 62, 63, 161T

S1 nuclease mapping 32, 33F
Salmonella typhimurium 56, 75, 80
sequence determination 40–6, 52, 162
sequence ladder 42, 43F, 45
sequence, non-coding 159
sequence, protein 46, 159
sequence, repeated 102–5, 126, 133

serine proteases 160
Shine–Delgarno sequence 61
shot-gun library 34–6
shuttle vector 51
sickle cell anaemia 37, 116
sigma (σ) 16, 59, 60, 61
signal peptide 99, 103
single-stranded binding protein 8, 56
snurp 95
somatomammotrophin 134–5
Southern blotting 37, 39
splicing 96, 98, 113, 114, 135, 146
steroid hormones 154
stem and loop structures 15F, 57, 61, 62, 63F, 78, 80, 83, 104, 152
sticky ends 34, 48–9
stringent response 84–5
supercoiling 6, 7F, 8, 56
SV40 (Simian virus 40) 55
switch 119, 126

TATAA box 60, 92, 106, 133, 151, 152T
template 14, 28, 31, 43, 59
termination 16, 27, 61, 113, 115
terminator 15F, 78, 79F, 80, 83
thalassaemia 95, 96, 113–15, 157
thymine 14
T_m 38
TMP (thymidine monophosphate) 1, 2F, 29
topoisomerase 6, 8, 56
transcription 8, 14, 16, 56, 59–62, 64, 66, 70, 82, 84, 87, 89, 90–2, 93, 111, 115, 134, 143, 147, 149, 153, 154, 155T, 156T, 157, 158

transfection 47, 50, 51
transition 11
translation 23–7, 61, 64, 66, 147, 149
translocation 12, 109, 110, 158
transplantation antigens 127, 128, 131
transposons 63–5, 104, 106, 158
transversion 11
triplet 21, 23
trp operon 15F, 77–80
trypanosomes 158

ubiquitin 90
ultra-violet radiation 11, 33
UMP (uridine monophosphate) 14, 15, 29
uracil 14
UTP (uridine triphosphate) 14

V region genes 12, 120, 121, 122, 125, 161T
variable region 118
vasopressin 138, 139
vector 29, 30, 31, 47–55, 126
virus 4, 30, 52, 55, 93, 158

wobble hypothesis 22

X-rays 11

yeast 101, 104, 152
yeast genes 23, 92, 98
yeast mitochondria 143–7
yeast plasmid 51

THE CAMBRIDGE
FIELD GUIDE TO
PREHISTORIC
LIFE

THE CAMBRIDGE FIELD GUIDE TO
PREHISTORIC LIFE

David Lambert
and the Diagram Group

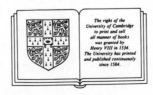

The right of the
University of Cambridge
to print and sell
all manner of books
was granted by
Henry VIII in 1534.
The University has printed
and published continuously
since 1584.

CAMBRIDGE UNIVERSITY PRESS

Cambridge

London   New York   New Rochelle   Melbourne   Sydney

Published by the Press Syndicate of the University of Cambridge
The Pitt Building, Trumpington Street, Cambridge, CB2 1RP
32 East 57th Street, New York, NY 10022, USA
10 Stamford Road, Oakleigh, Melbourne 3166, Australia

First published 1985

Printed in Great Britain at the University Press, Cambridge

British Library Cataloguing in Publication Data:

Lambert, David, *1932–*
  The Cambridge field guide to prehistoric life
  1. Palaeontology
  I. Title   II. Diagram Group
  560   QE711.2

ISBN 0 521 26685 8
ISBN 0 521 31299 X

The Diagram Group

Art director Mark Evans
Artists Graham Rosewarne, and
Alastair Burnside, Brian Hewson, Richard Hummerstone,
Karen Johnson, Alison Jones, Pavel Kostal,
Jerry Watkiss
Art assistant Robert Jones

Editor Ruth Midgley
Indexer Mary Ling

Consultants Dr R. A. Fortey, British Museum (Natural History),
London
Dr Angela Milner, British Museum (Natural
History), London
Dr Ralph E. Molnar, Queensland Museum,
Queensland, Australia
Professor R. J. G. Savage, University of Bristol,
England
Mr C. A. Walker, British Museum (Natural History),
London

Fossil hunters and fossils in
Wirksworth Cave, Derbyshire;
from an engraving of 1851.

FOREWORD

This book is a concise key to prehistoric animals, plants, and other organisms. It is the first to use field-guide techniques to picture and describe all the major forms of prehistoric life, giving precise details of what was what and what lived when. Large, labelled, life-like restorations, reconstructed skeletons, marginal "field-guide" illustrations, diagrams, and family trees – all integrated with the text – help readers to grasp important information at a glance. All this, and a use of both popular and scientific terms, will make the guide accessible to anyone from the inquiring eleven-year-old to the budding scientist.

There are ten chapters. Each has a brief explanatory introduction, followed by topics arranged under bold headings.

Chapter 1 (Fossil Clues to Prehistoric Life) briefly shows what fossils are and what they tell us of past life forms, their evolution, classification, and extinction.

Chapter 2 (Fossil Plants) describes briefly the major groups of plants known from the fossil record.

Chapter 3 (Fossil Invertebrates) ranges through all major groups of prehistoric animals that lacked a backbone.

Chapter 4 (Fossil Fishes) covers the world's earliest great group of backboned animals.

Chapter 5 (Fossil Amphibians) deals with the first group of backboned animals designed to walk on land.

Chapter 6 (Fossil Reptiles) describes vertebrates that dominated life on land for over two hundred million years.

Chapter 7 (Fossil Birds) gives a brief account of all known orders.

Chapter 8 (Fossil Mammals) gives examples from all major branches of the backboned group that replaced reptiles as masters of the land.

Chapter 9 (Records in the Rocks) shows which plants and creatures thrived through different periods and epochs on our ever-changing planet.

Chapter 10 (Fossil Hunting) tells how fossil hunters work and use what they discover. The chapter outlines achievements of famous palaeontologists and gives a worldwide list of fossil collections.

Lastly there is a list of books for further reading, and an index.

The producers of this guide consulted many works, but owe a special debt to these writers of books listed under Further Reading: U. Lehmann and G. Hillmer (invertebrates), and Edwin Colbert, Alan Feduccia, and Alfred Sherwood Romer (vertebrates). The author is answerable for all facts here presented, but thanks those named and unnamed experts whose advice has helped to make this book more accurate and up to date.

CONTENTS

Chapter 1
FOSSIL CLUES TO PREHISTORIC LIFE

12 Introduction
14 What fossils are
16 Fossils in the rocks
18 Fossil clues to evolution
20 How living things are classified
22 The tree of life
24 How ancient organisms lived
26 What lived where?
28 Mass extinctions
30 The oldest fossils

Chapter 2
FOSSIL PLANTS

32 Introduction
34 Simple plants
36 Protophytes
38 Vascular plants
40 Ferns and gymnosperms
42 Flowering plants

Chapter 3
FOSSIL INVERTEBRATES

44 Introduction
46 About invertebrates
48 Protozoans
50 Parazoans
52 Coelenterates
54 Molluscs
58 Worms
60 Arthropods
68 Tentaculates
70 Echinoderms and branchiotremes

Chapter 4
FOSSIL FISHES

72   Introduction
74   About fishes
76   Jawless fishes
78   Spiny fishes
80   Placoderms
82   Sharks and their kin
84   Bony fishes

Chapter 5
FOSSIL AMPHIBIANS

88   Introduction
90   About amphibians
92   Labyrinthodonts
96   Lepospondyls
97   Modern amphibians

Chapter 6
FOSSIL REPTILES

98   Introduction
100   About reptiles
102   Reptile pioneers
104   Turtles and tortoises
106   Early euryapsids
108   Plesiosaurs
110   Ichthyosaurs
112   Eosuchians
114   Lizards and snakes
115   Rhynchocephalians
116   Thecodonts
118   Crocodilians
120   Pterosaurs
122   Saurischian dinosaurs
126   Ornithischian dinosaurs
130   Mammal-like reptiles

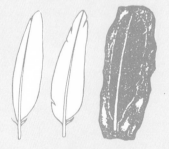

Chapter 7
FOSSIL BIRDS

136 Introduction
138 About birds
140 Early birds
142 Flightless birds
144 Water birds
146 Land birds

Chapter 8
FOSSIL MAMMALS

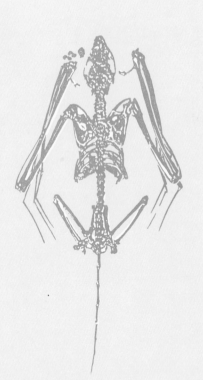

148 Introduction
150 About mammals
152 Early mammals
154 Pouched mammals
156 Placental pioneers
158 Primates
162 Creodonts
164 Modern carnivores
168 Condylarths
170 Amblypods
172 Subungulates
174 Proboscideans
178 South American ungulates
180 Horses
182 Brontotheres and chalicotheres
184 Tapirs and rhinoceroses
186 Early even-toed ungulates
188 Camels and their kin
190 Primitive pecorans
192 Deer and giraffes
194 Cattle and their kin
196 "Toothless" mammals
198 Whales
200 Rodents and rabbits

Chapter 9
RECORDS IN THE ROCKS

202 Introduction
204 Precambrian time
205 Cambrian Period
206 Ordovician Period
207 Silurian Period
208 Devonian Period
210 Carboniferous Period
212 Permian Period
214 Triassic Period
216 Jurassic Period
218 Cretaceous Period
220 Palaeocene Epoch
221 Eocene Epoch
222 Oligocene Epoch
223 Miocene Epoch
224 Pliocene Epoch
225 Pleistocene Epoch

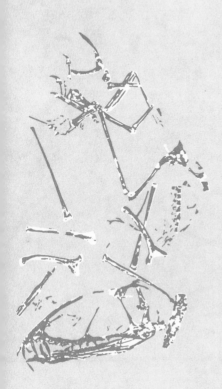

Chapter 10
FOSSIL HUNTING

226 Introduction
228 Finding fossils
230 Extracting fossils
232 Cleaning and repairing fossils
234 Fossils for display
236 Using fossils
238 Famous fossil hunters
242 Museum displays

248 Further reading
249 Index

Chapter 1

FOSSIL CLUES TO PREHISTORIC LIFE

This chapter explains fossils as keys that help us to unlock the puzzle of past life.

We show how parts of prehistoric organisms have survived as fossils in layered rocks, and what those fossils reveal – about the ages of the rocks themselves, the evolution (and extinction) of past life forms, and how and where prehistoric plants and creatures lived.

The chapter ends with a glimpse of Earth's oldest organisms – precursors of those plants and animals which, group by group, fill later pages of this book.

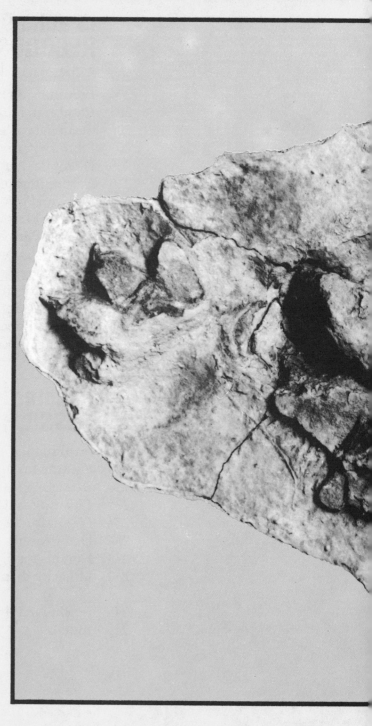

Fossil footprints tell us much about the shape, size, stance, and speed of otherwise unknown prehistoric animals. This print from Late Triassic rocks in Cheshire, England, came from *Cheirotherium* ("mammal hand"), a beast with a huge, thumb-like little finger. Narrow trackways show it walked on all fours, with limbs held well below the body. The feet would have resembled those of the thecodont reptile *Euparkeria*.

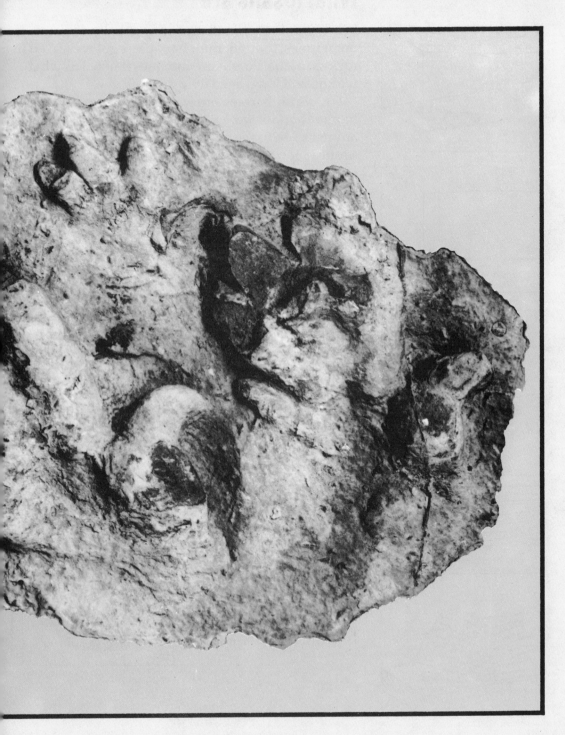

What fossils are

When plants or animals die, they usually decay. Sometimes, though, their hard parts get preserved in rock as fossils. Fossils are the clues that tell us what we know of long-dead living things.

Fossils form in several ways. The process usually happens under water. First, a newly dead plant or animal sinks to the bottom of a lake, sea, or river. Soft tissues soon rot. But before bone or wood decay, sand or mud may cover them, shutting out the oxygen needed by bacteria that cause decay. Later, water saturated with dissolved minerals seeps into tiny holes in the bone or wood. Inside these tiny pipes the water sheds some of its load of minerals. So layers of substances such as calcite, iron sulphide, opal, or quartz gradually fill the holes. This strengthens the bone or wood and helps it to survive the weight of sand or mud above. Sometimes bits of bone or wood dissolve, leaving hollows that preserve their shapes – fossils known as "moulds". If minerals fill a mould they form a "cast".

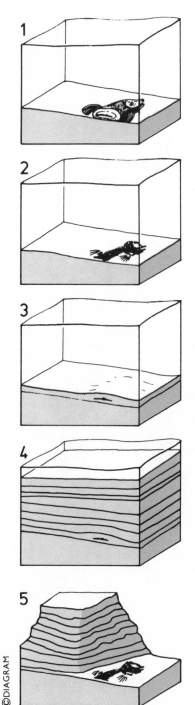

A fossil's story (left)
1 A fish that has just died lies on a sea bed.
2 Flesh rots, revealing bones.
3 Mud or sand covers the bones, preventing decay.
4 Layers of mud and sand bury the bones, now reinforced and fossilized by minerals.
5 Weather exposes the fossil bones by eroding the layered sediments above, long since hardened into stone and raised by uplift of the Earth's crust.

Mould and cast (right)
A A shell in rock dissolved to leave this shell-shaped hollow, called a mould.
B Minerals later filled the mould to form a cast.

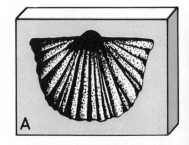

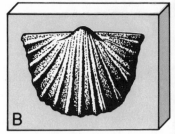

©DIAGRAM

14

Not only wood and bone become preserved. Skin, leaves, burrows, footprints, other tracks, and even droppings may form fossils. But soft-bodied creatures such as worms form fossils only in the finest fine-grained rock.

As a fossil hardens under water, layers of mud or sand grow above it. Their crushing weight and any natural cements that they contain may change thick layers of sand or soft mud into thin beds of hard rock. Millions of years later, great movements of the Earth's crust might heave up these beds to build mountains. Rain, frost, and running water slowly wear them down. In time, weather bares the mountains' inner layers and their fossils. Many fossils develop as we have described. But some form under sands piled up by desert winds, while amber, frozen mud, and tar preserve some ancient organisms whole.

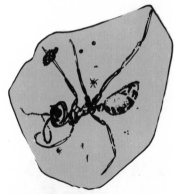

Ant in amber
One hundred million years ago resin leaking from a tree trunk trapped this worker ant. The resin hardened into amber, preserving all but the ant's soft internal organs.

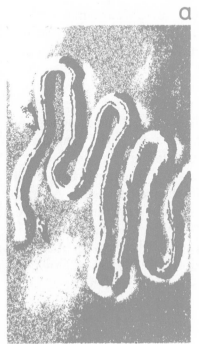

a   b

Fossil tracks
Two examples show different types of fossil tracks (not drawn to scale).
a Trail made probably by a snail-like creature, and preserved in Pennsylvanian (Late Carboniferous) rock.
b Footprints and beak marks left in mud by *Presbyornis*, an Eocene wading bird of North America. (Illustration after Erickson.)

Fossils in the rocks

Rocks of North America
a Precambrian rocks: older than 600 million years
b Palaeozoic rocks: 600–225 million years old
c Mesozoic rocks: 225–65 million years old
d Cenozoic rocks: less than 65 million years old

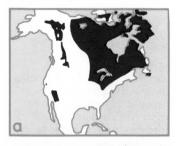

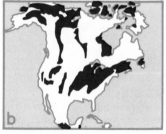

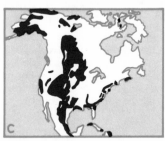

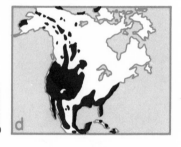

Fossils hold clues to the ages of the rocks containing them – sedimentary rocks such as limestones, mudstones, and sandstones. Each rock layer or "stratum" includes fossils special to that layer – remains of organisms living when the stratum formed. A cliff, cutting, or quarry may reveal many strata, like layers in a slice of cake. Usually the lowest layers are the oldest. Fossil organisms in these layers may have lived many million years before organisms in the top strata.

Geologists group rock strata in "systems" that give their names to geological "periods" in which the rocks formed. Groups of periods form larger time units called "eras". A period can be divided into Early, Middle, and Late, corresponding with the Lower, Middle, and Upper strata of the system that it represents. Some periods are divided into named "epochs", each corresponding with a "series" of rocks. And each series can be subdivided into "stages" or "formations".

Different kinds of prehistoric organisms lived at different times. Finding similar fossils in different continents helps geologists to correlate the ages of rocks around the world. This process is vital to the study called stratigraphy. Fossils show only the relative ages of the rocks. Radiometric dating gives rock ages more precisely. This method depends on radioactive elements inside some rocks. Half the substance in a radioactive element decays into another element in a known time called a half-life. In the next half-life, half of what remains decays, and so on. So certain key ingredients in ancient rocks reveal their ages to within a few million years. At first, radiometric dating worked well only with once-molten rocks in which there are no fossils. Now, radiometric techniques can help us date fossil-bearing sedimentary rocks as well.

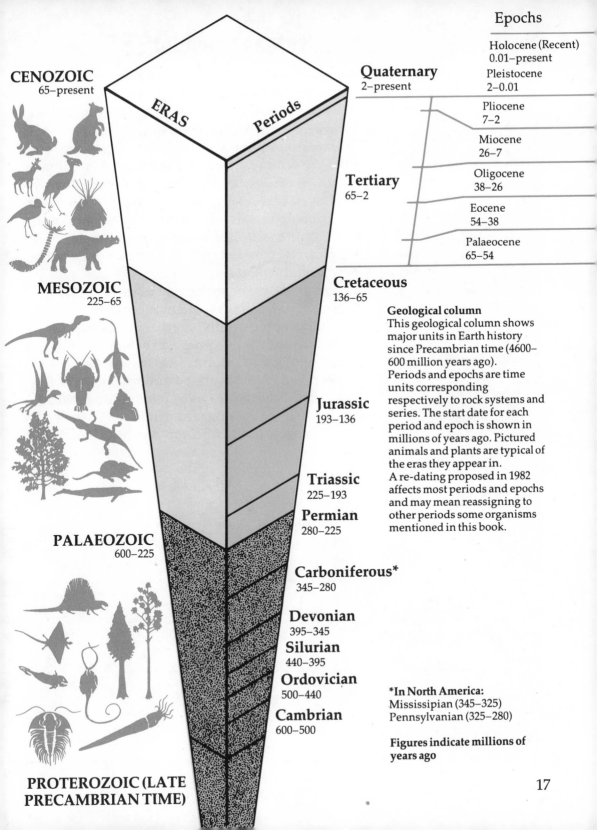

Epochs

Quaternary 2–present	Holocene (Recent) 0.01–present
	Pleistocene 2–0.01
Tertiary 65–2	Pliocene 7–2
	Miocene 26–7
	Oligocene 38–26
	Eocene 54–38
	Palaeocene 65–54

ERAS

Periods

CENOZOIC 65–present

MESOZOIC 225–65

PALAEOZOIC 600–225

PROTEROZOIC (LATE PRECAMBRIAN TIME)

Cretaceous 136–65

Jurassic 193–136

Triassic 225–193

Permian 280–225

Carboniferous* 345–280

Devonian 395–345

Silurian 440–395

Ordovician 500–440

Cambrian 600–500

Geological column
This geological column shows major units in Earth history since Precambrian time (4600–600 million years ago). Periods and epochs are time units corresponding respectively to rock systems and series. The start date for each period and epoch is shown in millions of years ago. Pictured animals and plants are typical of the eras they appear in. A re-dating proposed in 1982 affects most periods and epochs and may mean reassigning to other periods some organisms mentioned in this book.

*In North America:
Mississipian (345–325)
Pennsylvanian (325–280)

Figures indicate millions of years ago

17

Fossil clues to evolution

Evolutionary advances
1 Organic compounds
2 Cells
3 Food-producing cells
4 Multi-celled animals
5 Limbs and shells
6 Vertebrate skeleton
7 Fishes evolve jaws and fins
8 Land plants
9 Arthropods invade land
10 Some fishes develop lungs
11 Amphibians: first vertebrates with legs
12 Reptiles: vertebrates with eggs that develop on dry land
13 Flying insects
14 Seed plants spread
15 Warm-blooded reptiles
16 Dinosaurs
17 Mammals
18 Birds
19 Flowering plants evolving
20 Placental mammals
21 Primates
22 Bats
23 Whales
24 Grass, ungulates evolving
25 Big modern carnivores
26 Early man

Fossils found in rocks of different ages show how living things evolved, or changed, through time. The first life forms were microscopically tiny organisms. Later came soft-bodied sea creatures. Some gave rise to animals with shells or inner skeletons. One group – the fishes – gave rise to amphibians. Amphibians led on to reptiles; reptiles separately to the birds and mammals. Body changes that produced each major group of organisms happened bit by bit. From one generation to another, slight shifts in inherited characters accumulated, in time producing brand new kinds of plant and animal.

Scientists can see such changes in the making by studying fossils in sequences of zones – biostratigraphic subdivisions of the rocks. "Key fossils" useful for this purpose include ammonites, brachiopods, and trilobites – fossil sea creatures widespread in rocks formed under ancient seas. (Most fossils were preserved in marine deposits.) Microfossils – fossil algae and other tiny fossils – are other valuable guides. So, too, are the minute fossil

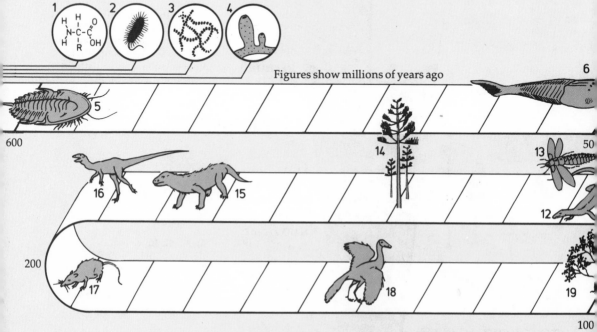

Figures show millions of years ago

18

spores and pollen grains produced by plants. Indeed, palynology – the study of fossil spores and pollen – is a special branch of fossil studies. Gaps blur the fossil record. Some organisms left no trace. Others have yet to be discovered. But enough remain for us to learn which organisms came from what – at least for many major groups. Wary palaeontologists watch out for homeomorphs: unrelated "lookalike" species similarly adapted for the same lifestyle.

Most major groups are many million years old. Within these, though, each kind of organism endured only as long as it could fend off enemies and rivals. New kinds of lethal enemy or harsh climatic changes wiped out unresistant species by the dozen. Fossils show that major evolutionary changes came in fits and starts. After mass extinctions (see pp. 28–29) new life forms sprang up and diversified explosively. New predators and herbivores soon populated habitats emptied of their old-established counterparts.

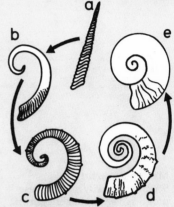

Evolving ammonites
These five fossils from successively younger rocks reveal one line of evolution among ammonites, molluscs with coiled shells.
a Shell straight
b Shell curved
c Shell strongly curved
d Shell loosely coiled
e Shell tightly coiled

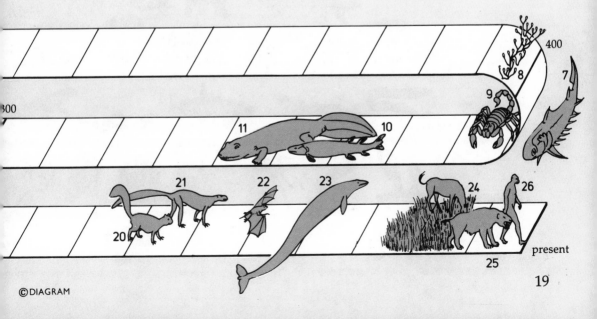

How living things are classified

Biologists classify or group all organisms, alive or extinct, according to how closely they resemble one another, or are related. Either way, those that can breed among themselves but not with others form one "species". Different species resembling one another more than they resemble other species form a "genus". Similar genera make up a "family".

Man's place in nature
Scientists classify man in these progressively higher categories and (not illustrated) subgroups.
1 Species: *Homo sapiens*
2 Genus: *Homo*
3 Family: Hominidae
  Superfamily: Hominoidea
4 Order: Primates
5 Class: Mammalia
  Subphylum: Vertebrata
6 Phylum: Chordata
7 Kingdom: Animalia
8 Superkingdom: Eukaryota

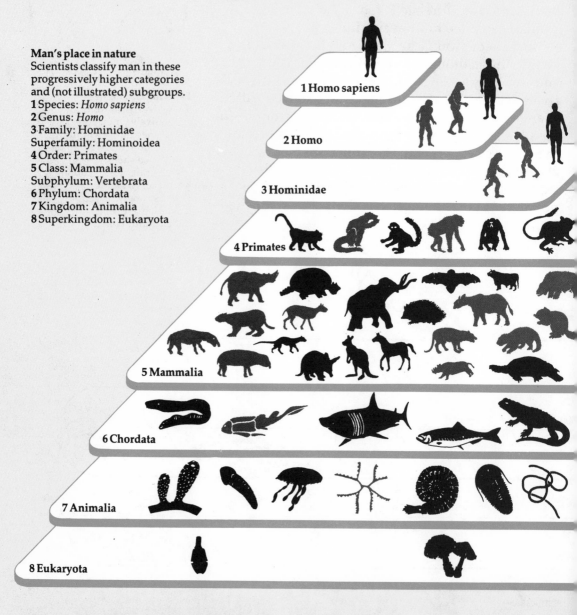

1 Homo sapiens
2 Homo
3 Hominidae
4 Primates
5 Mammalia
6 Chordata
7 Animalia
8 Eukaryota

Similar families form an "order"; similar orders make a "class"; similar classes make a "phylum" or, if plants, a "division". Similar phyla or divisions form a "kingdom".

Various experts disagree about how some organisms should be classified. "Lumpers" lump together in one group organisms that "splitters" would split up among several groups.

Scientific names of living things come at least partly from Greek or Latin. (Animal family names end in -idae, plant family names usually end in -ceae.) Scientists of all nations use the same scientific name for a given kind of animal or plant. This helps prevent confusion.

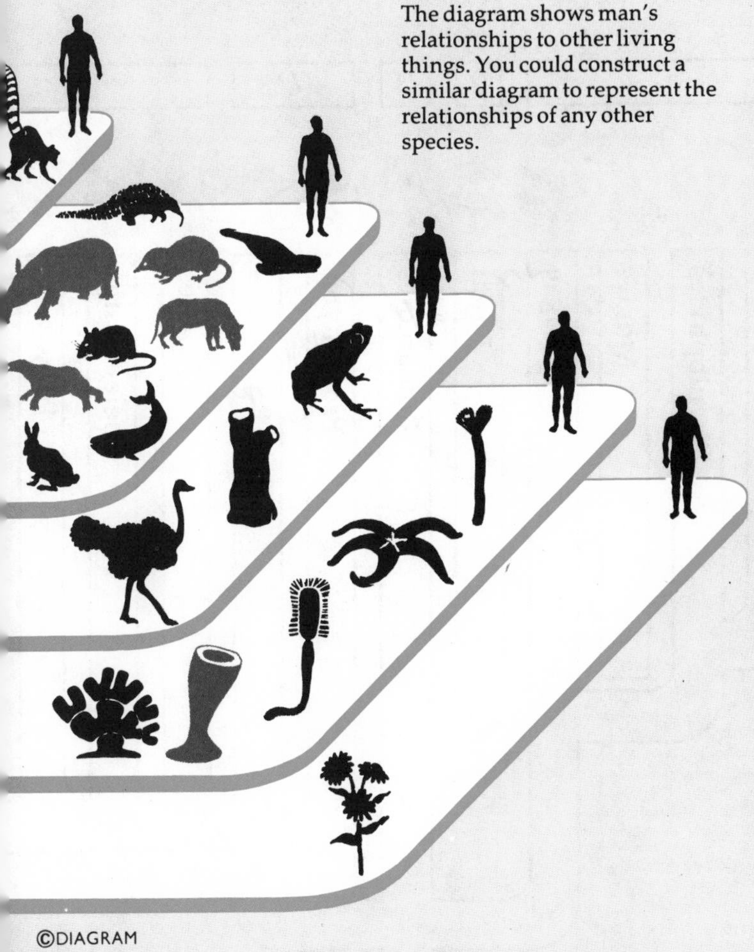

Man's fellow creatures
The diagram shows man's relationships to other living things. You could construct a similar diagram to represent the relationships of any other species.

Each level shows a grouping comprising lesser groups with the status of the group above. Individual creatures stand for groups. For instance, each creature shown for the phylum Chordata stands for a different class of animal. Each creature shown for the kingdom Animalia stands for a different phylum, and so on. Animals shown as pale shapes stand for groups that are extinct. Some groups are not depicted.
1 Species *Homo sapiens*
2 Genus *Homo*, containing three species
3 Family Hominidae, containing three genera
4 Order Primates, containing maybe 18 families
5 Class Mammalia, containing more than 30 orders
6 Phylum Chordata, containing more than a dozen classes, nine of which are vertebrates (backboned animals)
7 Kingdom Animalia, containing 19 phyla, all but one of them invertebrates (animals without a backbone)
8 Superkingdom Eukaryota, containing four kingdoms (animals, plants, fungi, and protists)

21

The tree of life

Scientists believe that all living things can be related
to ancestors originating in minute one-celled
organisms. First, lightning and ultraviolet radiation
acting on a primeval atmosphere formed organic
compounds from simple chemicals. Next, organic
compounds organized in self-replicating
committees formed simple, one-celled organisms –
bacteria and blue-green algae – called collectively
prokaryotes. Committees of prokaryotes arguably
then produced more complex organisms – protists,
fungi, plants, and animals – called collectively
eukaryotes. If the superkingdom Prokaryota
represents the main trunk of the evolutionary tree of
life, Eukaryota – other living things – provides the
branches.

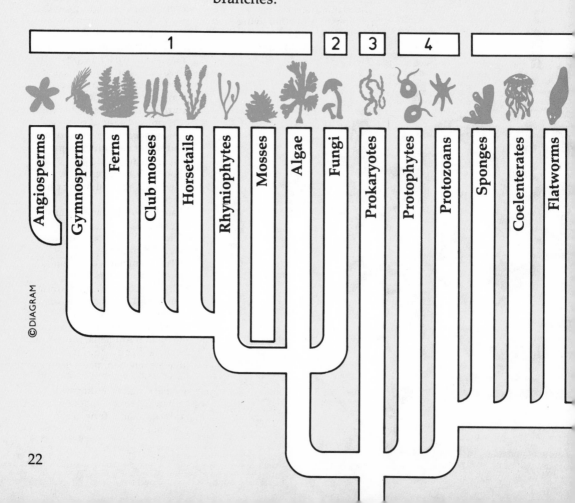

©DIAGRAM

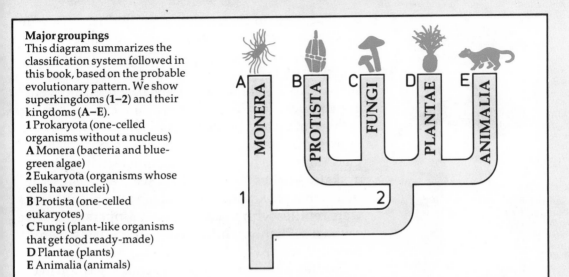

Major groupings
This diagram summarizes the classification system followed in this book, based on the probable evolutionary pattern. We show superkingdoms (**1–2**) and their kingdoms (**A–E**).
1 Prokaryota (one-celled organisms without a nucleus)
A Monera (bacteria and blue-green algae)
2 Eukaryota (organisms whose cells have nuclei)
B Protista (one-celled eukaryotes)
C Fungi (plant-like organisms that get food ready-made)
D Plantae (plants)
E Animalia (animals)

MONERA A
PROTISTA B
FUNGI C
PLANTAE D
ANIMALIA E

1   2

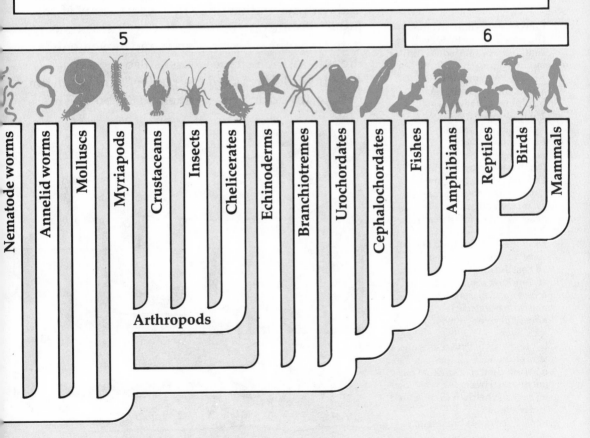

5   6

Nematode worms
Annelid worms
Molluscs
Myriapods
Crustaceans
Insects
Chelicerates
Echinoderms
Branchiotremes
Urochordates
Cephalochordates
Fishes
Amphibians
Reptiles
Birds
Mammals

Arthropods

How ancient organisms lived

Fossils reveal much about how individual organisms lived. Sometimes we can even learn how prehistoric plants and creatures interacted, as prey or predators.

Just by looking at a fossil creature, experts may be able to work out how it ate, moved, sensed, or even grew. Clues often lie in jaws or other features similar to those of certain living animals with habits that are known. For instance, prehistoric sea reptiles called placodonts had broad, flat crushing teeth like those of skates – fishes that crunch up sea-bed shellfish. It seems that placodonts ate shellfish too. Ichthyosaurs were reptiles with paddle-shaped flippers rather like a whale's. Plainly, ichthyosaurs, like whales, were accomplished swimmers, unable to walk on land. Trilobites were sea creatures resembling wood lice and with eyes rather like a fly's. Close study of one fossil trilobite proves that it saw in all directions, detecting other creatures' sizes and movements, though not their shapes. Fossils also show that growing trilobites shed their outer "skin" from time to time, to add an extra segment to the body. This

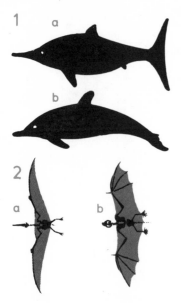

Convergent evolution (above)
Unrelated individuals evolved similar shapes suited for similar modes of life.
1 Fish-like swimmers:
a ichthyosaur (a reptile)
b dolphin (a mammal)
2 Skin-winged fliers:
a pterosaur (a reptile)
b bat (a mammal)

Death assemblage (right)
This diagram shows how the remains of living creatures can get mixed and broken after death to form a fossil death assemblage.
1A Prehistoric land and sea communities of organisms
1B Fossils from these sources and another source, mixed by erosion and undersea currents, and incorporated in rock
a Fossil derived from an older rock layer
b, c Remains of land plants and a land animal
d Soft-bodied creature that has left no fossil trace
e Shellfish shell halves separated and realigned
f Fragile crinoids broken up

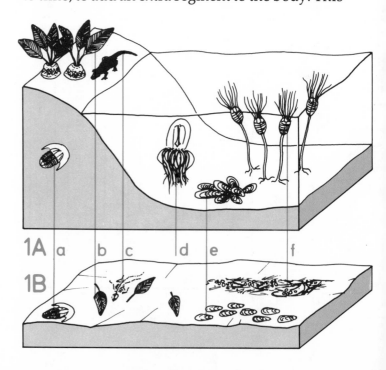

gives a rough idea of individuals' ages. (Other clues to fossil organisms' ages include tooth wear in mammals, and growth rings in tree trunks, fish scales, and mollusc shells.)

Where many different fossils crop up together in the same rock zone, palaeontologists may be able to identify plants, plant-eating animals, and the predators that preyed upon these herbivores. In this way an expert can work out a prehistoric food chain whose links consist of eaters and eaten – even perhaps a food web made up of interlinking chains.

Working out links is easy when fossils form a life assemblage – a group of organisms preserved as they once lived. Unfortunately many fossil groups are death assemblages. Such groups can include "outsiders" washed in by floods from other habitats. Then, too, predators, winds, currents, or chemicals destroy fragile bones and shells, or carry them away. So certain species from a given habitat survive only as broken scraps, or disappear completely.

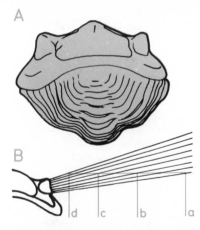

Trilobite abilities (above)
A This fossil trilobite was found curled up for protection.
B Study of the many lenses in its eyes shows that lenses at ever higher levels detected the advance (from **a** to **d**) of objects of a certain height.

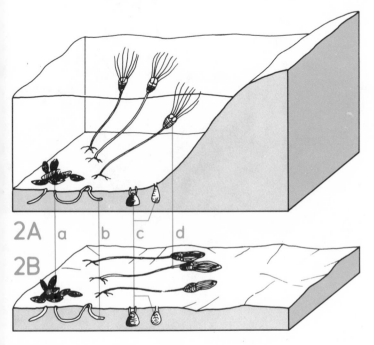

Life assemblage (left)
This shows how creatures' hard parts and tracks may be preserved intact in the positions where they lived.
2A Prehistoric sea-bed community of organisms
2B Fossils of the community undisturbed by erosion or undersea currents before being incorporated in rock
a Shellfish shells intact and aligned as during life
b Tracks of a burrowing worm preserved in rock
c Burrowing bivalve and sea urchin, with hollows their soft parts had made in sand
d Fragile crinoids unbroken

©DIAGRAM

25

What lived where?

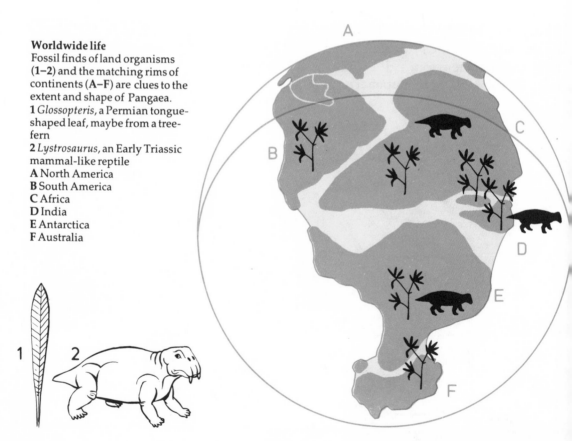

Fossils tell us much about the surface of the world in ancient times. Then, as now, different species were designed for life in different habitats – for instance forest, grassland, desert, swamp, or river. Fossils therefore tell us indirectly about the kind of place they lived in – and about its climate. Widespread finds of fossil desert animals dating from a given time hint that desert climates were widespread too. Desert fossils crop up in Permian rocks as far apart as the United States, the USSR, and South Africa.

Fossils of lush tropical vegetation show that much of North America and Europe had a warm, wet climate in later Carboniferous times, when northern

Supercontinent (above)
This global view shows the supercontinent Pangaea in Permian times.

Worldwide life
Fossil finds of land organisms
(1–2) and the matching rims of continents (A–F) are clues to the extent and shape of Pangaea.
1 *Glossopteris*, a Permian tongue-shaped leaf, maybe from a tree-fern
2 *Lystrosaurus*, an Early Triassic mammal-like reptile
A North America
B South America
C Africa
D India
E Antarctica
F Australia

lands lay close to the Equator. Fossils have done much to prove that continents have drifted from their old positions. For instance, the Permian fossil plant *Glossopteris* occurs in all southern continents, now widely separated by oceans. Plainly, when *Glossopteris* grew, all southern continents lay locked together. Geologists believe they have been tugged apart by currents in the molten rock beneath the oceans, where Earth's crust of solid rock is thin and weak. But finds of certain fossil species only in specific regions suggest that natural barriers such as mountains, seas, or temperature boundaries stopped those organisms spreading.

Isolated life (below)
Shown on this map (**1–8**) are mammals that evolved in isolation after Pangaea broke up and continents (**A–F**) drifted apart, some split by seas.
1 Uintatheres
2 Opossum rats
3 Pyrotheres
4 Aardvarks
5 Embrithopods
6 Lemurs
7 Insectivores
8 Spiny anteaters
A North America
B South America
C Africa
D Europe
E Asia
F Australia

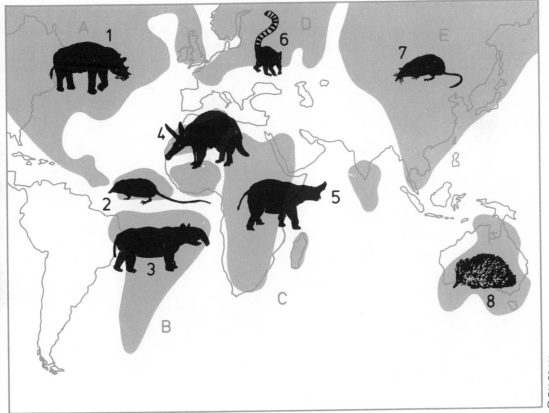

Mass extinctions

Plants and animals alive today account for most known species. Yet species that became extinct must have outnumbered these by far. One calculation suggests there are 4.5 million living species, but that 980 million species evolved in the last 600 million years. This estimate supposes each species lasted on average for only 2.7 million years.

In fact some forms were far more durable. The Australian lungfish may have survived for about 100 million years. Sea creatures like the king "crab", coelacanth (a fish), and *Neopilina* (a mollusc) are close kin of beasts with even longer histories.

Somehow, such living fossils resisted changes that wiped out many other organisms. Sometimes, disasters struck down many groups together. The greatest mass deaths marked the ends of eras. Thus many sea creatures and the great group of reptiles called pelycosaurs died out as the Palaeozoic Era ended. Major reptile groups including dinosaurs, and those once-abundant molluscs ammonites, all vanished as the Mesozoic Era closed.

Lifelines
Thick and thin horizontal bands show the expansion and collapse of certain groups of organisms through prehistoric time. (Band thicknesses are not to scale.) Thick vertical lines show mass collapses changing the make-up of marine communities. Widespread land and sea extinctions marked the end of the Cretaceous period.
1 Archaeocyathids
2 Reef-forming stromatolites
3 Tabulate corals
4 Trilobites
5 Bivalves
6 Ammonites
7 Archosaurian reptiles (including dinosaurs)

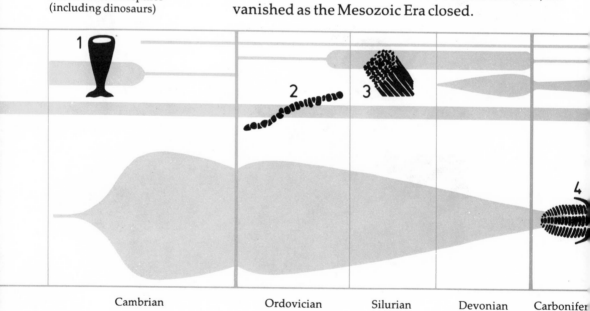

Cambrian   Ordovician   Silurian   Devonian   Carbonifer

Experts disagree about what caused these mass extinctions. Perhaps cosmic rays from an exploding star deformed unborn young. Maybe deadly rays from space poured down when the Earth's magnetic field reversed. Or perhaps changes in the positions and levels of the continents caused killing periods of cold.

Some scientists blame huge lumps of rock that could have crashed on Earth from space like massive bombs. The impact of a rock 10 kilometres across would hurl enough dust and moisture into the air to darken skies around the world for months. Plants and plant-eating creatures would die in droves. The world could briefly freeze. But once dust settled, moisture still up in the sky would trap the Sun's incoming heat. Then overheating could destroy creatures unable to control their body temperature.

Yet no mass extinction theory convincingly explains why many groups of living things survived.

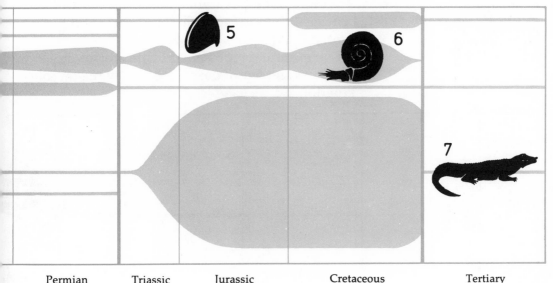

Permian   Triassic   Jurassic   Cretaceous   Tertiary

©DIAGRAM

29

The oldest fossils

The earliest identifiable living things were bacteria and blue-green algae – members of the superkingdom Prokaryota, tiny one-celled organisms whose genetic blueprints did not lie in a single nucleus. Prokaryotes probably evolved more than 3600 million years ago from combined amino acids created in the early ocean – a chemical-rich "test tube" shrouded by a methane-rich atmosphere intensively bombarded by solar radiation and electric storms.

Pioneer prokaryotes consumed ready-made amino acids. Later came blue-green algae containing the green pigment chlorophyll, which enabled them to use the energy in sunshine to build their own food compounds from carbon dioxide and water. This process, photosynthesis, yielded free oxygen as waste.

In time, free oxygen screened the Earth's surface from the Sun's harmful ultraviolet rays and formed a rich new source of energy. This was tapped by new, more complex life forms, members of the only other superkingdom: Eukaryota. Eukaryotes, with a central nucleus to each cell, comprise the fungi, protists, plants, and animals.

Our examples represent the sole prokaryote kingdom and one of the most primitive eukaryote kingdoms.

Accelerating evolution
Prokaryotes played a crucial role in evolution, as we show here. The thicker the line, the more diversified the life forms that it represents.
A Chemical evolution: amino acids, proteins, sugars, etc formed in water.
B Slow organic evolution: photosynthesizing prokaryotes enriched the atmosphere with oxygen; some formed ozone.
C Rapid evolution, mainly of eukaryotes deriving energy from oxygen and screened from harmful solar rays by ozone in the atmosphere. Evolution accelerated 1000–700 million years ago as atmospheric oxygen reached more than 1% of its present level.

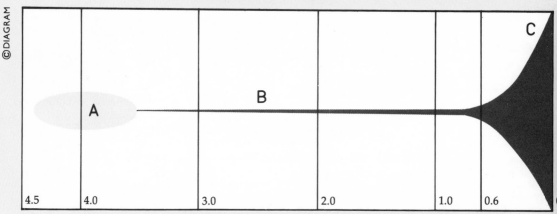

| 4.5 | 4.0 | 3.0 | 2.0 | 1.0 | 0.6 |

billion years ago

©DIAGRAM

1 **Kakabekia** resembled a microscopic umbrella with a clubbed stalk. It might have been a bacterium dividing into two. Size: most bacteria are under 1 micron (0.001mm) across. Time: Precambrian (2000 million years ago). Place: Canada. Kingdom: Monera (bacteria and blue-green algae). Subkingdom: Bacteria (probably the earliest, most primitive living organisms).

2 **Nostoc,** a blue-green alga, forms a necklace-like mass. It probably evolved from one-celled Precambrian ancestors. Kingdom: Monera. Subkingdom: Cyanophyta (blue-green algae, mostly tiny plants under 25 microns across).

3 **Aspergillus** is a mould, producing mildew on fruit, leather, or walls. Its growing, thread-like hyphae produce "side shoots" bearing spores from which new moulds grow. Size: minute. Kingdom: Fungi (plant-like organisms that lack chlorophyll and feed on dead or living organisms). Fossil hyphae have been found in Precambrian rocks, but claims for a 3800-million-year-old fossil fungus found in Greenland proved mistaken.

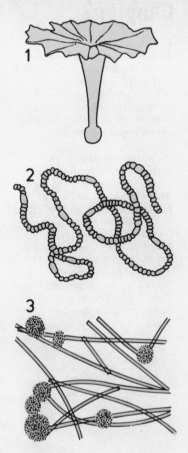

 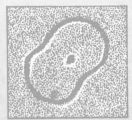  

Early complex cells
Fossils of cells multiplying by division hint at early complex life 1000 million years ago. Shown here much magnified are one-celled eukaryotic algae fossilized in chert at various stages of division. Genes within cell nuclei determine how each daughter cell develops. Most multicellular organisms undergo sexual reproduction, in which a fertile egg cell receives genes from both parents. This innovation increased the chance of genetic variation and so of evolution.

Chapter 2

FOSSIL PLANTS

These pages give a brief evolutionary overview of the plant kingdom (Plantae) – that vast group of organisms on which all animals depend directly or indirectly for their food. We describe key features of major groups, and give examples.

The chapter starts with divisions of those primitive plants collectively called algae. For convenience we include protophytes: tiny, one-celled algae now usually ranked (with protozoans) in a separate kingdom from true plants (see pp. 20–23 for classification).

Most of the chapter deals in turn with increasingly advanced divisions, of which some of the names and groupings are disputed.

This artist's reconstruction shows a coal forest of Late Carboniferous (Pennsylvanian) Europe or North America. Huge horsetails, scale trees, ferns and tree ferns sprout from rich swamp mud and jostle one another for a share of light.

33

Simple plants

True plants (members of the kingdom Plantae) consist of cells surrounded by a cellulose cell wall, not just a membrane like that around animal cells. Plants contain the green pigment chlorophyll, enabling them to manufacture food from carbon dioxide and water.

The simplest plants are green, red and brown algae. (Other algae are classed as prokaryotes and protists.) Most green algae are freshwater plants. Green, red and brown algae comprise the seaweeds. The simplest kinds of land plants are bryophytes, including liverworts and mosses. Some have leaves but none has roots. All reproduce by spores. "True" algae evolved (perhaps from blue-green algae) at least 1000 million years ago. Green algae were probably ancestral to all higher plants. Bryophytes appeared at least 350 million years ago.

A brown alga
Here we show features of the inter-tidal seaweed *Fucus vesiculosus* (bladder wrack).
a Receptacles (swollen branch tips containing mucilage-filled conceptacles where sex cells develop)
b Air bladder, for buoyancy
c Blade
d Stipe
e Holdfast

Sexual reproduction
Fucus vesiculosus has the life cycle shown in this diagram.
A Male conceptacle
B Female conceptacle
1 Antheridium
2 Male sex cells being released
3 Oogonium
4 Female sex cells being released
5 Male sex cells surrounding a female sex cell – one male cell fertilizes the female

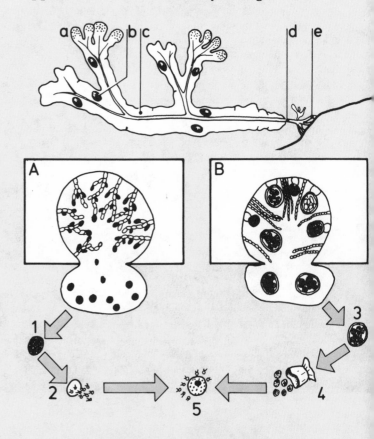

1 **Stephanochara,** an Oligocene stonewort, produced this female sex organ with a wall composed of helically arranged cells, shown magnified about 20 times. Division: Charophyta, an algal group that dates back to Devonian times.

2 **Ulva,** "sea lettuce", is a green, shallow-water seaweed. Size: 10–46cm (4–18in). Division: Chlorophyta (green algae), known since at least as early as Ordovician times.

3 **Fucus** is a brown shore seaweed. Height: 13–91cm (5in–3ft). Division: Phaeophyta (brown algae), perhaps dating from Devonian times.

4 **Ceramium** is a red seaweed, banded and with forked, inturned tips. Height: 2.5–30.5cm (1in–1ft). Division: Rhodophyta (red algae), dating from Ordovician times.

5 **Marchantia** is a liverwort. Size: 5cm (2in). Division: Bryophyta (liverworts and mosses), dating from Devonian times.

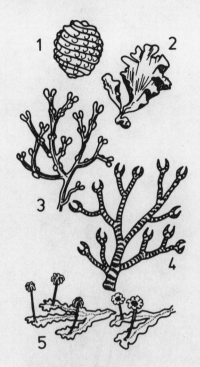

Colour and depth (left)
a–f Green seaweeds need red light which only penetrates shallow water.
g–l Brown seaweeds, whose brown pigments mask their green chlorophyll, live in inter-tidal zones.
m–o Red seaweeds, with a masking red pigment, can use blue light, which penetrates 100m (330ft) deep.

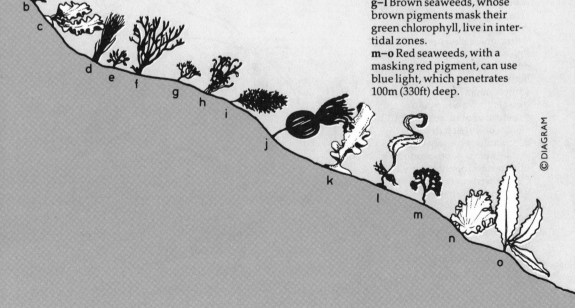

Protophytes

Described here are certain one-celled algae that possess a nucleus and can move about like animals. Scientists now usually rank such algae in the subkingdom Protophyta ("first plants") of the kingdom Protista. Protists evolved by 1200 million years ago, and probably came from simple cells that swallowed bacteria and even blue-green algae which then lived on inside them. The new "committee" micro-organisms reproduced by dividing, like bacteria, but most of their ingredients split too, in complex ways. Of our protophyte examples, number 1 is in the phylum Pyrrophyta (algae with yellowish or brownish food stores). Numbers 2–4 are in the Chrysophyta (algae with golden-brown food stores). Prehistoric protophytes mostly drifted in the sea. Their fossils help scientists date rocks formed in the last 200 million years.

1 **Gonyaulacysta** is known only from fossils of its resting form: a hard-walled cyst. Class: Dinophyceae (dinoflagellates), sea-surface micro-organisms

Food factory
This diagram shows (much magnified) structures in a tiny, living, one-celled protophyte, *Prymnesium*. Even such microscopic scraps of life are highly organized.
a Haptonema
b Flagellum (a whip-like structure used in locomotion)
c Golgi body (rich in fat)
d Pyrenoid (a protein body)
e Chloroplast (food-producing unit containing chlorophyll)
f Fat (stored energy supply)
g Nucleus (control centre essential for the cell's life and reproduction)
h Leucosin vesicle, containing food reserves

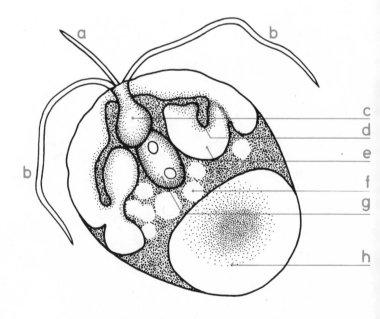

0.005–2mm across. Each has a groove around the
body and lashes itself along with two cilia ("whips").
Time: Silurian onward (guide fossils since Triassic
times).

2 **A silicoflagellate** These have one or two "whips"
and a tube- or rod-like opal skeleton, often fossilized.
Silicoflagellates are organisms 0.02–0.1mm across,
dating from Cretaceous times.

3 **Cymbella** has a two-valved skeleton like a glass
box with an overlapping lid. Length: 0.03mm. Time:
Pliocene onward. Class: Bacillariophyceae (diatoms)
– oblong, round, and other micro-organisms dating
from Cretaceous times. Accumulations of their shells
form diatomaceous earth.

4 **Coccolithus** is a ball-shaped sea micro-organism
covered in limy plates. Time: Pliocene onward.
Class: Coccolithophyceae – important rock formers
since Jurassic times. Their plates 0.002–0.01mm
across sink on death, building layers of chalk or
chalky mud.

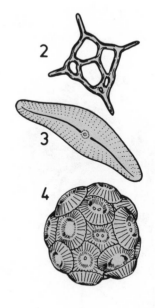

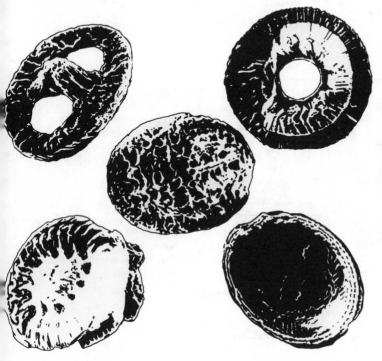

Coccoliths
Shown here much enlarged are
examples of coccoliths
(literally "seed stones"). Many of
these limy platelets, measuring
less than one hundredth of a
millimetre across, go to cover
each *Coccolithus* and its relatives.
Distinctive shapes and fast
evolution make coccoliths good
guides to the ages of such rocks
as chalk.

Vascular plants

Vascular plants – plants with internal channels for transporting liquids – include ferns, conifers, and flowering plants. Algal ancestors gained support and nourishment in water but vascular plants evolved for life on land. Roots give anchorage and obtain water from soil. Stems raise the food-producing leaves up to the light. A waterproof cuticle prevents desiccation. Stomata – holes that can be closed – let gases in and out for food production and respiration. Above all, a "plumbing" system of tiny internal tubes transports water and salts up from the soil and dissolved food down from the leaves.

Vascular plants appeared about 400 million years ago. Pioneers had short, bare stems and neither roots nor leaves. Tall plants with woody stems became the first true trees. Among these were the giant club mosses and horsetails of swampy Carboniferous forests. Their compressed carbonaceous remains formed coal.

Here we describe prehistoric examples from three spore-producing divisions: Rhyniophyta, Lycophyta, and Sphenophyta.

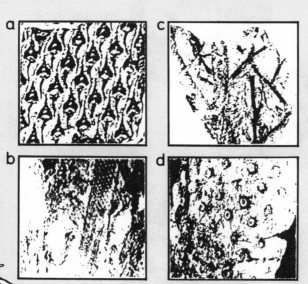

Lepidodendron restored
Fossil finds (left) enabled scientists to reconstruct the tree they came from (far left).
a Piece of trunk scarred by shedding old leaves
b Piece of grooved, mid-level trunk
c Branches with leaves
d Surface of root-like anchoring structure
e Whole tree, with a man to show its size. (Many other plant fossils left such fragmentary remains that we cannot even guess the shape and size of the whole plant.)

1 **Cooksonia**, the first known vascular plant, had simply forked, leafless stems, each ending in a spore-filled cap. Height: about 5cm (2in). Time: Late Silurian. Place: Wales. Division: Rhyniophyta.

2 **Asteroxylon** had a main stem with forked branches bearing tightly packed, tiny leaflets. Height: up to 1m (3ft 3in). Time: Early Devonian. Place: Scotland. Division: Lycophyta (club mosses).

3 **Lepidodendron** was a giant club moss with a root-like anchoring organ divided into four main branches; a broad, tall, bare trunk crowned by two sets of repeatedly forked branches bearing narrow leaves; and small and large spores – the latter borne in cone-like structures. Height: 30m (100ft). Time: Carboniferous. Place: Europe and North America. Division: Lycophyta.

4 **Calamites** had a tall, jointed stem and upswept branches bearing rings of narrow leaves. Height: up to 30m (100ft). Time: Carboniferous. Place: Europe and North America. Division: Sphenophyta (horsetails).

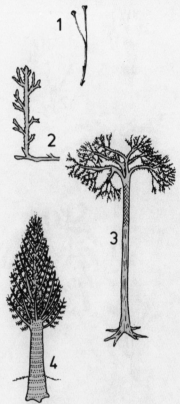

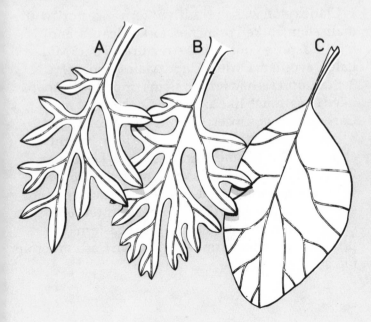

Leaves evolving *(left)*
These three illustrations retrace the likely evolution of compound leaves from stems.
A Branching stems of certain early plants probably had broad flat rims. These increased the surface area exposed to sunlight. Such stems trapped more light than others and so produced more food.
B Later plants produced more branches than their ancestors, so increasing the surface area still further.
C Later still came plants whose many, wide-rimmed branches fused to form broad plates that faced the Sun. These plates were leaves.

©DIAGRAM

Ferns and gymnosperms

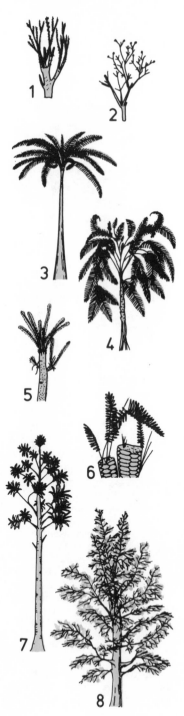

Ferns and gymnosperms probably arose from plants like *Cooksonia* more than 350 million years ago. Many ferns have feathery or strap-like leaves called fronds. They lack flowers or seeds and disperse by spores shed from the fronds' undersides. A spore falling on damp soil grows into a minute "breeding" plant producing male and female cells. Male cells fertilize female cells which then develop into normal fern plants.

Gymnosperms ("naked-seed" plants) include seed-ferns and conifers – flowerless plants that reproduce by seeds not spores. Seeds develop when wind-blown pollen fertilizes egg cells still on the parent plant. The ripe seeds have food supplies and waterproof coats. So fallen seeds survive drought, sprouting when rain soaks the soil. Ferns and seed-ferns flourished in warm, wet Carboniferous forests. Gymnosperms became increasingly varied during Mesozoic time. These pages show examples of the division Polypodiophyta or ferns (items 1–3), and gymnosperms, an artificial grouping (items 4–9).

1 **Cladoxylon** was a primitive "fern ancestor" with a main stem, forked branches, forked leaves, and fan-like spore-containing structures. Time: Mid-Late Devonian. Order: Cladoxylales (extinct).
2 **Stauropteris** had forked stems tipped with spore-filled caps, much like ancestral *Cooksonia*. Time: Carboniferous. Order: Coenopteridales (extinct).
3 **Psaronius** was a prehistoric tree fern. Height: about 7m (23ft). Time: Carboniferous-Permian. Order: Marattiales (still surviving in the Tropics).
4 **Medullosa** was a seed-fern: producing seeds not spores. Height: 4.6m (15ft). Time: Carboniferous. Order: "Pteridospermales" – the seed-ferns, primitive gymnosperms abundant in Carboniferous forests.

5 **Williamsonia** sprouted stiff, palm-like fronds from a trunk with scars where stems of old, dead leaves had fallen off. Height: 3m (10ft). Time: Jurassic. Order: Bennettitales (once widespread but long extinct).

6 **Zamia** has leathery fronds rising from a short, fat, pithy stem. It is a survivor of the abundant Mesozoic order Cycadales (cycads).

7 **Cordaites** had a tall, slim, straight main trunk with a crown of branches bearing long, strap-shaped leaves. Seeds grew on stalks from cone-like buds. Height: up to 30m (100ft). Time: Carboniferous. Order: Cordaitales (long extinct).

8 **Ginkgo**, a living fossil, has fan-shaped leaves. Height: up to 30m (100ft). Time: Permian onwards, widespread in the Jurassic. Order: Ginkgoales ("maidenhair trees").

9 **Araucaria** (monkey puzzle) is an evergreen, cone-bearing tree with stiff, flat, pointed leaves. Height: up to 45m (150ft). Time: Jurassic onwards. Order: Coniferales (conifers – pines, firs, etc).

9

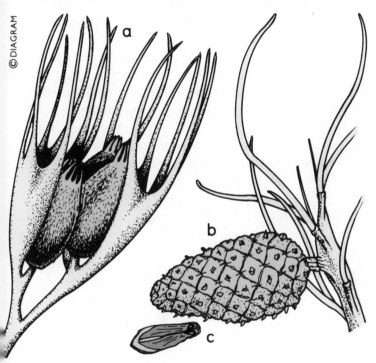

© DIAGRAM

Seeds and cones
These two examples contrast the seeds and seed-containing structures of an early and a modern gymnosperm.
a Forked branches form the "bracts" which only loosely surround the seeds in a Late Devonian plant called *Archaeosperma*.
b A woody cone surrounds many individual seeds in a modern conifer. The cone opens when the seeds are ripe, to let them fall.
c A conifer seed has a built-in, wing-like blade. Ripe seeds flutter lightly down, dispersing on the wind.

41

Flowering plants

Flowering plants are among the most successful land plants ever. They thrive from high latitudes to the Equator, from seashore to mountaintop. They include most garden plants, farm crops, and broadleaved trees, and range in size from 100-metre (330ft) *Eucalyptus* trees to tiny duckweeds.

Reasons for success included a carpel, a protective covering that earns this major group of vascular plants its scientific name: Angiospermae ("enclosed seed plants"). Angiosperm seeds develop in greater safety than the "naked" seeds of gymnosperms, their likely ancestors. Finds of fossil leaves and pollen hint that flowering plants evolved about 120 million years ago. By 65 million years ago over 90 per cent of known fossil plants were angiosperms. There were beeches, birches, maples, poplars, walnuts, and many other familiar kinds. Since Tertiary times the main change has been in distribution, as warmth-loving species tended to retreat from cooling polar regions.

Our examples represent both subclasses: the Monocotyledonae (monocots, with one seed-leaf) and Dicotyledonae (dicots, with two seed-leaves). Dicots form the vast majority of angiosperms.

1 **Buchloe,** the living buffalo grass, represents the grasses – monocots that first became plentiful in the Miocene. Order: Graminales (grasses and sedges).

2 **Magnolia** is a genus of dicot trees and shrubs with large leaves and showy flowers regarded as primitive in structure. Time: Early Cretaceous onwards. Place: Asia and the Americas. Order: Ranales, among the earliest of all known flowering plants.

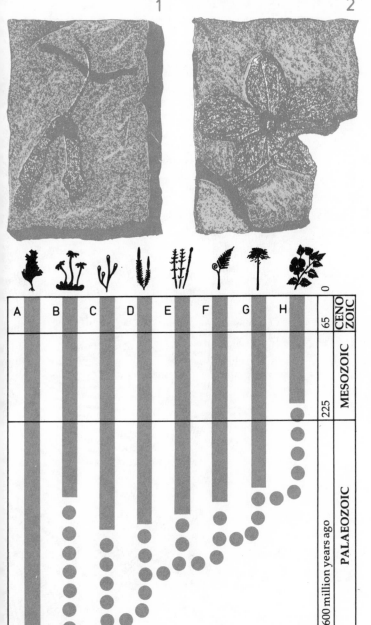

Fossil angiosperms
These Miocene plant fossil remains come from Switzerland. Fossil seeds and flowers are far rarer than fossil leaves but most fossil angiosperms closely resemble living kinds.
1 Winged seeds of a Miocene maple, *Acer trilobatum*
2 The blossom of *Porana oeningensis*

Plants family tree
This shows important groups of true plants. Groups B, C, D, E, F and H represent divisions. A and G are artificial groupings.
A Algae (four divisions, some older than others)
B Bryophyta (liverworts and mosses)
C Rhyniophyta (*Cooksonia*)
D Lycophyta (club mosses)
E Sphenophyta (horsetails)
F Polypodiophyta (ferns)
G Gymnosperms
H Angiospermae (flowering plants)

43

Chapter 3

FOSSIL INVERTEBRATES

Most phyla (the major groups of animals) are invertebrates – creatures with no backbone. According to some experts, invertebrates account for 18 of 19 phyla comprising the kingdom Animalia. (Even the nineteenth phylum includes some groups without a backbone.) This chapter surveys the chief invertebrate groups significant as fossils, from simple sponges to complex echinoderms. For convenience, we start with the tiny protozoans, now usually ranked (with protophytes) in a separate kingdom from true animals (see pp. 20–23 for classification).

Early, soft-bodied invertebrates left few fossil traces. But most phyla may have appeared in the sea between 800 and 600 million years ago. This upsurge in evolution became possible as Earth's atmosphere became breathable, and ozone shielded the surface of the sea from lethal ultraviolet radiation.

All invertebrate phyla that evolved survive today, but through the ages many subgroups disappeared.

Clymenia sedgwickii here represents that huge, long-lived group of invertebrates, the ammonoid cephalopods. Part of its outer shell has been removed to show the distinctive suture lines of its internal chambers. Sutures help to identify this as a goniatite ammonoid. Clymeniids thrived in Late Devonian seas. (The Mansell Collection.)

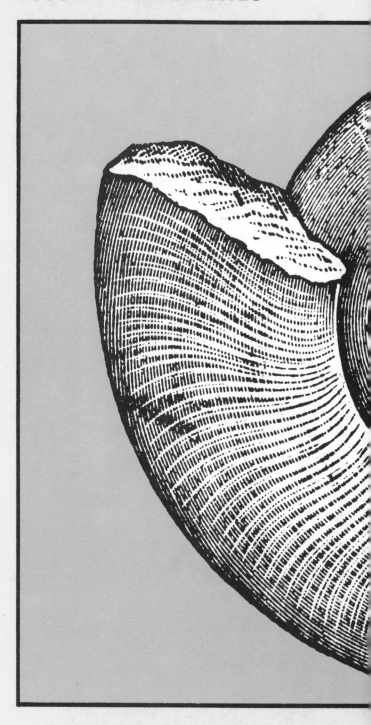

44

About invertebrates

Invertebrates is a name popularly used for the larger, lowlier section of the animal kingdom. This huge group – including insects, worms, snails, etc – comprises living things with a distinct body shape and size that develop from an embryo or larva formed when male and female sex cells meet. Most animals move about and, unlike plants, all feed upon organic matter, for they cannot manufacture food. Zoologists put animals in two subkingdoms: the primitive Parazoa (sponges and close kin), and the more advanced Metazoa (animals composed of many cells of different types, specialized for performing different tasks). All parazoans and most metazoans are invertebrate: they lack a notochord ("backstring") or vertebrae (bones making up a backbone). For convenience, this chapter includes those tiny one-celled organisms protozoans, although these rank as a subkingdom in the Protista, a kingdom that also includes microscopic plant-like organisms (see pages 36–37).

Borehole tracks from rocks in Zambia suggest that protozoans had given rise to parazoans and metazoans over 1000 million years ago. By 700 million years ago, jellyfishes, worms, and other lowly metazoans lived in the sea. This chapter traces different groups of fossil invertebrates, from minute protozoans through the progressively more complex sponges, corals, molluscs, worms, and arthropods, on to echinoderms – indirect ancestors of backboned animals. Most major invertebrate groups evolved in Palaeozoic times, yet still endure. So living species often help us to deduce what their fossil forebears looked like, and how these lived and moved about.

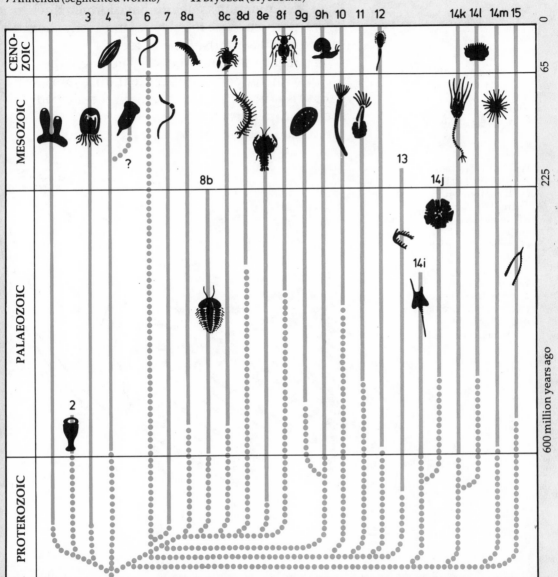

Invertebrate groups
This shows most major groups
significant as fossils and
featured in this chapter.
Numbers indicate phyla, letters,
subphyla.
1 Porifera (sponges)
2 Archaeocyatha
(archaeocyathines)
3 Cnidaria (jellyfishes and corals)
4 Rhizopoda (rhizopods)
5 Ciliata (ciliates)
6 Aschelminthes (roundworms)
7 Annelida (segmented worms)

8 Arthropoda (arthropods)
a Onychophora (velvet worms)
b Trilobitomorpha (trilobites)
c Chelicerata (chelicerates –
scorpions, spiders etc)
d Myriapoda (myriapods –
millipedes etc)
e Crustacea (crustaceans)
f Insecta (insects)
9 Mollusca (molluscs)
g Amphineura (chitons)
h Conchifera (snails etc)
10 Phoronida (phoronids)
11 Bryozoa (bryozoans)

12 Brachiopoda (brachiopods)
13 Conodontophorida
(conodonts)
14 Echinodermata
i Homalozoa (carpoids)
j Blastozoa (blastoids etc)
k Crinoidea (sea lilies)
l Asterozoa (sea anemones)
m Echinozoa (sea urchins etc)
15 Branchiotremata
(branchiotremes)

Protozoans

Protozoans ("first animals") form a subkingdom in the kingdom of microscopic one-celled organisms, Protista. Unlike their plant-like counterparts (see pages 36–37), protozoans tend to feed on other organisms or organic substances.

Scientists group protozoans in different phyla according to how they move about. Only prehistoric kinds with hard parts have survived as fossils. The earliest date back over 700 million years. Billions lived, and live, on sea floors or in surface waters. Their accumulating skeletons have formed thick rock layers in which some fossil types indirectly show the saltiness and warmth of ancient seas.

Of our examples, numbers 1–3 belong to the phylum Rhizopoda; number 4 to the phylum Ciliata.
1 **Fusulina** had a large, chalky, spindle-shaped skeleton tapered at both ends. Size: 6cm (2.4in). Time: Late Carboniferous. Class: Foraminifera

1

Fusulinid features
Here the big foraminiferan *Fusulina* is shown enlarged.
a Large test (skeleton)
b Spindle (sometimes spherical) shape
c Composition: calcium carbonate
d Three- or four-layered spirotheca (wall)
e Concentric chambers
f Fluted septa (divisions)
g Septal pores
h Foramen (opening)

(foraminiferans – micro-organisms usually known as fossils from their calcareous shells pierced by tiny holes. From these, in life, thread-like "false feet" projected for locomotion and seizing particles of food). Order: Fusulinida (fusulines).

2 **Nummulites** belonged to a group of giant coin-shaped foraminiferans with limy, perforated shells comprising many chambers. Size: 1–6cm (0.4–2.4in). Time: Palaeocene–Oligocene. Order: Rotaliida.

3 **Cryptoprora** resembles a lacy, pointed hat with ribbons hanging from the rim. Size: 0.1–1mm (0.004–0.04in). Time: Eocene onward. Class: Actinopoda. Subclass: Radiolaria (protozoans with glassy, perforated skeletons shaped like hats, urns, spheres, etc).

4 **Tintinnopsis** has a bell-shaped skeleton with a pointed "tail". Size: 0.1–0.2mm (0.004–0.008in). Time: Recent. Phylum: Ciliata (ciliates), fringed by fine hairs (cilia) whose rhythmic beating helps them swim.

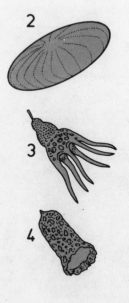

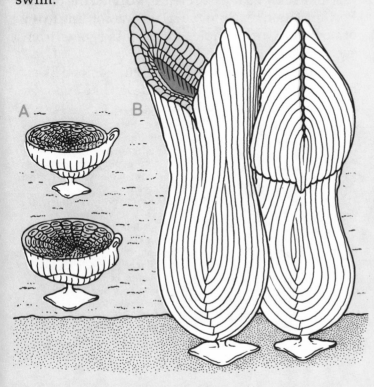

Problematic organisms
Some fossil organisms fit no known group of protists, plants, or animals. They include the tiny aquatic acritarchs and chitinozoans, also the larger, many-celled Petalonamae, shown here.
A Juvenile specimens
B Reconstructed group (smaller than life size) on a sandy sea floor, in late Precambrian times. Bodies comprised funnel-shaped colonies of branching tubes. The central cavity and scattered "needles" in the skeleton show that the Petalonamae had certain features found in sponges.

49

Parazoans

Sponges and the sponge-like archaeocyathids make up the animal subkingdom Parazoa. Parazoans are many-celled animals that lack true tissues or organs and can look misleadingly like plants. Sponges have simple bag-like bodies open at one end and anchored to the sea bed at the other. Mineral spicules (tiny rods) reinforce the body wall, which is pierced by tiny holes. The body cavity is lined with cells equipped with whips that draw in water through the holes. A sponge extracts food particles and oxygen from this water, and then "whips" drive the used water out through the main body opening, the osculum.

Some sponges live singly, others form colonies measuring from under 1 centimetre to over 1 metre across. Sponges grow as "vases", branches, or rock-encrusting blobs. Fossil skeletal remains suggest that sponges evolved from protozoans 700 million years ago. Some prehistoric sponges built reef-like sea-bed rocks. Scientists divide the phylum Porifera (sponges) into four classes according to type of skeleton. Our examples show a fossil genus from each class.

Archaeocyathids were usually cup-shaped like a fossil coral, but with a perforated wall like a sponge

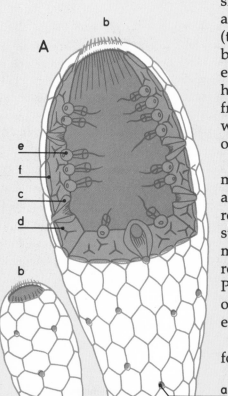

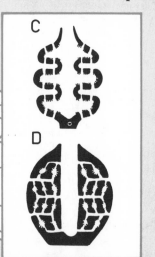

Sponge features
Shown here are different types of sponge, with features found in living specimens.
A Simple sponge: cutaway view of part of a colony, shown much enlarged
a Incurrent pores, letting water enter
b Osculum, letting water out
c Pore cell
d Spicule
e Collar cell
f Covering cell
B Simple sponge, actual size
C More advanced sponge, with a folded wall (shown in section).
D Complex sponge, with many canals and chambers (section).

(in fact most had a double wall). Experts put them in their own phylum, Archaeocyatha, with two classes. Archaeocyathids evidently lived on sea beds, much as sponges do, and flourished in warm, clear seas, worldwide – but only in Early–Mid Cambrian times.

1 **Siphonia,** about 1cm (0.4in) long, has persisted since Cretaceous times. Class: Demospongea (horny sponges largely reinforced by networks of horny fibres).

2 **Coeloptychium** was a mushroom-shaped sponge common in Late Cretaceous Europe. Diameter: 7.5cm (3in). Class: Hyalospongea (glass sponges with skeletons made up of six-rayed spicules).

3 **Polytholosia** resembled a string of pearls 5cm (2in) long. Time: Triassic. Class: Calcispongea (sponges with two- to four-rayed limy spicules).

4 **Chaetetopsis,** 4cm (1.6in) long, lived from Ordovician to Tertiary times. Class: Sclerospongea (coralline sponges with a skeleton of pin-shaped spicules).

5 **Tabellaecyathus,** was a cone-shaped archaeocyathid, 4cm (1.6in) long. Time: Cambrian. Class: Irregulares (cup- and disc-shaped forms, often irregularly shaped).

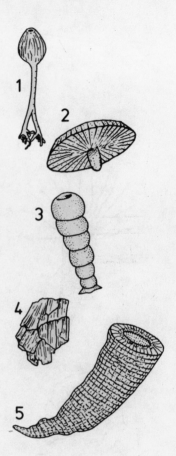

Spicules and fibres
Scattered spicules are often the only clues to the fossil sponges in a rock layer. Here we show spicules from two classes of sponge, and horny fibres from a third.
a Calcareous spicules belonging to the Calcispongea
b Siliceous spicules belonging to the Hyalospongea
c Horny fibres belonging to the Demospongea, and rare as fossils because they readily dissolve

©DIAGRAM

Coelenterates

Coelenterates or "hollow gutted animals" are mostly jelly-like sea creatures. Each consists of many cells organized as tissues to produce a central body cavity with a mouth surrounded by tentacles and stinging cells for catching prey. Coelenterates are less primitive than sponges but have no central nervous system or systems for breathing, circulation, or excretion.

Cnidaria is the only phylum known from fossils. Cnidarians include jellyfishes, hydrozoans, sea anemones, and corals. Free-swimming cnidarians (medusas) produce polyps "rooted" to the sea floor. Polyps produce medusas, and so on. Cnidarians evolved about 700 million years ago. Limy cups enclosing colonial corals and hydrozoans in time built limestone reefs that formed rock layers.

Fossil cnidarians include these five examples. Numbers 3–5 are from the class Anthozoa (corals and sea anemones).

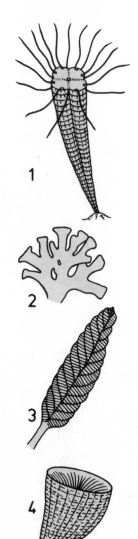

1 **Conularia** produced a polyp in a tall, hard, hollow "pyramid" inverted on the sea floor. Tentacles probably surrounded the broad mouth end. Length: 6–10cm (2.4–4in). Time: Cambrian–Triassic. Class: Scyphozoa (scyphozoans) – jellyfishes etc.

2 **Millepora** comes in colonies of feeding polyps in central tubes, surrounded by protector polyps in other tubes. Size: 7cm (2¾in). Time: Recent. Class: Hydrozoa (hydrozoans).

3 **Charnia,** a sea pen, formed colonies of "feathers" growing from the sea bed. Height: 14cm (5½in). Time: Precambrian (about 700 million years ago). Subclass: Octocorallia (Alcyonaria), soft corals.

4 **Streptelasma** was a rugose ("wrinkled") coral, named from horizontal wrinkles on the skeleton's outer wall. Height: about 2.5cm (1in). Time: Ordovician–Silurian. Subclass: Zoantharia (stony corals). Order: Pterocorallia (rugose corals), mostly solitary, now extinct.

5 **Halysites,** a "chain coral", produced chains of tubes stuck together at the sides. Size: 5cm (2in) across. Time: Ordovician–Silurian. Subclass: Zoantharia. Order: Tabulata (tabulate corals, named from internal horizontal plates, or tabulae), extinct since Permian times.

6 **Acropora** produces a light, branching skeleton about 30.5cm (1ft) long. Time: Eocene–Recent. Subclass: Zoantharia. Order: Cyclocorallia (Scleractinia), the major modern reef-builders, probably evolved from naked sea anemones.

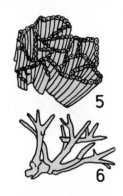

Scarce survivors
This fossil "jellyfish" came from Carboniferous rocks in Belgium. Fossils of soft-bodied beasts like this are rare. They survive only in rocks such as shales or limestones formed from fine-grained particles of mud.

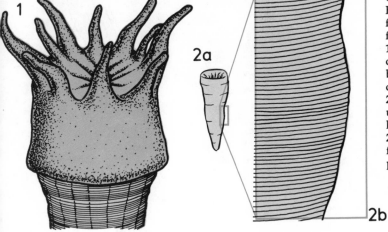

Growth of coral
Here we show how coral polyps form cups that have survived as fossils.
1 A living coral polyp in its coral cup. The creature builds its cup by adding a daily layer of calcium carbonate to its top.
2a Growth bands plainly show up on this Silurian coral cup, real height less than 2.5cm (1 inch).
2b This magnified view shows a few of the daily growth rings that produced the cup.

Molluscs 1

Molluscs ("soft" animals) include snails, clams, squids, and other creatures with a shell and muscular foot, or parts derived from these, and a digestive tract. Most live in the sea: grazing on algae, filtering food particles from mud or water, or hunting. Molluscs probably evolved in Precambrian times from animals like flatworms. By Carboniferous times, some molluscs invaded land and fresh water. No group of animals except the arthropods have diversified more. The phylum Mollusca produced many thousands of species, grouped into two subphyla with nine known fossil classes. Our examples all lived in the sea.

1 **Chiton** (coat-of mail shell). Chitons resemble flattened wood lice guarded by a shell of seven or eight overlapping plates. They cling to rocks and graze on algae. Chitons date from Late Cambrian times. Only known fossil class: Polyplacophora.

2 **Scenella** was limpet like, with a cap-shaped single shell. Size: 1cm (½in) across. Time: Cambrian. Class: Monoplacophora (maybe ancestral to the squids, octopuses, and other cephalopods).

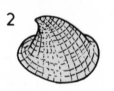

1

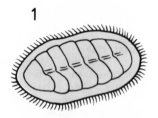

2

Four types of mollusc
Shown here are sections through four of the main types of mollusc described on pages 54–57.
A Chiton (coat-of-mail shell)
B Clam (bivalve mollusc)
C Snail (gastropod)
D Squid (cephalopod)
Despite differences, all share the same basic body plan, with the following common features.
a Shell (usually the only part preserved in fossils)
b Foot
c Digestive tract

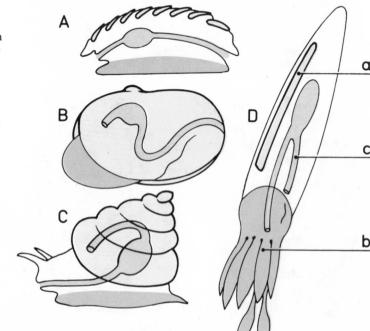

3 **Pleurotomaria** was a primitive snail with a spiral, pointed shell about 5cm (2in) high. Time: Jurassic–Cretaceous. Class: Gastropoda (molluscs with head, foot, and a mantle that secretes a bowl-shaped or spiral shell). They feed with a radula – a horny ribbon, armed with rows of rasping "teeth".

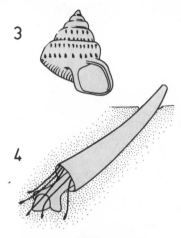

4 **Dentalium** lives in a slim, tusk-like shell and burrows head-first in soft sediment, extending thread-like tentacles to catch small organisms. Shell length: up to 12cm (4.7in). Class: Scaphopoda (tusk shells), known since Ordovician times.

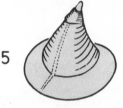

5 **Conocardium** had a shell about 6cm (2.4in) across. Time: Ordovician–Permian. Class: Rostroconchia – extinct molluscs with a fused two-part shell. Rostroconchs perhaps gave rise to bivalves.

6 **Mytilus,** the common mussel, is a bivalve, with a hinged, two-part shell about 5cm (2in) long. Time: Triassic–Recent. Class: Lamellibranchia (bivalves) – headless molluscs with a hinged, two-valved shell.

Unlikely bivalves
This restoration shows a sea-bed colony of rudists, 100 million years ago. In these bivalves one valve was a lid that opened to let in food and water. Rudist shells built reefs before these molluscs died out at the end of the Cretaceous Period.

©DIAGRAM

55

Molluscs 2

Squids, octopuses, and their kin form the cephalopods ("head footed"): the most highly developed class of molluscs. These big-brained, keen-eyed sea beasts can swim backward fast by squirting water forward. Tentacles at the head end seize prey and feed it to the beak-like jaws. Some cephalopods have an outside shell, some an inner shell, others none at all. Cephalopods range from a species only millimetres long to the giant squid – at up to 22 metres (72ft) long, the largest living invertebrate. Cephalopods probably evolved in Cambrian times, from gastropod-like molluscs. Our examples come from the four cephalopod subclasses: nautiloids, bactritoids, ammonoids, and coleoids. Fossils show that extinct ammonoids and (coleoid) belemnites were abundant in certain Mesozoic times. Besides cephalopods, there were two more mollusc classes: coniconchs and calyptoptomatids. These were small beasts with straight, cone-shaped shells. The last died out in Permian times.

A living nautiloid
Clues to fossil nautiloids come from the surviving *Nautilus*.
A Head-end, revealing the body in the outer chamber
B Side view, with shell cut open to reveal its structure
a Tentacles
b Jaws
c Gut
d Heart
e Gill
f Valve controlling water used for jet propulsion
g Siphuncle – tube connected to the inner chambers
h Chambers, filled with gas or air (*Nautilus* pumps water in or out to alter buoyancy, and so sink or rise)

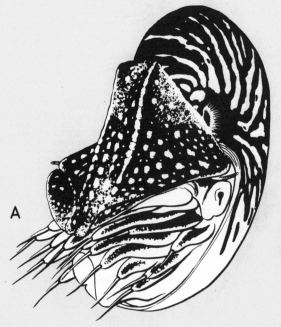

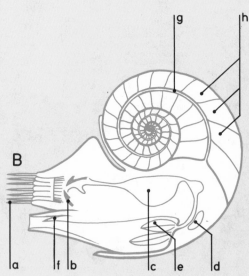

1 **Orthoceras** had a straight shell 15cm (6in) long. Time: Ordovician–Triassic. Subclass: Nautiloidea (nautiloids) – cephalopods with a straight or curved shell divided into chambers. Nautiloids persisted from Cambrian times to today.

2 **Cyrtobactrites** had a tusk-shaped shell. Length: 4.5cm (1.8in). Time: Early Devonian. Subclass: Bactritoidea (ancestors of ammonoids).

3 **Anetoceras** had a loosely coiled shell about 4cm (1.6in) across with zig-zag growth lines. Time: Early Devonian. It belonged to the goniatites, early members of the subclass Ammonoidea. Ammonoids resembled nautiloids but had shells with wrinkled growth lines between chambers.

4 **Stephanoceras** had a coiled, ribbed, disc-shaped shell 20cm (8in) across. Time: Mid Jurassic. It belonged to the ammonoids called ammonites. Thousands of species teemed in Mesozoic seas.

5 **Gonioteuthis** was a belemnite, with a long, squid-like body known only from the rear end's fossilized, bullet-shaped internal guard. Guard length: 7cm (2.8in). Time: Late Cretaceous. Order: Belemnitida, in the subclass Coleoidea (squids, cuttlefish, octopuses, and belemnites).

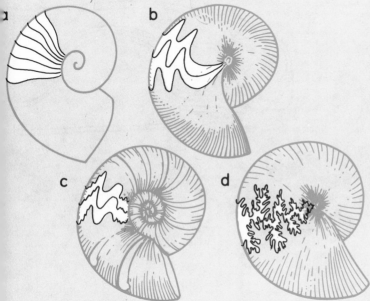

Shell sutures
Nautiloids, goniatites, ceratites and ammonites respectively evolved ever more complex sutures – lines formed where internal partitions met the outside shell. These four cut-open shells show stages in this trend.
a *Nautilus* (simple sutures)
b *Goniatites* (zig-zag sutures)
c *Ceratites* (wavy sutures)
d *Phylloceras* (very complex sutures)
Complex sutures might have made shells strong enough to endure water pressure deep down. Many ammonites were arguably bottom feeders.

Worms

Soft-bodied beasts such as worms are seldom fossilized, but we know of prehistoric kinds from tracks, tunnels, tubes, jaw remains, and even impressions of their bodies left in fine-grained rocks. Such clues reveal that worms were crawling on and in, and swimming just above, the floors of shallow coastal waters some 700 million years ago, where South Australia's Ediacara Hills now stand. Southwest Canada's famous Burgess Shales reveal that many sea worms had evolved 550 million years ago. Of our examples, number 1 comes from the phylum Aschelminthes (roundworms, built on simple lines), 2–5 come from the Annelida (worms divided into many segments, often with complicated jaws, and tentacles projecting from the head). Annelids probably gave rise to millipedes and other arthropods.

1 **Gordius,** a "hair worm", resembles a living thread up to 15cm (6in) long. Adults wriggle in ponds and ditches worldwide. Young live inside aquatic insect larvae; adults invade land insects, then return to water to breed. Time: Eocene on. Phylum: Aschelminthes. Class: Nematomorpha.

Durable mouth parts
Prehistoric polychaete worms are known mostly from their mouth parts – the only hard parts of their bodies, and so the only structures tending to be fossilized. Such fossils are called scolecodonts.
A Enlarged section through a polychaete's head (items **b–e** are the hard parts likely to be fossilized)
a Mouth
b Mandible
c Maxilla
d Dental plate
e "Pincer"
B Five Permian scolecodonts 15 times larger than life

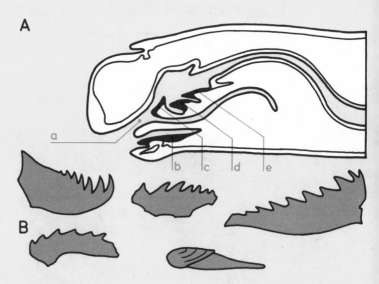

2 **Spriggina,** one of the first known worms, was probably an annelid. It was slim and flexible, with a long, strong, curved head shield, and up to 80 "limbs" with spiny ends. In Late Precambrian times it swam offshore in what is now Australia. Length: up to 4.5cm (1.8in).

3 **Dickinsonia** had a broad, flat, oval body crossed by 20 to 550 ridges. Length: 0.25–60cm (0.1in–2ft). It lived in a Late Precambrian sea in what is now Australia. It resembled some flatworms but was probably an annelid.

4 **Canadia** was a marine annelid with bristles, and long "legs" projecting from its sides. Time: Mid Cambrian. Place: south-west Canada. Class: Polychaeta (segmented worms with "legs" and bristles). Order: Errantia (mainly active polychaetes, with a well-defined head and jaws).

5 **Serpula** lives in a white, limy tube glued to an undersea shell or stone. Tentacles projecting from the head catch passing particles of food. Tube length: about 8cm (3in). Time: Silurian to present. Class: Polychaeta. Order: Sedentaria (jawless worms living in a tube or tunnel).

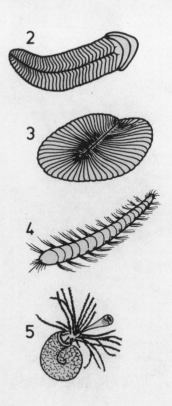

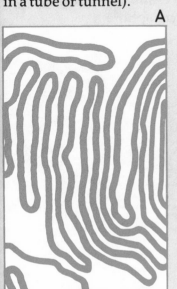

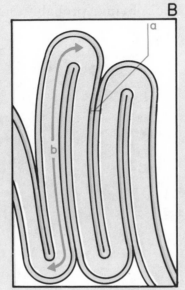

A

B

Worm trails
Some rocks preserve worm tracks made in soft mud that later hardened into stone.
A An unknown worm-like beast left this fossil track, seen from above. The creature ploughed through underwater sediments. Most likely it was feeding on organic particles or tiny organisms.
B This reconstruction shows how track **A** took shape.
a The width of churned up sediment fixed the distance of adjoining trails.
b The lengths of straight sections hint at the length of the worm that formed the U-turns seen in the track.

©DIAGRAM

Arthropods 1

Most known kinds of animals belong to the phylum Arthropoda: "jointed legged" animals. Arthropods include the extinct trilobites, as well as insects, spiders, crabs, centipedes, and relatives. All have jointed limbs, and an exoskeleton shed from time to time as they grow. Arthropods probably evolved from annelid seaworms. Early arthropods lived in the sea but some were colonizing land by 400 million years ago. These two pages give fossil examples of major groups based mostly on land. Onychophorans ("velvet worms") may be a primitive link with the arthropods' annelid worm ancestors. Many-legged myriapods (centipedes and millipedes) were among the first land arthropods. From early, wingless insects came the bees and butterflies – winged pollinators evolving with, and feeding on, the flowering plants. Most fossil insects date from the last 60 million years. Some of the best specimens occur in coal and amber.

1 **Aysheaia** looked like a worm with a pair of stubby legs on each body segment and a pair of "feelers". It was arguably related to the living velvet worms but lived below the sea, in Mid Cambrian times. Subphylum: probably Onychophora.

Arthropleura
At 1.8m (6ft) *Arthropleura* was the largest-ever land arthropod. This huge, flat "millipede" munched rotting vegetation on the floors of Carboniferous forests.

2 Latzelia has been called an early centipede. It had poisoned fangs, a flat back and a pair of walking legs on each of its many body segments. It roamed damp forest floors, hunting worms and soft insects. Time: Carboniferous. Subphylum: Myriapoda (centipedes, millipedes).

3 Rhyniella, the first known (wingless) insect, was a springtail, about 1cm (½in) long. Springtails live in soil, browse on decaying plants, and flip into the air if scared. Time: Devonian, about 370 million years ago. Subphylum: Insecta (insects) – small invertebrates with six legs and a three-part body (head, thorax, abdomen). Many undergo great body changes (metamorphosis) as they grow.

4 Meganeura, the largest-known winged insect, had a wing span of up to 70cm (27½in). Time: Late Carboniferous. *Meganeura* belonged to the Megasecoptera, primitive winged insects unable to fold back wings held at rest.

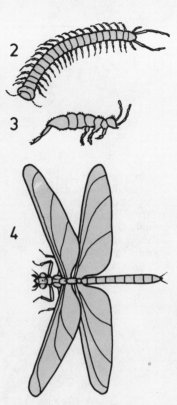

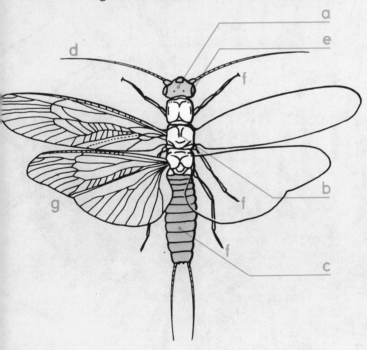

Insect features
This stonefly illustrates typical insect features, such as the three-part body. Early flying insects had two pairs of wings of equal size. They could flap each pair separately, but only up and down; they could not fold their wings back at rest. In such later insects as butterflies, both pairs are coupled. In flies, hind wings are tiny and only serve as balancers.
a Head
b Thorax
c Abdomen
d One pair of antennae
e Compound eyes
f Three pairs of legs
g Wings reinforced by "veins"

Arthropods 2

Trilobites ("three lobed") were marine arthropods resembling wood lice. Their name comes from a central raised ridge along the back, flanked by flattish side lobes. There was a head shield and an armoured thorax and tail, both divided into many segments. Each of these sprouted a pair of limbs designed for walking, swimming, breathing, and handling food. Trilobites crawled or swam, and many could curl up if threatened. Scientists know of several thousand genera and perhaps 10,000 species, from less than 4mm (0.1in) to 70cm (28in) long. Trilobites lived from about 600 to 250 million years ago. They dominated shallow seas in the Cambrian Period. Ordovician genera included many specialized species with bizarre spines or knobs. These pages show a few contrasting kinds.

Trilobite body plan
Calymene illustrates the basic body plan of trilobites.
A Cephalon, a head with a central hump
B Thorax, consisting of a number of jointed segments
C Pygidium, a tail of fused segments articulated with the thorax

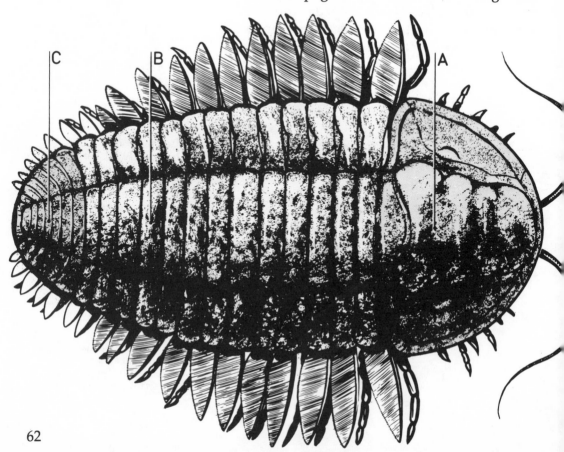

1 **Ampyx** was lightly built and filter fed. Length: 4cm (1.6in). Time: Ordovician.
2 **Encrinurus** had a relatively large, heavy head with eyes on stalks, and was a bottom dweller. Length: 5cm (2in). Time: Ordovician-Silurian.
3 **Lonchodomas** had a long spike jutting forward from the head. Length: about 5cm (2in). Time: Ordovician.
4 **Pliomera** Tooth-like ridges jutting from the front of the head interlocked with the tail tip when this animal rolled up. Length: about 4cm (1.6in). Time: Ordovician.
5 **Trimerus** was elongate and lived by burrowing in mud. Length: 8cm (3in). Time: Silurian-Devonian.

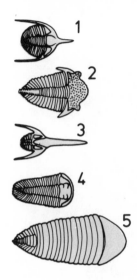

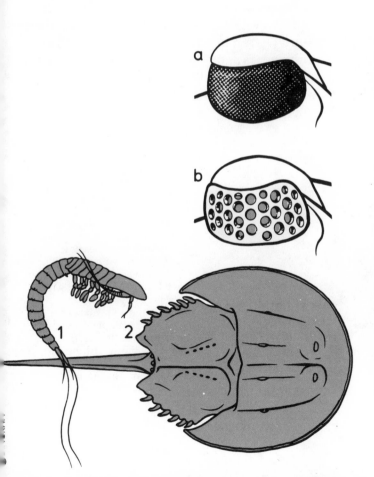

Amazing eyes
Trilobites were among the first known animals with efficient eyes. These had many calcite-crystal lenses fixed at different angles to register movement and light from different directions.
a Compound or holochroal eyes consisted of 100 to 15,000 closely packed hexagonal lenses and resembled insects' eyes.
b Schizochroal eyes featured groups of lenses, relatively few in number.

Living fossils
Trilobites died out more than 230 million years ago, but we show two kinds of living arthropods which have been regarded as close kin of these creatures.
1 Cephalocarids are tiny, primitive, shrimp-like animals with segmented bodies much like trilobites'. First found in 1955, cephalocarids may be their nearest living relatives.
2 The horseshoe crab looks a bit like a trilobite and has a larval stage reminiscent of a young trilobite. This type of "crab" has lived for over 300 million years.

63

Arthropods 3

Spiders, scorpions, horseshoe crabs, and the extinct sea scorpions make up the subphylum Chelicerata. Prehistoric chelicerates included arthropods longer than a man – fish-eating terrors of ancient seas.

Chelicerates get their name from the two chelae ("biting claws") in front of the mouth. Behind these come a pair of pedipalps ("foot feelers"), used by horseshoe crabs as legs, by scorpions for seizing prey, by male spiders to grip when mating. Then come four pairs of legs. The two-part body's limbless abdomen often ends in a flat or spiky tail. Horseshoe crabs and sea scorpions had gills and breathed under water. Modern spiders and scorpions breathe atmospheric air with help from "lung-books" or air holes called tracheae.

Chelicerates appeared in the sea over 560 million years ago. They gave rise to sea scorpions soon after 500 million years ago. True scorpions evolved about 440 million years ago. The first known spiders were catching insects 370 million years ago. Most chelicerates grow a rather soft body covering, so few survive as fossils.

Spider in amber
Above: sticky sap trapped a spider 30 million years ago. Below: the sap hardened into amber which still preserves the spider's life-like body.

©DIAGRAM

Eurypterid body plan (right)
Shown here are two life-size views of the Silurian "sea scorpion" *Eurypterus*.
A View of underside:
a Chelicerae (appendages with jointed pincers – long and formidable in beasts like *Pterygotus* which was capable of catching fishes)
b Four pairs of walking legs
c Large "paddles"
d Genital appendages
B View of upper side:
a Prosoma (forepart)
b Opisthosoma (hind part, comprising 12 articulated segments) including:
c Telson ("tail")

1 **Palaeolimulus** had a broad, horseshoe-shaped forepart with two large, many-faceted eyes. The narrow abdomen ended in a spiky "tail". Length: 6cm (2.4in). Time: Permian. Place: shallow seas. Class: Merostomata. Subclass: Xiphosura (horseshoe crabs).

2 **Pterygotus,** one of the largest-ever arthropods, was a formidable predator. Its forepart had two long, strong pincers, eight legs, and two large, broad paddles for swimming. Twelve segments and a tail formed the long hind end. Length: up to 2.3m (7ft 4in). Time: Silurian. Place: seas. Class: Merostomata. Subclass: Eurypterida (eurypterids or sea scorpions).

3 **Palaeophonus,** a late Silurian scorpion, had big pincers and a large sting on its tail. It might have been the first land animal. Class: Arachnida (spiders and scorpions). Order: Scorpionida (scorpions).

4 **Arthrolycosa** was a large, long-limbed early spider. It had eight legs and eight eyes and probably attacked insects with help from poisoned chelae ("fangs"). Time: Carboniferous. Class: Arachnida. Order: Araneae (spiders).

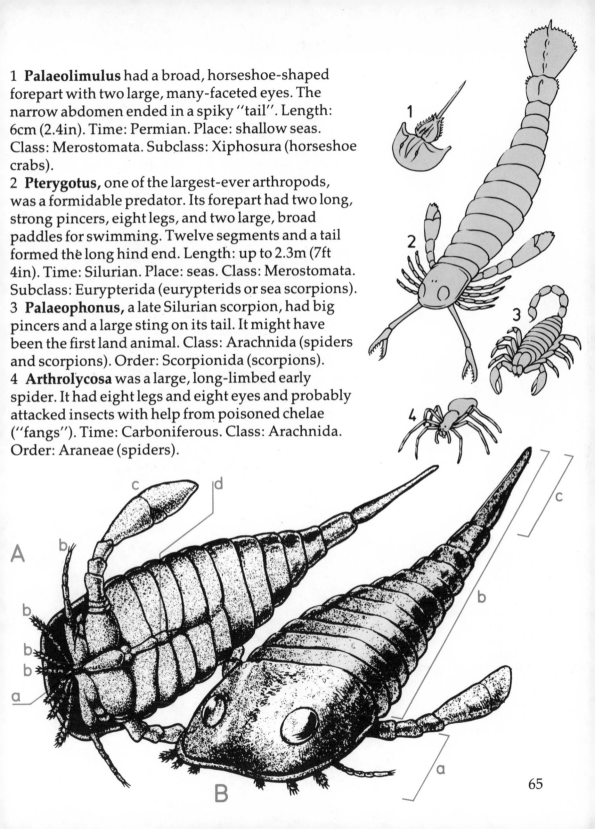

65

Arthropods 4

Crustaceans include crabs, barnacles, and relatives with flexible shells. Members of their subphylum, the Crustacea, have three main body parts (the first two often look like one), two pairs of antennae, and many more appendages. The limbs are forked. Crustaceans evolved by 650 million years ago. From early forms arose nine classes, all mostly found in water. Three produced a rich variety of fossils. These classes were the tiny ostracods, the cirripedes (barnacles and kin), and the malacostracans (including lobsters, crabs, and shrimps). Lobster-like crustaceans date from Triassic times and gave rise to crabs in the Jurassic Period.

1 **Cypridea** was a freshwater crustacean resembling a microscopic clam. Its body lay inside a two-valved shell with straight top and bottom edges and a knobbly surface. Time: Mid Jurassic–Early Cretaceous. Class: Ostracoda – ostracods (mostly marine organisms, valuable as guide fossils from Cambrian times onward).

Ostracod body plan
Experts identify fossil ostracods from the shapes and patterns of their shells. In life these small crustaceans (none larger than a bean) looked like the living specimen seen here enlarged. Such animals are found in seas and ponds.
a Hinged shell, shown in section; when shut it protected all parts of the body
b Short body
c Antennae (used as limbs)
d Mandibles (mouthparts)
e Maxillae (mouthparts)
f Trunk appendages
g Furca ("tail")

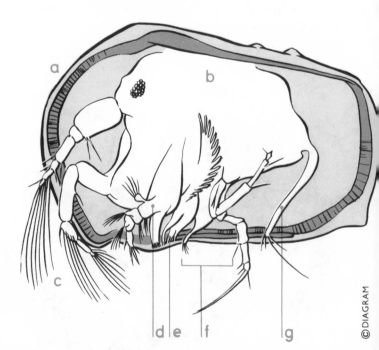

©DIAGRAM

2 **Balanus,** an acorn barnacle, lives inside a dome-like shell of six white limy plates, built on a tidal rock. At high tide curved, feathery appendages move in and out between shell plates to pull in scraps of food, and water. Size: about 1cm (½in) across. Time: Oligocene onward. Class: Cirripedia (barnacles) – crustaceans that fix their heads to rocks or other solid objects.

3 **Aeger** was a long-tailed ten-legged crustacean with a long bill-like rostrum and long antennae. Length: about 12cm (4.7in) excluding rostrum. Time: Late Triassic–Late Jurassic. Class: Malacostraca (advanced crustaceans with stalked eyes and a "shell" usually covering the head and thorax). Order: Decapoda (ten-legged crustaceans, including shrimps, crabs, and lobsters).

4 **Eryon** had a big, broad, flattened, crab-like cephalothorax (fused head and thorax) but a longish abdomen. Crabs evolved from beasts like these that tucked the abdomen beneath the forepart of the body. Length: 10cm (4in). Time: Mid Jurassic–Early Cretaceous. Order: Decapoda.

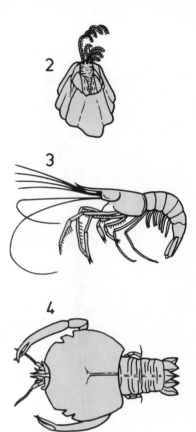

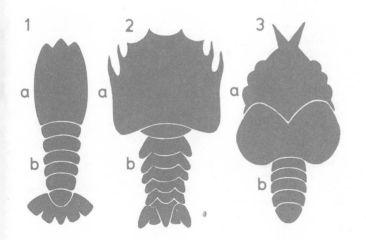

Crabs in the making
Shown here are the bodies of three prehistoric crustaceans (without their limbs).
1 *Eryma*
a Narrow cephalothorax (head and thorax)
b Abdomen as long as the cephalothorax
2 *Eryon*
a Broad, flat cephalothorax
b Long abdomen
3 *Palaeinachus*
a Big, broad cephalothorax
b Short, narrow abdomen

Tentaculates

This name is used for four great groups of lowly sea beasts. Tentaculates have a U-shaped gut and a mouth surrounded by a ring of tentacles, used for pulling tiny particles of food and water into the mouth. Most live rooted to the sea bed. Phoronids are worm-like creatures living in a hard protective tube. Bryozoans are tiny, soft-bodied beasts in hard limy or horny cases. Colonies form mats or miniature "trees" on underwater rocks or shells. Brachiopods ("arm footed" animals) resemble bivalve molluscs sprouting from a fleshy stalk, but shell valves cover the body from above and below, not from side to side. Conodonts ("cone teeth") are brown or greyish tooth-like microfossils from eel-like creatures.

Conodonts flourished from Cambrian to Triassic times. Brachiopods date from the Early Cambrian, bryozoans from the Ordovician, and phoronids from the Cambrian Period. All groups but conodonts have living representatives.

Conodont types (right)
Conodonts are preserved hard parts of soft-bodied animals resembling elvers (young eels), only recently discovered. These may not have been tentaculates at all. Their tiny tooth- and saw-like fossils came in several types. Three kinds appear here, much enlarged.
A Compound conodont with many "teeth"
B Simple conodonts
C Platform conodont

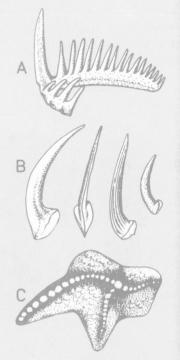

Evolved from worms (left)
Lingula's long worm-like body hints at the likely origin of brachiopods. Often called lampshells, they probably evolved from a tube-shaped worm that grew a pair of flat, protective shells.

1 **Phoronis** is a worm-like tube dweller about 20cm (8in) long. It lives buried in sand offshore. Tiny whip-like cilia on its tentacles drive water and food into its mouth. Time: maybe Cretaceous onward. Phylum: Phoronida (phoronids).

2 **Multisparsa** was a bryozoan forming tree-like colonies of tubes with touching sides. Colony size: 2cm (0.8in) across. Time: Mid Jurassic. Phylum: Bryozoa (sea mats).

3 **Lingula** lives inside a long, tongue-shaped shell growing from a strong stalk. It inhabits a burrow in soft undersea sediment. Diameter: 10cm (4in). Time: Ordovician onward (it is about the oldest living fossil). Phylum: Brachiopoda. Class: Inarticulata (brachiopods whose shell lacks a hinge).

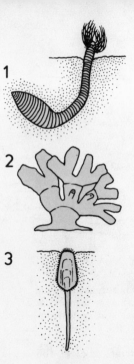

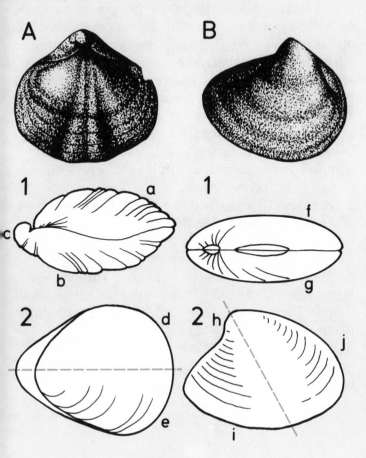

Brachiopods and bivalves

At first glance you might mistake a fossil brachiopod for a bivalve mollusc, another creature with a two-valved shell. These illustrations show how to tell both apart.

A Brachiopod: both valves seen edgewise (**1**), and one valve seen from above (**2**)

a,b Shell valves differ in size and curvature

c Pointed beak at hinge end

d,e Each half of one valve is a mirror image of the other

B Bivalve: both valves seen edgewise (**1**) and one seen from above (**2**)

f,g Shell valves alike in size and curvature

h No beak at hinge

i,j Valves not symmetrical

©DIAGRAM

Echinoderms and branchiotremes

Echinoderms ("spiny skinned" animals) include starfishes, sea urchins, and their relatives. These tend to have a five-rayed body with a skeleton of chalky plates just below the skin. Some sprout protective spines. Water flows through tubes inside the body, and pumps up many tiny tube feet tipped with suckers and used for walking, gripping, or as breathing aids. Echinoderms lack a normal "head", but have a well-developed nervous system. They probably gave rise to backboned animals by way of branchiotremes.

These small, soft-bodied, worm-like creatures with internal "gill baskets" include the pterobranchs and long-extinct but once abundant graptolites. Branchiotremes produced the beginnings of a notochord or "backstring" – the precursor of a backbone. Some experts rank them as a phylum, others consider them a subphylum of the Chordata, which includes all backboned animals. Echinoderms and many branchiotremes live on the sea bed. Branchiotremes use tiny tentacles to catch passing scraps of food. Echinoderms include hunters, grazers, scavengers, and filter feeders. Both phyla go back to Cambrian times. Examples 1–5 represent echinoderm subphyla. Examples 6–7 represent two classes of branchiotremes.

1 **Dendrocystites** looked like a double-ended thorn and probably laid flat on the sea bed, eating tiny organisms. Length: 9cm (3.5in). Time: Ordovician. Subphylum: Homalozoa (carpoids), extinct.
2 **Pleurocystites** sprouted two short, unbranched arms from a bud-shaped "cup" on a long stalk. Stalk height: 2cm (0.8in). Time: Ordovician. Subphylum: Blastozoa (blastoids and cystoids), extinct.
3 **Botryocrinus** was a flower-like sea-bed dweller with a stalk, crown, and feathery arms. Height: 15cm (6in). Time: Silurian. Subphylum: Crinoidea ("sea lilies"), Cambrian onward.

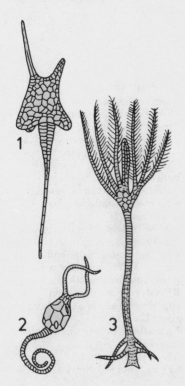

How tube feet work (above) This shows a section along a starfish arm. Water from a pipe inside fills bottles. Muscles squeeze them, driving water down into tube feet. The feet lengthen. As water flows back, each foot tip forms a sucker.

4 **Crateraster** had a broad central disc and five slim arms. Size: 10cm (4in) across. Time: Cretaceous. Subphylum: Asteroidea (starfishes and brittle stars), dating from Ordovician times.

5 **Bothriocidaris** resembled a small, prickly pincushion. Size: about 2cm (0.8in) across including spines. Time: Ordovician. Subphylum: Echinozoa (sea urchins and sea cucumbers), Ordovician on.

6 **Rhabdopleura** is a soft-bodied worm-like beast living in colonies in tubes on the sea bed. Size: 2–3mm (0.08–0.12in). Phylum: Branchiotremata (branchiotremes). Class: Pterobranchia, dating from Ordovician times.

7 **Dichograptus** was an eight-branched graptolite – a colony of tiny worm-like creatures, known from their flattened, "saw-edged" fossil tubes. Colony diameter: 6cm (2.4in). Time: Ordovician. Phylum: Branchiotremata. Class: Graptolithina (graptolites), extinct, dating from Cambrian to Carboniferous.

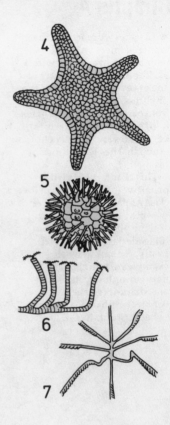

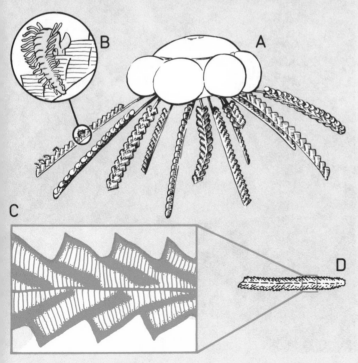

About graptolites
A Single-branched graptolites called diplograptids formed a colony buoyed up by a central float. Such colonies drifted at the surface of Early Palaeozoic seas. (Some others sprouted from the sea floor.)
B Each individual of a colony lived inside a cup-like theca.
C Much magnified view of fossil *Diplograptus* thecae. Infrared light shows details of structure poorly visible in ordinary light.
D *Diplograptus* colony, shown actual size.

Chapter 4

FOSSIL FISHES

With fishes we reach the first of five chapters on fossil vertebrates, or animals with a backbone. Vertebrates are just one of several subphyla in the phylum Chordata (chordates). But apart from hemichordates (also called branchiotremes), vertebrates are the only chordate group that has left important fossils.

This chapter starts with primitive jawless fishes in the class Agnatha. We briefly review the spiny fishes (Acanthodii) and the placoderms (Placodermi), both extinct. Lastly, we look at fossil examples of two classes flourishing today: the sharks and shark-like fishes (Chondrichthyes), and bony fishes (Osteichthyes).

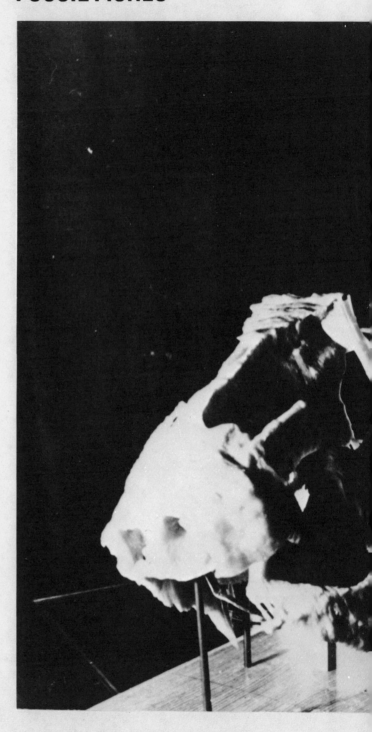

Swedish palaeontologist Erik Stensiö builds a three-dimensional model of a prehistoric fish's brain. In the 1920s, Stensiö's techniques of dissection and reconstruction of fossil brains revolutionized our understanding of some early fishes. (The Mansell Collection.)

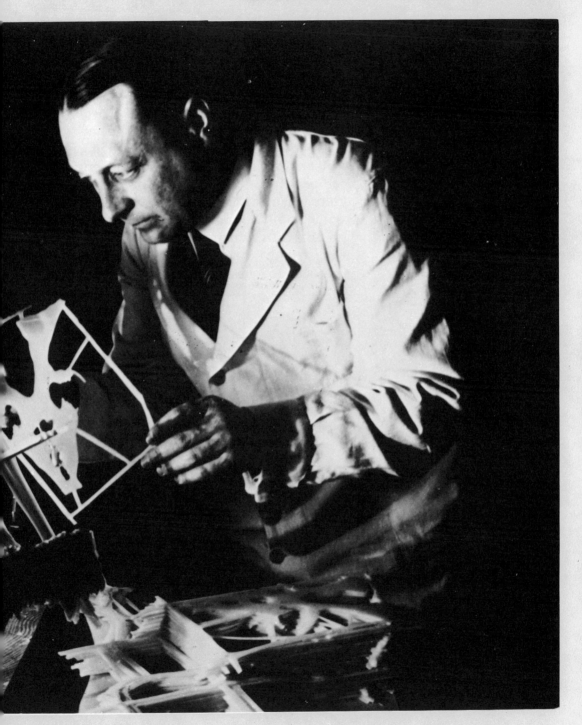

About fishes

Fishes are cold-blooded, backboned animals that live in water, breathing oxygen through gills. To swim forward most thrust water backward by waggling the tail and body.

Defined strictly, fishes have jaws, paired fins, and median fins (fins along the midline of the body). But this chapter includes creatures without jaws or true fins. Appearing over 500 million years ago, they were the first known backboned animals or vertebrates – creatures with a brain housed in a skull, and vertebrae (bones making up a backbone). Vertebrae provide bodily support and a protective channel for the spinal cord, which contains nerves connecting the brain with other parts of the body. From jawless fishes or close relatives there evolved four other

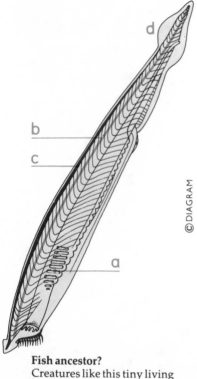

© DIAGRAM

Fish ancestor?
Creatures like this tiny living lancelet might have been the immediate ancestor of fishes. (A fossil lancelet 550 million years old is known from Canada.) Lancelets lack head, jaws, vertebrae, and paired fins, but have these fish-like features:
a Gills
b Notochord (flexible rod foreshadowing the backbone)
c Nerve cord
d Tail fin
(All animals with a notochord or backbone are chordates: members of the phylum Chordata. Lancelets belong to the subphylum Cephalochordata.)

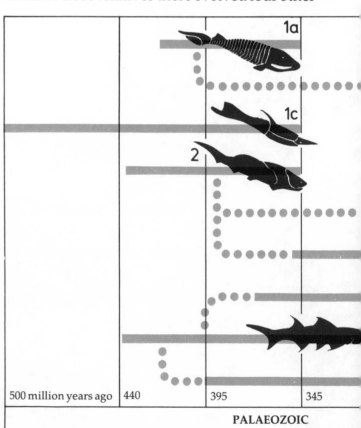

| 500 million years ago | 440 | 395 | 345 |

PALAEOZOIC

74

major groups of fishes, two now extinct. Bony fishes, the most successful group, evolved a more flexible and mobile body than the fishes they replaced.

Fishes probably originated in the sea indirectly from echinoderms (see pp. 70–71), but most early kinds inhabited fresh water. Prehistoric species ranged from creatures a few centimetres long to a 14-metre-long (46ft) relative of the man-eating great white shark alive today.

Fossil fishes have been found in every continent. Appropriate rocks yield isolated bony plates, spines, scales, vertebrae, and teeth. But whole skeletons are rare, and many early fishes had gristly skeletons that seldom formed good fossils.

Family tree of fishes
In this family tree of all fishes, numbers show classes and letters show subclasses.
1 Agnatha (jawless fishes)
a Monorhina (with a nostril between the eyes)
b Cyclostomata (living jawless fishes)
c Diplorhina (no dorsal nostril)
2 Placodermi (placoderms)
3 Chondrichthyes (fishes with a gristly skeleton)
d Holocephali (shark-like fishes)
e Elasmobranchii (sharks)
4 Acanthodii (spiny fishes)
5 Osteichthyes (bony fishes)
f Actinopterygii (ray-fins)
g Sarcopterygii (fleshy-finned fishes)

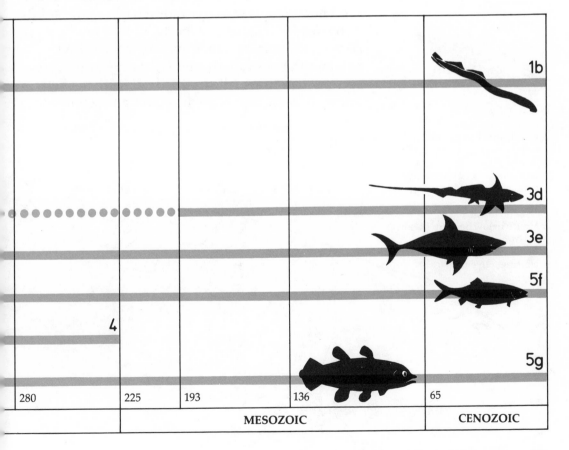

280		225	193	136		65	
				MESOZOIC		CENOZOIC	

75

Jawless fishes

Agnathans ("jawless fishes") included the first backboned animals. Body armour earns most extinct agnathans the collective name ostracoderms ("shell skins"). Armour probably saved some from being eaten by eurypterids (pp. 64–65). Ostracoderms flourished mainly in rivers and lakes about 510–350 million years ago. Most were small and lacked paired fins. Their armoured heads had jawless slits or holes for mouths, through which they sucked in water containing particles of food. Some scavenged in mud, others guzzled tiny organisms at the surface.

Ostracoderm fossils are plentiful in Late Silurian and Early Devonian rocks of Europe and North America.

Here is one example from each ostracoderm order.
1 **Hemicyclaspis** was an armoured fish with a solid bony head shield, usually backswept "horns", sensory fields in the head, eyes on top of the head and mouth below, body plated, triangular in cross section, tapering to the uptilted tail, with flat belly, one nostril, and many openings for gills. Length:

Cephalaspid body plan
These features include hints of the bottom-dwelling life of *Hemicyclaspis* and its relatives.
a Solid bony shield guarding head
b Eyes on top of head
c Mouth below head for sucking food from mud
d Sensory fields – aids for detecting movement

e Armour plates allowing side-to-side body movements
f Uptilted tail keeping head down
g Lateral fins helping balance
h Median fin stopping body rolling
i Flat belly

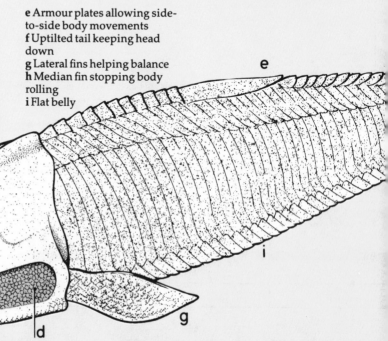

13cm (5in). Time: Late Silurian–Devonian. Place: northern continents. Order: Cephalaspida.

2 **Birkenia** was a small, deep-bodied ostracoderm, with small bony plates on the head, eyes at sides of the head, spines on the back, and tail angled downward. It had a fin along each side. Length: 10cm (4in). Time: Mid Silurian–Early Devonian. Place: Europe. Order: Anaspida.

3 **Pteraspis** was a fish with a long, rounded head shield, no visible nostril, long sharp "beak", slit-shaped mouth below, one gill opening per side, long spine on the back, no paired fins, and a downward-angled tail. Length 5.7cm (2¼in). Time: Early Devonian. Place: northern continents. Order: Pteraspida.

4 **Thelodus** was a flat fish covered with small, tooth-like denticles not flat plates. It had a long upper tail lobe and eyes on the sides of the head. Length: 18cm (7in). Time: Mid Silurian–Early Devonian. Place: Europe, North America. Order: Coelolepida.

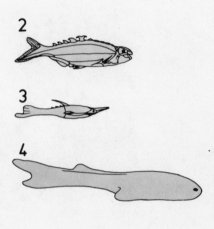

2

3

4

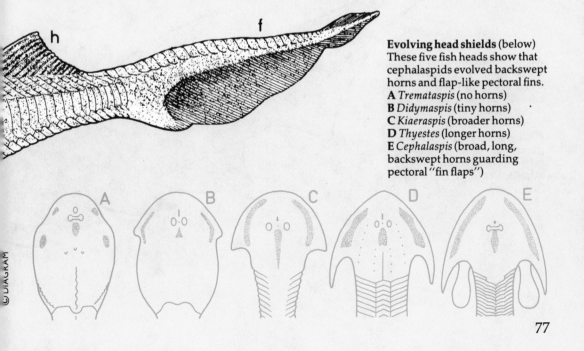

h

f

Evolving head shields (below) These five fish heads show that cephalaspids evolved backswept horns and flap-like pectoral fins.
A *Tremataspis* (no horns)
B *Didymaspis* (tiny horns)
C *Kiaeraspis* (broader horns)
D *Thyestes* (longer horns)
E *Cephalaspis* (broad, long, backswept horns guarding pectoral "fin flaps")

A   B   C   D   E

Spiny fishes

Nicknamed "spiny sharks" or "spiny fishes", the acanthodians had stout spines along the leading edges of their fins but were neither sharks, nor bony fishes of a modern kind. These small freshwater species lived about 400–230 million years ago, and were the first known vertebrates with jaws. They had a blunt head, small, stud-like body scales, and a long, tapered upper tail lobe. Acanthodians swam at mid and surface levels. Some probably ate small, freshwater invertebrates. Others preyed on jawless fishes.

Fossils crop up in Late Silurian to Permian rocks, and occur in almost every continent. Most are crushed flat in shale slabs. Early fossils are just spines and scales.

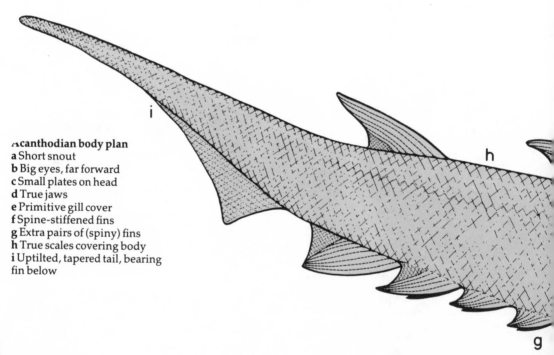

Acanthodian body plan
a Short snout
b Big eyes, far forward
c Small plates on head
d True jaws
e Primitive gill cover
f Spine-stiffened fins
g Extra pairs of (spiny) fins
h True scales covering body
i Uptilted, tapered tail, bearing fin below

Below are species from the three acanthodian orders: Climatiformes (primitive types); Ischnacanthiformes (types with reduced spines); Acanthodiformes (the last group, some degenerate).

1 **Climatius** had a short, deep body, and five pairs of extra fins below its belly. Length: 7.6cm (3in). Time: Late Silurian–Early Devonian. Place: northern continents. Order: Climatiformes.

2 **Ischnacanthus** was more advanced than *Climatius*, with fewer and slimmer but relatively longer and more deeply embedded spines. Time: Early–Mid Devonian. Place: Europe. Order: Ischnacanthiformes.

3 **Acanthodes** looked more eel-like than earlier acanthodians: it was partly scaleless, with fewer fins and spines and no teeth. Length: 30.5cm (1ft). Time: Late Devonian–Early Permian. Place: northern continents and Australia. Order: Acanthodiformes.

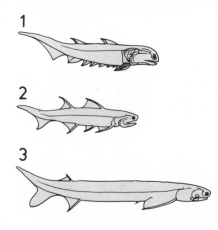

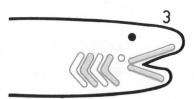

Jaws evolving (above)
Three diagrams show stages in the evolution of a fish's jaws, enabling it to open or close its mouth.
1 Jawless fish: bony gill supports (shown tinted) alternate with gills.
2 The first pair of gills shrinks to form a spiracle, a tiny hole for drawing in mud-free water.
3 The first pair of gill supports can now expand, and evolve into jaws.

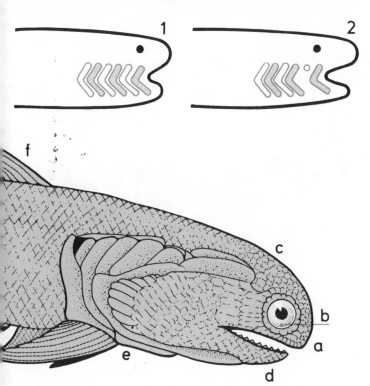

79

Placoderms

The placoderms or "plated skins" were among the first fishes with jaws and paired fins. Bony armour covered the head and forepart of the body. In many, a movable joint between head and body armour let the head rock back to open the mouth wide. The primitive jaws had jagged bony edges that served as teeth. The tail end usually lacked protection, even scales. Placoderms mostly swam with eel-like movements. Many lived on the sea bed. Some were the largest, most formidable creatures of their day. The group appeared in Silurian times, dominated Devonian seas (395–345 million years ago), and then died out under competition from sharks and bony fishes.

Placoderms are divided into seven orders. Here are examples of six of them, all Devonian.

1 **Gemuendina** was flat and broad. Length: 23cm (9in). Place: Central Europe. Order: Rhenanida (placoderms resembling skates).

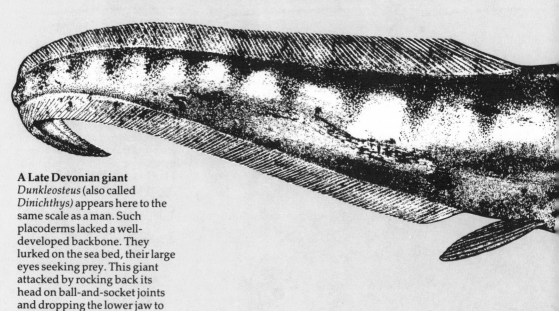

A Late Devonian giant
Dunkleosteus (also called *Dinichthys*) appears here to the same scale as a man. Such placoderms lacked a well-developed backbone. They lurked on the sea bed, their large eyes seeking prey. This giant attacked by rocking back its head on ball-and-socket joints and dropping the lower jaw to expose the bony cutting edges serving as its teeth.

2 **Lunaspis** had armoured skin all over and curved, bony shoulder spines. Length: 27cm (10½in). Place: Europe. Order: Petalichthyida.

3 **Dunkleosteus** (*Dinichthys*), a huge predator, could kill large fishes. Length: up to 9m (30ft). Place: North America and Europe. Order: Arthrodira (armoured placoderms with jointed necks).

4 **Rhamphodopsis** had grinding jaw plates and a big shoulder spine. Length: 10cm (4in). Place: Europe. Order: Ptyctodontida (small, armoured fishes).

5 **Phyllolepis** was flat, with little head armour. Length: 12.7cm (5in). Place: Australia, Europe, North America. Order: Phyllolepida (mostly flat and heavily plated placoderms).

6 **Bothriolepis** had a weak mouth, eyes on top of the head, and crab-like arms encasing the front fins. Length: up to 30cm (1ft). Place: found in most continents. Order: Antiarchi (small fishes with jointed, movable, spiny front fins).

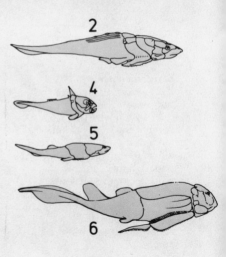

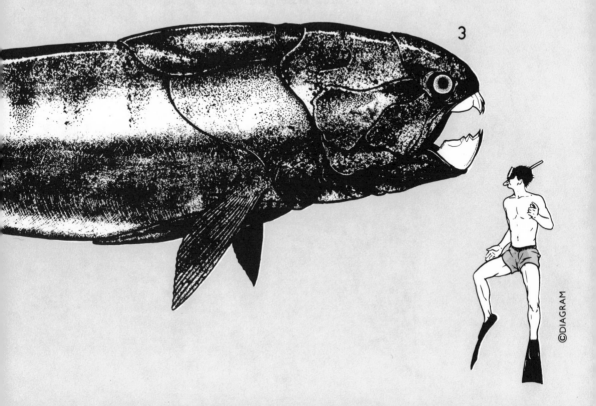

©DIAGRAM

Sharks and their kin

Sharks and shark-like fishes make up the Chondrichthyes – one of two classes of so-called higher fishes. Chondrichthyans have skeletons of tough, gristly cartilage, not bone, and tiny tooth-like scales. They have paired fins but lack gill covers, or swim bladders to adjust buoyancy.

Some are ferocious, streamlined killers with razor-sharp teeth. But skates, rays, and rat fishes include bottom-dwellers with low, broad teeth for crunching shellfish. Most kinds live in the sea.

Sharks probably evolved from placoderms 390 million years ago. They evolved fast, but many kinds died out. Shark fossils occur worldwide. Many are just teeth, fin spines, or tooth-like denticles from skin. The soft skeletons have mostly rotted. But fine-grained Late Devonian Cleveland shales preserve fine specimens of early sharks along Lake Erie's southern shore.

These fishes represent five chondrichthyan orders.
1 **Cladoselache** had a torpedo-shaped body, short snout, big eyes, broad-based fins, and long upper tail lobe. Some individuals had a spine on the back. Length: 50cm–1.2m (1ft 8in–4ft). Time: Late Devonian. Place: Europe and North America. Order: Cladoselachiformes (extinct ancestral sharks).

Early and modern
Here we compare some primitive features of *Cladoselache* with (in parentheses) those of fully modern sharks.
a Jaws at front of head (jaws on underside of head)
b Upper jaw fixed to braincase at back and front (upper jaw fixed to braincase at back only, allowing mouth to gape wide)
c Snout short and rounded (head pointed)
d Broad-based triangular fins (more mobile fins with narrow bases)
e Horizontal fin "rudders" near tail (no such fins)
f Torpedo-like body (similar)
g No claspers on pelvic fins (claspers on males' pelvic fins grip females during mating)

1

e   g

2 **Xenacanthus** had a long dorsal fin, long tail ending in a point, and a spine jutting back from its head. Length: 76cm (2ft 6in). Time: Late Devonian–Mid Permian. Place: Americas, Europe, and Australia. Order: Pleuracanthiformes (early freshwater sharks).

3 **Hybodus** had narrow-based, manoeuvrable fins, and a small anal fin. Length: over 2m (6ft 6in). Time: Late Permian–Early Cretaceous. Place: worldwide. Order: Selachii (modern sharks and close kin).

4 **Aellopos** was a flat fish with wing-like fins and whiplash tail. Length: 1.5m (5ft). Time: Late Jurassic. Place: Europe. Order: Batoidea (skates and rays).

5 **Ischyodus** had a stout dorsal spine, wing-like pectoral fins, and whip-like tail. Length: 1.5m (5ft). Time: Mid Jurassic–Palaeocene. Place: worldwide. Order: Chimaeriformes ("rat fishes").

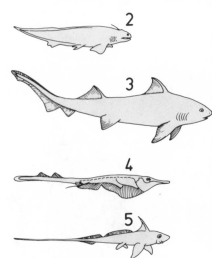

Tooth replacement (bottom) This diagram shows how new teeth continuously grow forward in a shark's jaw to replace old teeth that drop out. Young sharks replace each tooth almost once a week. Hard dentine and enamel make sharks' teeth durable. Vast numbers survive as fossils.

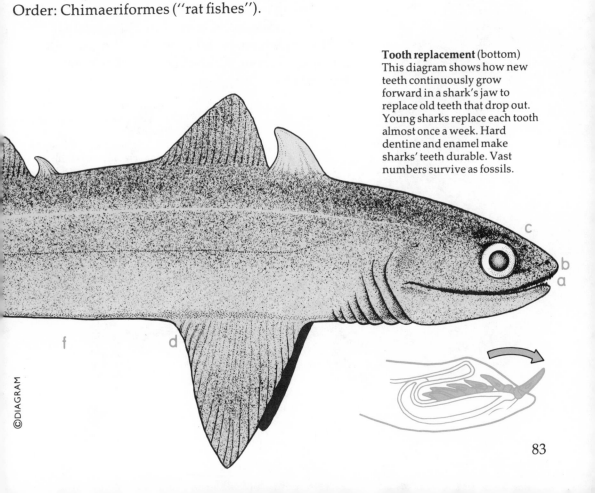

©DIAGRAM

83

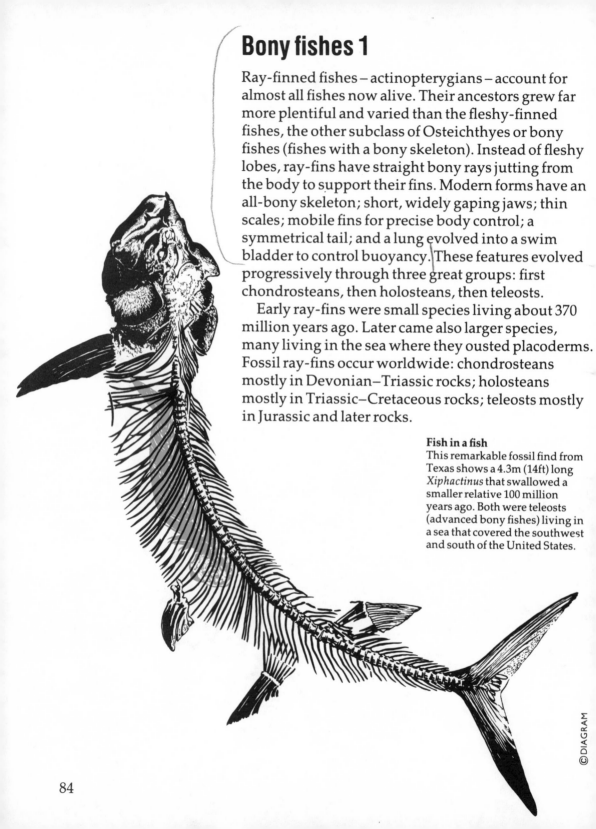

Bony fishes 1

Ray-finned fishes – actinopterygians – account for almost all fishes now alive. Their ancestors grew far more plentiful and varied than the fleshy-finned fishes, the other subclass of Osteichthyes or bony fishes (fishes with a bony skeleton). Instead of fleshy lobes, ray-fins have straight bony rays jutting from the body to support their fins. Modern forms have an all-bony skeleton; short, widely gaping jaws; thin scales; mobile fins for precise body control; a symmetrical tail; and a lung evolved into a swim bladder to control buoyancy. These features evolved progressively through three great groups: first chondrosteans, then holosteans, then teleosts.

Early ray-fins were small species living about 370 million years ago. Later came also larger species, many living in the sea where they ousted placoderms. Fossil ray-fins occur worldwide: chondrosteans mostly in Devonian–Triassic rocks; holosteans mostly in Triassic–Cretaceous rocks; teleosts mostly in Jurassic and later rocks.

Fish in a fish
This remarkable fossil find from Texas shows a 4.3m (14ft) long *Xiphactinus* that swallowed a smaller relative 100 million years ago. Both were teleosts (advanced bony fishes) living in a sea that covered the southwest and south of the United States.

©DIAGRAM

These three fishes represent evolutionary trends in the subclass Actinopterygii (ray-fins).

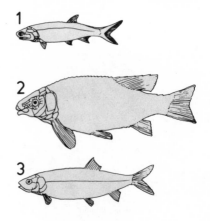

1 **Palaeoniscum** had a long upper tail lobe, thick scales, bony head armour, long jaws hinged far back, and a kind of lung. Length: 30cm (1ft). Time: mainly Permian. Place: worldwide. Infraclass: Chondrostei.

2 **Lepidotes** differed from *Palaeoniscum* in its short upper tail lobe, thinner scales, swim bladder, deeper body, more manoeuvrable paired fins, and shorter jaws. Length: up to 1.2m (4ft). Time: mainly Jurassic. Place: worldwide. Infraclass: Holostei.

3 **Leptolepis** had a herring-like shape. It differed from *Palaeoniscum* amd *Lepidotes* in its symmetrical tail, thinner scales, shortened jaws with wide gape, and fewer skull bones. Length: 23cm (9in). Time: mainly Jurassic. Place: seas worldwide. Infraclass: Teleostei.

Evolving tails and heads
Here we show evolutionary trends in bony fishes, from chondrosteans (**a,d**) through holosteans (**b,e**) to teleosts (**c,f**).
a Heterocercal tail: the backbone's upturned end produces a long lower tail lobe that tends to drive the head down.
b Abbreviated heterocercal tail: upper lobe shortened; lift comes from swim bladder.
c Homocercal tail: tail lobes seem equal, but most rays sprout from the backbone's still upturned end.
d Jaws work as a snap trap.
e Jaws are shortened but gape wide to suck in food.
f Jaws are shortened further but protrude when opened to create a suction tube.

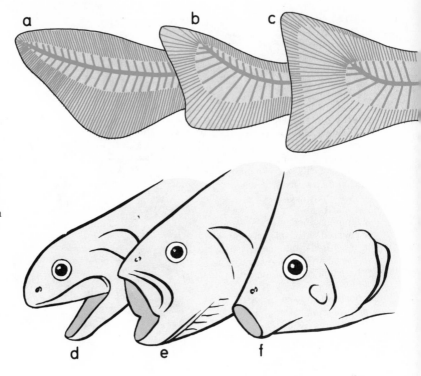

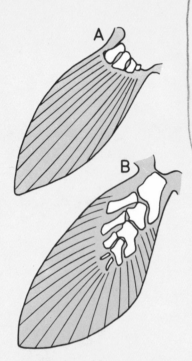

Bony fishes 2

The Sarcopterygian subclass of bony fishes had paired fins borne on scaly lobes containing bones and muscles. Such fleshy fins gave rise to the limbs of backboned animals that live on land. Fleshy-finned fishes appeared perhaps 390 million years ago. Three main groups evolved: rhipidistians, coelacanths, and lungfishes. The last two still survive.

Rhipidistians were long-bodied flesh-eaters that lurked in shallow waters – fresh and salt. They could use fins as legs and breathed air with lungs if hot weather made their water foul.

Coelacanths were mostly deep-bodied and lived in oceans. Their lungs became swim bladders that regulated buoyancy. All were thought extinct for over 60 million years until a fisherman caught one off south-east Africa in 1938.

Lungfishes had weaker limbs and a flimsier skeleton than other fleshy-finned fishes. Some could (and can) breathe atmospheric air if their ponds or rivers dry up.

Two types of fin (above)
A Ray-fin's fin: rays spring from bones at the base.
B Fleshy-finned fish's fin: rays spring from bones along the centre of the fin itself.

Fleshy-finned features (right)
Here we show similarities and differences between *Osteolepis* (**A**), an early fleshy-finned fish, and *Cheirolepis* (**B**), an early ray-finned fish. Both date from Mid Devonian times.
Similarities:
a tapered at both ends;
b covered with heavy scales;
c primitive, uptilted tail;
d paired fins similarly spaced;
e bony plates covering skull.
Differences:
f position and size of eyes;
g proportions of skull bones;
h number of dorsal fins;
i fin design.

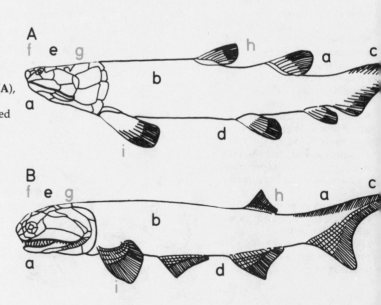

The animals below represent three main groups of fleshy-finned fishes. The first two were in the order Crossopterygii, the third in the order Dipnoi.

1 **Eusthenopteron** was a long-bodied, carnivorous freshwater fish with paired fins and a "three-pronged" tail fin. Skull, backbone, and limb bones resembled those of early amphibians. Nostrils opened into the mouth. Length: 30–60cm (1–2ft). Time: Late Devonian. Place: Europe and North America. Suborder: Rhipidistia.

2 **Macropoma** was a deep-bodied coelacanth with a short, deep skull, three-pronged tail fin, and fan-shaped dorsal and anal fins. Nostrils did not open into the mouth. Length: 56cm (22in). Time: Late Cretaceous. Place: oceans; fossils come from Europe. Suborder: Coelacanthini.

3 **Dipterus** had a long body tapered at both ends, paired, leaf-shaped fins, an uptilted tail, big, thick scales, and a braincase largely made of gristle. Length: 36cm (14in). Time: Middle Devonian. Place: North America and Europe. Order: Dipnoi (lungfishes).

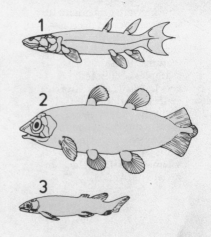

Finny adventurers
Agile young of *Eusthenopteron* might have flipped ashore to dodge cannibal adults. Some found a food supply and stayed awhile. Breathing posed no problem, for these fishes were equipped with lungs.

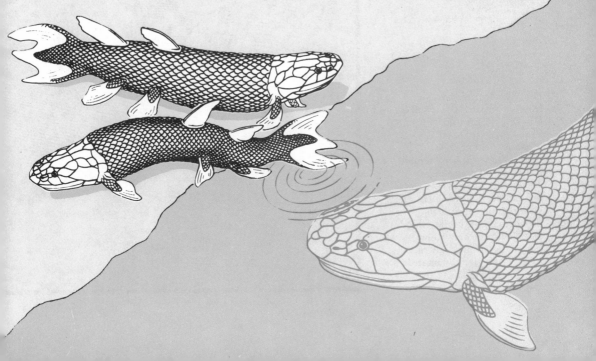

Chapter 5

FOSSIL AMPHIBIANS

The first vertebrates with legs were members of the class Amphibia. In this chapter we summarize their origins and key features, and show fossil examples of their four subclasses. These comprise the extinct labyrinthodonts (Labyrinthodontia) and lepospondyls (Lepospondyli), and the surviving frogs and toads (Anura), newts and salamanders (Urodela), and caecilians (Apoda).

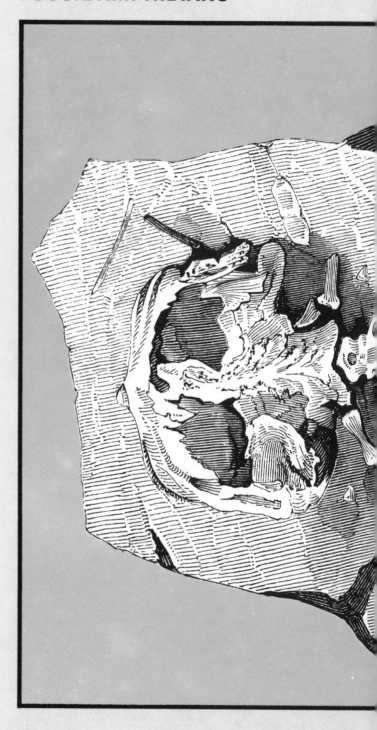

Andrias scheuchzerii was a Miocene salamander about 60cm (2ft) long. Impressed by its man-like form, Swiss scientist Johannes Scheuchzer in 1726 named the fossil *Homo diluvii testis* ("a man who had witnessed the flood"). Scheuchzer believed all fossils were remains of animals drowned by the biblical flood. Scientists now know that *Andrias scheuchzerii* was closely related to the living giant salamanders of China and Japan. (The Mansell Collection.)

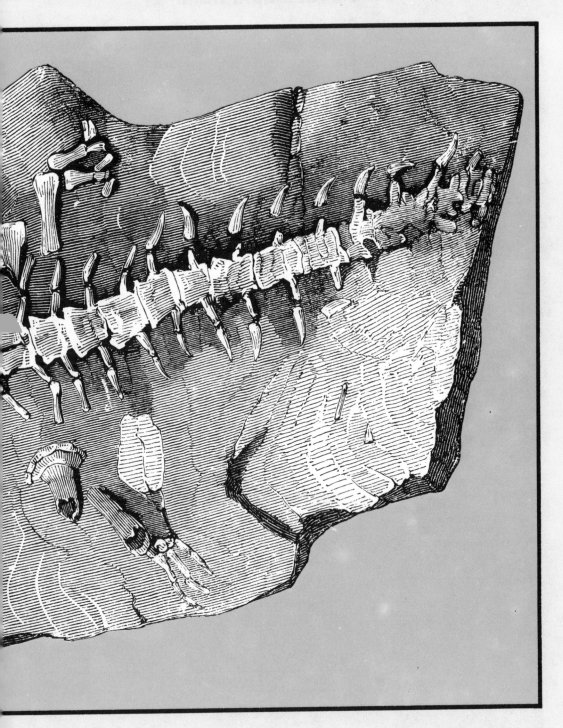

About amphibians

Rhipidistian fishes (p. 86) gave rise to amphibians about 350 million years ago. Amphibians were the first backboned animals with limbs designed for use on land. Like their descendants, frogs and newts, early amphibians had to lay their shell-less eggs in water to prevent them drying up. The eggs hatched into tadpoles which breathed through gills that later usually shrank. Some adults grew as big as any crocodile. Many had heavy skeletons, with powerful sprawling legs. They breathed through lungs and had a covering of fish-like scales or tough skin. (In contrast, modern amphibians are mostly small, with light skeletons and soft, moist skin; adults get more oxygen through skin than lungs.) Early amphibians lived mainly in or near fresh water, hunting fishes, insects, or early reptiles. They dominated swamps that covered much of coastal North America and

Amphibian family tree
In this family tree of the class Amphibia (amphibians) numbers show subclasses and superorders and letters show orders.
1 Labyrinthodontia (labyrinthodonts)
a Ichthyostegalia (ichthyostegids)
b Batrachosauria (batrachosaurs)
c Temnospondyli (temnospondyls)
2 Lepospondyli (lepospondyls)
d Nectridea (nectrideans)
e Aistopoda (aistopods)
f Microsauria (microsaurs)
3 Apoda (caecilians)
4 Anura (frogs and toads)
5 Urodela (newts and salamanders)

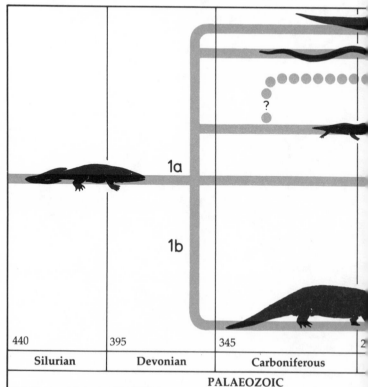

440	395	345	2
Silurian	Devonian	Carboniferous	
PALAEOZOIC			

Europe in Late Carboniferous (Pennsylvanian) and Early Permian times. Fossils of extinct amphibian groups come mostly from rocks that formed then.

Both main groups of early amphibians died out by 160 million years ago. Meanwhile an unknown species of amphibian had given rise to reptiles and so, indirectly, to all other backboned land animals.

Limbs from fins

This comparison between bones of a fish's fin and an early amphibian's limb reveals that limb bones correspond to and evolved from fishes' fin bones.
1 Bones supporting a pelvic fin of a Devonian rhipidistian fish, *Eusthenopteron*
2 Corresponding bones in a hind limb of the Permian amphibian *Trematops*
a Pelvis (hip region)
b Femur (thigh bone in land vertebrates)
c Tibia and fibula (leg bones in land vertebrates)
d Pes (foot): the small bones in the rhipidistian fish's fins evolved into the toes and fingers of amphibians and their descendants
(Diagram after Colbert.)

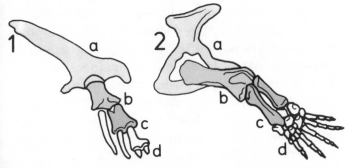

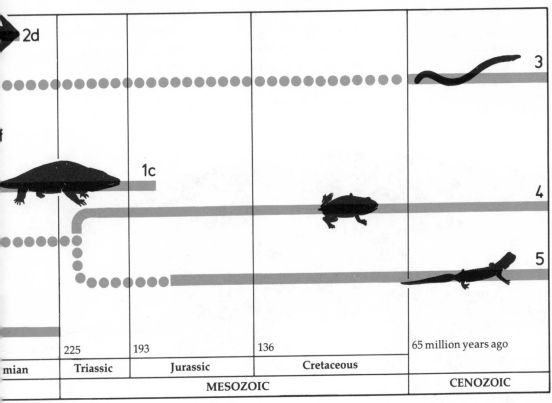

	225	193		136		65 million years ago	
mian	Triassic	Jurassic		Cretaceous		CENOZOIC	
		MESOZOIC					

Labyrinthodonts 1

Labyrinthodonts formed the largest subclass of prehistoric amphibians. Their name comes from the labyrinth-like structure of their teeth.

Labyrinthodonts had solid skulls and complex spinal bones. Some evolved strong backbones and strong, sprawling limbs. These became the first land-living vertebrates. Others had weaker skeletons and eel-like bodies, and lived in water.

Labyrinthodonts ranged from a few centimetres long to 9m (30ft). They lived about 350–180 million years ago and spread worldwide. There were three orders: ichthyostegids, batrachosaurs (notably anthracosaurs), and (pp. 94–95) temnospondyls.

Ichthyostegids were sprawling amphibians with some fish-like features. Maybe they were ancestral to all later labyrinthodonts. Fossils come from Late Devonian rocks. (See example 1.)

Anthracosaurs were Late Palaeozoic amphibians. Some evolved like reptiles and might have given rise to these, but the anthracosaurs' middle ear design makes this unlikely. (See examples 2–6.)

1 **Ichthyostega** had well-developed limbs but traces of a fish's tail and scales. Length: 1m (3ft 3in). Time: Late Devonian. Place: Greenland.

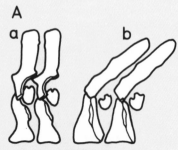

An early amphibian
Vertebrae (above) and skeleton and restoration (below) show features of *Ichthyostega*.
Fish-like features are:
A bones of vertebrae (a) matching those found in the fish *Eusthenopteron* (b);
B skull roof still solid;
C fish-like tail;
D fish-like scales.
Amphibian innovations are:
E strong shoulder girdle;
F strengthened spine;
G strong ribs;
H strengthened hip girdle;
I fully formed limbs.

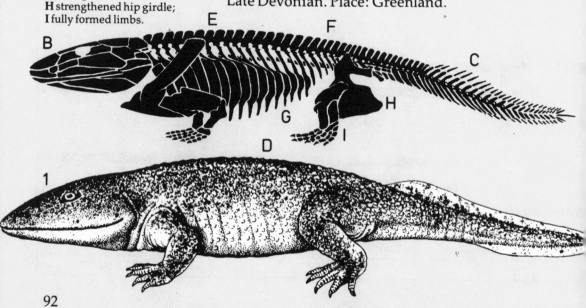

2 **Protogyrinus,** a very early anthracosaur, had a rather high skull and sturdy limbs. Length: 1–1.5m (3ft 3in–4ft 11in). Time: Late Carboniferous. Place: West Virginia, USA. Infraorder: Embolomeri.

3 **Gephyrostegus** had a small head and sturdy, sprawling limbs. Length: 45cm (18in). Time: Late Carboniferous. Place: Czechoslovakia. Infraorder: Gephyrostegoidea (terrestrial anthracosaurs).

4 **Eogyrinus** had a long, eel-like body and tail, weak limbs, and a crocodile-like skull. It lived in water. Length: 4.6m (15ft). Time: Late Carboniferous. Place: Europe. Infraorder: Embolomeri ("typical" aquatic anthracosaurs).

5 **Seymouria** had longer, stronger limbs than the first amphibians and lived on land. Length: 60cm (2ft). Time: Early Permian. Place: Texas, USA. Infraorder: Seymouriamorpha (reptile-like anthracosaurs).

6 **Diadectes,** the earliest known plant-eating vertebrate, had heavy bones and shortened jaws with blunt teeth. Length: 3m (10ft). Time: Early Permian. Place: Texas, USA. Infraorder: Seymouriamorpha.

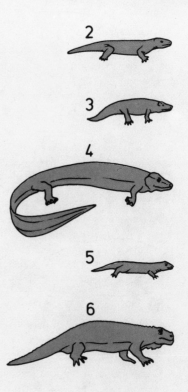

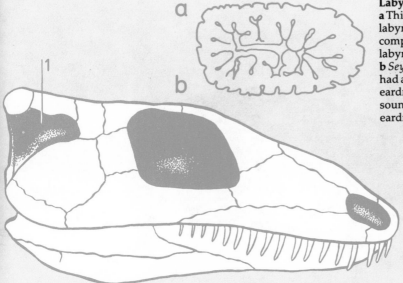

Labyrinthodont features
a This cross-section through a labyrinthodont tooth shows the complex folds that earned labyrinthodonts their name.
b *Seymouria's* skull (actual size) had an otic notch (**1**) to take an eardrum sensitive to airborne sounds. Its fish ancestors lacked eardrums.

© DIAGRAM

Labyrinthodonts 2

Temnospondyl labyrinthodonts had distinctive vertebrae and other features that distinguish them from anthracosaurs. Like those, some temnospondyls lived on land, others in water. They persisted from Carboniferous to Jurassic times and became the most abundant amphibians.

The temnospondyl order held three suborders: rhachitomes, stereospondyls, and plagiosaurs. Rhachitomes, the basic stock, formed a large, varied group: dominant amphibians of Permian times.

1 **Eryops** was heavy with a strong skeleton, short, strong limbs, and big, broad skull. It might have lived in water and on land, like crocodiles. Length: 1.5m (5ft). Time: Early Permian. Place: Texas, USA.

2 **Trimerorhachis** had a body protected by overlapping "fish scales". It probably swam in pools and streams. Length: 60cm (2ft). Time: Early Permian. Place: Texas, USA.

3 **Aphaneramma** had a head one third of its total length. It swam in seas and caught fishes with its long, slim, sharp-toothed jaws. Length: 60cm (2ft). Time: Early Triassic. Place: worldwide. Fossils occur as far apart as Australia and Spitsbergen.

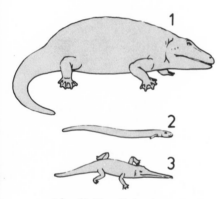

A landlubber (below)
The Permian temnospondyl *Cacops* was one of the amphibians best designed for life on land. It had these features.
a Length 40cm (16in)
b Sturdy limbs
c Large eardrum
d Armoured skin on the back
e Large head with long jaws

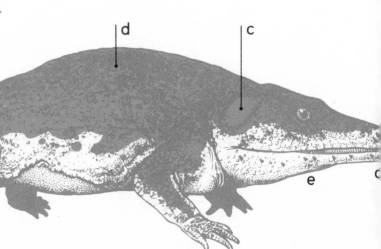

Stereospondyls were swimmers with degenerate skeletons. They did not need strong bony scaffolding like beasts that live on land and must resist the tug of gravity. Some developed broad, flat bodies and huge heads. Stereospondyls included the largest amphibians ever. The group dominated inland waters in Triassic times, then all died out.

Plagiosaurs were even more grotesque aquatic beasts, of Permian to Late Triassic times.

1 **Cyclotosaurus,** a stereospondyl, was as large as a crocodile but it had very small, weak legs and needed the support of water. Length: 4.3m (14ft). Time: Late Triassic. Place: Europe.

2 **Gerrothorax** was a plagiosaur with a short, wide head with gills; a flat, broad, armoured body; short tail; and tiny limbs. Length: 1m (3ft 3in). Time: Late Triassic. Place: southern Germany.

Built for water life (below)
Paracyclotosaurus, an aquatic Triassic stereospondyl, was modified for life in water.
a Length 2.25m (7ft 5in)
b Short, rather weak limbs
c Somewhat flattened body
d Flattened skull
e Mouth opened by raising the skull instead of dropping the lower jaw

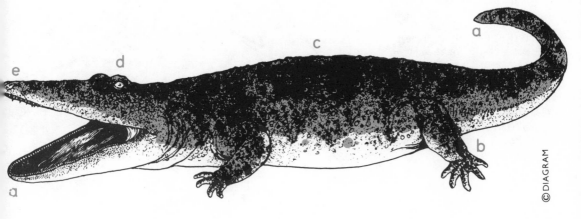

Lepospondyls

Lepospondyls formed a mixed subclass of small amphibians that thrived in swamps about 320–230 million years ago, dying out before Triassic times. They had simpler teeth and fewer skull bones than labyrinthodonts, and a different type of vertebra. This had a bony cylinder firmly joined to an arch and pierced by a hole to take a notochord. Below are examples of the three lepospondyl orders: aistopods, nectrideans, and microsaurs.

1 **Ophiderpeton** was a typical aistopod: aquatic, limbless, and snake-like, with forked ribs and 200 vertebrae. Length: 70cm (27.5in). Time: Late Carboniferous. Place: Europe and North America.

2 **Diplocaulus** had a flat body, weak limbs, and a grotesque head like a cocked hat. Length: 1m (3ft 3in). Time: Early Permian. Place: Texas, USA. Order: Nectridea (newt-like or snake-like).

3 **Pantylus** had a heavy body, small limbs, and a big, deep head. Length: 26cm (10in). Time: Early Permian. Place: Texas, USA. Order: Microsauria (mostly sturdy, land-based insect-eaters).

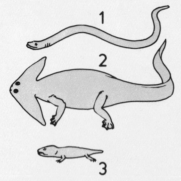

A puzzling head
Diplocaulus's strange head poses questions of design.
A Growth stages show that "horns" grew relatively longer as the head enlarged. This occurred because some skull bones outpaced others.
B Two illustrations suggest possible uses for a head with backswept "horns".
1 "Horns" might have acted as a hydrofoil, helping to give lift to raise the animal through the water.
2 "Horns" might have made it difficult for predators to swallow *Diplocaulus*.

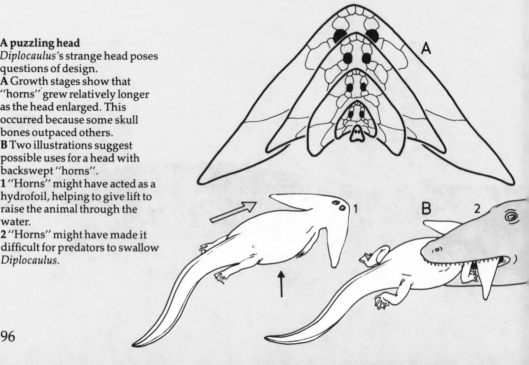

Modern amphibians

Frogs and toads, tailed amphibians, and caecilians are often lumped together in one subclass: Lissamphibia. Few known ancestors of modern amphibians date back to the last labyrinthodonts but similarities in skull design show that frogs, toads and urodeles came from a group of temnospondyl labyrinthodonts (the dissorophids). Below are early examples of each group.

1 **Triadobatrachus,** the first known fossil frog, lived in Early Triassic Madagascar. Length: 10cm (4in). Superorder: Anura.

2 **Karaurus,** the oldest known complete salamander skeleton, had a broad skull with sculptured bones. Length: 19cm (7.5in). Time: Late Jurassic. Place: Kazakhstan, USSR. Superorder: Urodela (newts and salamanders).

3 **Apodops,** an early apodan, is known from one vertebra. Time: Palaeocene. Place: Brazil. Superorder: Apoda (caecilians – worm-like burrowers).

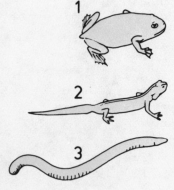

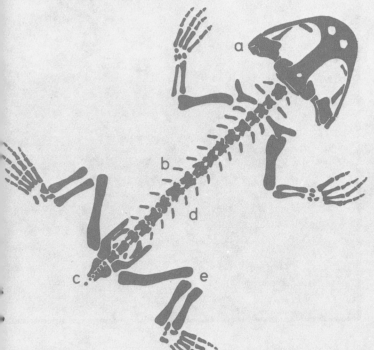

A frog ancestor
Here we show evolutionary changes that produced *Triadobatrachus* from long-bodied amphibian ancestors.
a Frog-like skull
b Shortened back with reduced spinal bones (modern frogs have fewer still)
c Shortened tail (modern adult frogs have none)
d Shortened ribs (modern frogs have none)
e Leg design still primitive (modern frogs have very long hind limbs, for jumping)

97

Chapter 6

FOSSIL REPTILES

The class Reptilia includes the first vertebrates designed to live and also to breed on land. Reptiles colonized the continents, and for 200 million years were masters over them.

These pages explore key features and fossil examples of orders belonging to the four reptile subclasses: the extinct euryapsids (Euryapsida) and mammal-like synapsids (Synapsida), and the surviving anapsids (Anapsida) and diapsids (Diapsida).

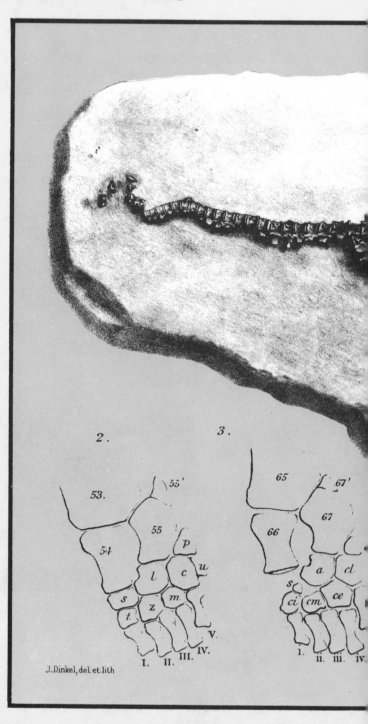

Limbs that evolved as flippers adapted *Plesiosaurus* for life in water. This large aquatic reptile flourished in the Mesozoic Era: the "Age of Reptiles" or "Age of Dinosaurs". Our book follows the practice of classifying plesiosaurs in one of four reptile subclasses determined by holes in the skull. A classification proposed in 1980 would restructure subclasses and regroup their contents. (Illustration from *A History of British Fossil Reptiles* by Sir Richard Owen.)

J. Dinkel, del. et lith

4.   5.   6.

PLESIOSAURUS RUGOSUS

About reptiles

By 300 million years ago amphibians had given rise to reptiles, the first backboned animals able to live entirely on dry land. Reptiles are cold-blooded animals with dry, scaly, waterproof skin. Their eggs are fertilized inside the females. Tough skin or a hard shell stops the eggs drying up after laying. These devices freed reptiles from the waterside. They colonized high, dry lands between river valleys and invaded the spreading deserts of Permian and Triassic times.

By 270 million years ago small, early, swampland reptiles had given rise to four great stocks named from the number and type of holes behind the eyes, on each side of the skull – holes that left space for strong jaw muscles to contract. Anapsids, without such holes, possibly include tortoises and turtles. Synapsids, with one low opening in the cheek, made

Reptile family tree
This shows major reptile groups as set out in this chapter. (New studies will lead to some revision.) Numbers represent subclasses: small letters represent orders.
1 Anapsida (anapsids)
a Cotylosauria (cotylosaurs)
b Mesosauria (mesosaurs)
c Chelonia (turtles)
2 Synapsida (synapsids)
d Pelycosauria (pelycosaurs)
e Therapsida (therapsids)
3 Euryapsida (euryapsids)
f Ichthyosauria (ichthyosaurs)
g Araeoscelidea (araeoscelids)
h Placodontia (placodonts)
i Sauropterygia (nothosaurs and plesiosaurs)
4 Diapsida (diapsids)
A Infraclass Lepidosauria (lepidosaurs)
j Eosuchia (eosuchians)
k Rhynchocephalia (rhynchocephalians)
l Squamata (lizards and snakes)
B Infraclass Archosauria (archosaurs)
m Thecodontia (thecodonts)
n Crocodilia (crocodilians)
o Saurischia (saurischian dinosaurs)
p Ornithischia (ornithischian dinosaurs)
q Pterosauria (pterosaurs)

345	280	225	1!
Carboniferous	Permian	Triassic	
PALAEOZOIC			

up the mammal-like reptiles. These dominated life on land for 70 million years and then became extinct. Meanwhile some gave rise to mammals. Euryapsids, with a high opening in the cheek, were mostly sea reptiles such as plesiosaurs; none survives. Diapsids, with two openings behind each eye, had two main subgroups: lepidosaurs and archosaurs. Lepidosaurs included snakes, lizards, and their ancestors. Archosaurs comprised the thecodonts, crocodiles, dinosaurs, and pterosaurs.

From Late Palaeozoic times all through the Mesozoic Era, big reptile predators and herbivores were the "lions" and "zebras" of their day. Others evolved wings or flippers, and took to air or water. Worldwide fossil finds bear witness to this Age of Reptiles. Such beasts no longer rule the Earth, but their heirs the birds and mammals do.

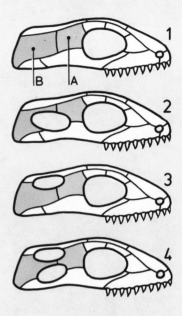

Four types of reptiles
Here reptiles are put in four subclasses according to their skull design (though some show exceptions to this rule and a 1980 reclassification proposes a different breakdown).
1 Anapsids show no hole between postorbital (**A**) and squamosal (**B**) bones.
2 Synapsids show one hole between and below these bones.
3 Euryapsids show one hole between and above these bones.
4 Diapsids show two holes between these bones.

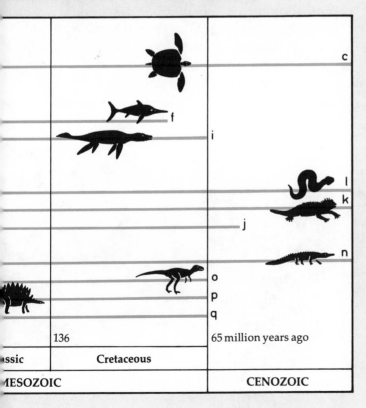

136   65 million years ago

ssic   Cretaceous

MESOZOIC   CENOZOIC

Reptile pioneers

Cotylosaurs ("stem reptiles") formed the first known reptile order. Cotylosaurs ranged from small, lizard-like beasts to big, bulky creatures 3m (10ft) long. They had sturdy, sprawling limbs and a solidly roofed skull, with eardrums just above the jaw hinge. Most inhabited swamps, 290–200 million years ago. Fossils crop up worldwide, mostly in Permian and Triassic rocks.

Mesosaurs were very early swimming reptiles that caught fishes in freshwater ponds and lakes. Fossils some 265 million years old come from Lower Permian rocks of Brazil and southern Africa. Cotylosaurs probably gave rise to mesosaurs and all the other reptile orders.

We show here a mesosaur and examples of both cotylosaur suborders: captorhinomorphs (small, carnivorous reptiles) and procolophonoids (the captorhinomorphs' mostly small descendants).
1 **Hylonomus,** one of the first reptiles, was a small, low captorhinomorph, with sprawling limbs, long tail, short neck, short, pointed snout, and sharp teeth. Length: 1m (3ft 3in). Time: Late Carboniferous (Lower Pennsylvanian). Place: Nova Scotia, Canada.

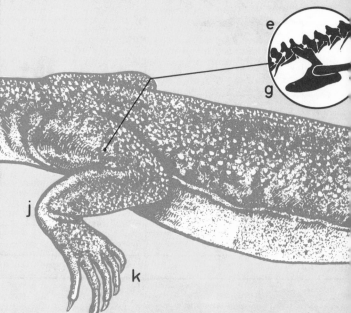

2 **Hypsognathus** was a lizard-like procolophonian, with broad cheek teeth, and spikes jutting back from its head. Length: 33cm (13in). Time: Late Triassic. Place: New Jersey, USA.

3 **Scutosaurus** belonged to the pareiasaurs: big, heavy, procolophonoid herbivores, that stood more upright than many reptiles. The small, saw-edged teeth in its broad head probably sliced up vegetation. Length: 2.4m (8ft). Time: Late Permian. Place: USSR.

4 **Millerosaurus** was a lizard-like reptile, with broad cheeks, short jaws, and a skull hole behind each eye. It might have given rise to the diapsids. Length: 1m (3ft 3in). Time: Late Permian. Place: South Africa.

5 **Mesosaurus,** a mesosaur, had a long, slim body, slender, sharp-toothed jaws, paddle-shaped limbs, and long, deep, swimmer's tail. Length: 71cm (28in). Time: Early Permian. Place: Brazil and South Africa.

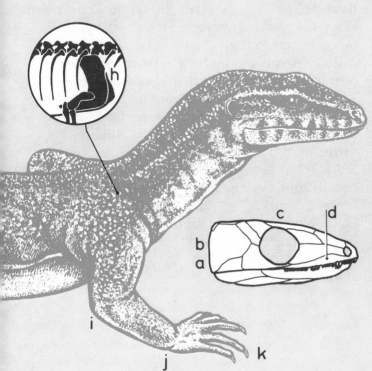

Reptilian features
Letters indicate features that help palaeontologists to identify fossil bones as those of a reptile, not an amphibian, though certain amphibians share some of these features.
a Relatively deep skull
b Usually no otic (ear) notch
c Small size or distinctive position of some skull bones
d Any teeth on palate small not tusk-like
e Two or more vertebrae join spine to hip girdle
f Pleurocentrum is the main part of each vertebra
g Enlarged ilium (a hip bone)
h Shoulder blade well developed but some bones in the shoulder girdle reduced
i Limb bones slimmer than in labyrinthodont amphibians
j Fewer wrist and ankle bones than in amphibians
k Distinctive numbers of toe and "finger" bones

©DIAGRAM

103

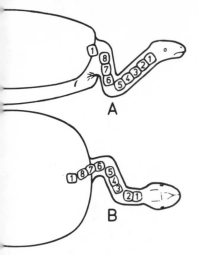

Turtles and tortoises

Chelonians (turtles and tortoises) form one of the oldest living reptile orders. Probably evolving from a cotylosaur, they appeared well developed by 205 million years ago.

Most had a broad, short body protected by a bony shell sheathed with horny scutes. Many could (and can) pull head, tail, and limbs inside the shell for protection. They evolved a toothless beak for slicing meat or vegetation. From the first chelonians came today's land tortoises, and turtles designed to swim and hunt in rivers, pools, or seas. Of four (perhaps five) suborders, two survive.

The earliest fossils come from Triassic Germany and Thailand. Later fossils occur worldwide.

We show three chelonians representing evolutionary trends: proganochelyids gave rise to amphichelyids (not shown) which led to cryptodires and pleurodires (both flourishing today).

1 **Proganochelys** had a well-developed shell but "old-fashioned" skull with teeth as well as beak. Probably it could not pull limbs, tail, or head inside its shell. Shell length: 61cm (2ft). Time: Late Triassic. Place: Germany. Suborder: Proganochelydia (ancestral chelonians).

2 **Podocnemis** was and is a freshwater turtle (terrapin) pulling the neck in sideways. Length: to 76cm (30in). Time: Late Cretaceous onward. Place: once widespread, now South America and Madagascar. Suborder: Pleurodira (side-neck turtles).

3 **Archelon** was a huge, early marine turtle. It had a broad, light, flattened shell and long paddle-like limbs. Length: 3.7m (12ft). Time: Late Cretaceous. Place: North America. Suborder: Cryptodira (vertical-neck turtles, including all living tortoises and most turtles).

Neck benders (above)
These diagrams reveal how two types of chelonian pull the head inside the shell.
A Cryptodires bend the neck down and back. Bending occurs mostly between cervical (neck) vertebrae 5 and 6, and between cervical vertebra 8 and dorsal (back) vertebra 1.
B Pleurodires bend the neck sideways. Bending occurs mostly between cervical vertebrae 2 and 3 and 5 and 6, and between cervical vertebra 8 and dorsal vertebra 1.

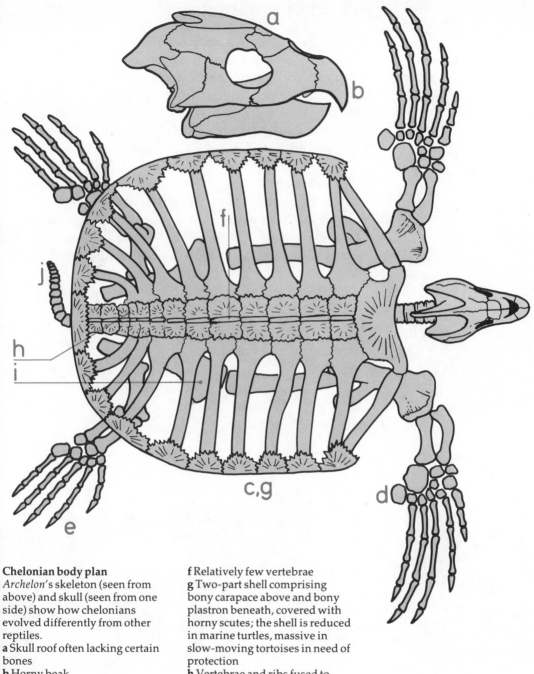

Chelonian body plan

Archelon's skeleton (seen from above) and skull (seen from one side) show how chelonians evolved differently from other reptiles.

a Skull roof often lacking certain bones
b Horny beak
c Short, broad body
d Heavy limbs projecting sideways
e Relatively few toe bones

f Relatively few vertebrae
g Two-part shell comprising bony carapace above and bony plastron beneath, covered with horny scutes; the shell is reduced in marine turtles, massive in slow-moving tortoises in need of protection
h Vertebrae and ribs fused to carapace
i Limb girdles and upper limb bones fitting inside ribs
j Short tail

Early euryapsids

Euryapsids were a subclass of mostly marine reptiles with a hole high in each temple of the skull. Three major early groups were the araeoscelids, placodonts, and nothosaurs. Araeoscelids were lizard-like and largely lived on land. Placodonts had short, stout, armoured bodies, paddle-like limbs, and blunt teeth. Nothosaurs were slimmer, with longer necks and bodies, and sharp teeth. Both groups hunted in the sea: placodonts for molluscs, nothosaurs for fishes. Araeoscelids seemingly evolved from primitive cotylosaur reptiles in early Permian times, 270 million years ago. Placodonts and nothosaurs flourished in Triassic times until wiped out by competition from bony fishes and new aquatic reptiles. Many fossils in these groups occur in Triassic rocks in parts of Europe, North Africa, and Asia that rimmed the ancient Tethys Sea.

Placodont body plan
The 2m (6ft 6in) *Placodus* from Mid Triassic Europe showed these typical placodont features.
a Powerful jaws
b Peg-like front teeth
c Flat, broad, crushing tooth plates at the back of the mouth
d Short, heavy, rounded body
e Short neck
f Extra set of (belly) ribs
g Small bones forming protective armour
h Limbs designed as paddles
i Skin joining toes and fingers
j Flattened tail (short in later placodonts)

1 **Araeoscelis** was small, lightly built, and rather lizard-like, but with long, slim shins and "forearms". Length: 66cm (26in). Time: Early Permian. Place: Texas, North America. Order: Araeoscelidea.

2 **Tanystrophaeus** had a grotesquely long neck, used maybe as a fishing rod. Adults lived in the sea, young on shore. Length: up to 6m (20ft). Time: Mid Triassic. Place: Central Europe and Israel. Order: Araeoscelidea, or maybe Eosuchia (p. 112).

3 **Henodus** shows how later placodonts evolved like marine turtles: with flippers, a broad, flat body protected by a bony shell, and a horny, toothless beak. Length: 1m (3ft 3in). Time: Late Triassic. Place: Germany. Order: Placodontia.

4 **Nothosaurus** had a long, slim neck and body, long forelimbs, and long, slim jaws bristling with sharp teeth shaped for catching fishes. Length: 3m (10ft). Time: Mid Triassic. Place: Central Europe, North Africa, South-West Asia, East Asia. Order: Sauropterygia. Suborder: Nothosauria.

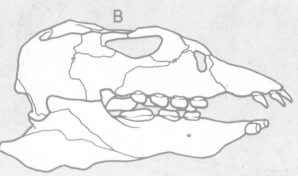

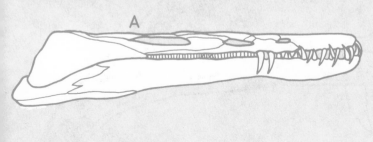

Skulls compared
These diagrams contrast the jaws of two euryapsids that ate different foods.
A *Nothosaurus* jaws were long and slim, with long, sharp teeth that interlocked to seize and grip slippery fishes.
B *Placodus* jaws were deep and strong, and closed by muscles with great crushing force. Jutting front teeth were pincers to pluck molluscs from the sea bed. Flat back teeth and bones in the skull roof and lower jaw crushed mollusc shells.

Plesiosaurs

Plesiosaurs ("near lizards") were bigger than the Triassic nothosaurs and better built for water life. They had a bulky, barrel-shaped body; broad ribs; "belly" ribs; four long, flat flippers; and a short tail. With ichthyosaurs they ruled Jurassic and Cretaceous seas and oceans, and then died off.

Their suborder held two superfamilies. Long-necked plesiosaurs were expert fishers at or near the surface. Short-necked plesiosaurs (pliosaurs) dived and preyed on ammonites. Both types swam with flippers, as marine turtles do. Flippers also helped haul them ashore to lay eggs.

Fossil plesiosaurs occur in Jurassic and Cretaceous clays and limestones – especially Liassic (early Jurassic) rocks in England and Germany, Late Cretaceous rocks in western North America, and Early Cretaceous rocks in Australia.

Plesiosaur body plan
Long-necked plesiosaurs had the following features.
a Bulky, rounded body
b Very long front flippers
c Hind flippers shorter than front flippers
d Very long neck, for darting down on fishes from above or snaking swiftly sideways through the water
e Small head
f Sharp, needle-like teeth for gripping fishes
g Short tail

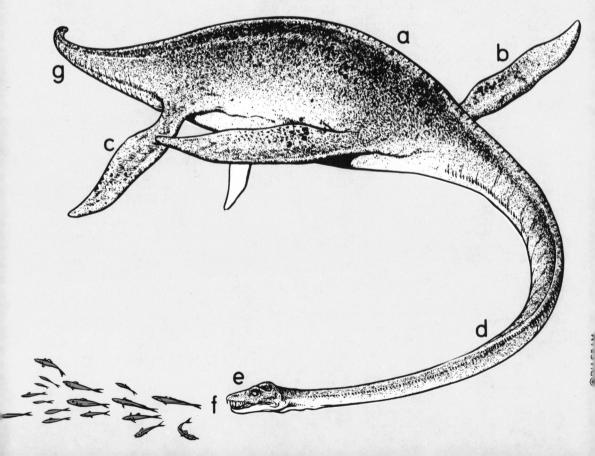

These four plesiosaurs represent evolutionary trends in both superfamilies.

1 **Peloneustes** was a small short-necked plesiosaur. Length: 3m (10ft). Time: Late Jurassic. Place: Western Europe. Superfamily: Pliosauroidea.

2 **Kronosaurus** was a huge short-necked plesiosaur with massive teeth. Length: up to 17m (56ft), one quarter of this being head. Time: Early Cretaceous. Place: Australia. Superfamily: Pliosauroidea.

3 **Thaumatosaurus,** an early long-necked plesiosaur, had a neck less than one quarter its total length. Length: 3.4m (11ft). Time: Early-Mid Jurassic. Place: Europe. Superfamily: Plesiosauroidea.

4 **Elasmosaurus** was more than half neck, with more than 70 neck vertebrae. Length: 12m (39ft) or longer. Time: Late Cretaceous. Place: North America. Superfamily: Plesiosauroidea.

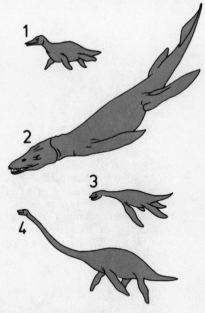

Plesiosaur skeleton
Seen from below, the skeleton of *Thaumatosaurus* shows the overall design of an early long-necked plesiosaur.
a Moderate length: 3.4m (11ft)
b Many neck bones
c Plate-like shoulder bones anchoring muscles that pull front flippers down and back
d Plate-like hip bones supporting muscles that operate hind flippers
e Short upper limb bones
f Long feet and "hands"
g Belly ribs

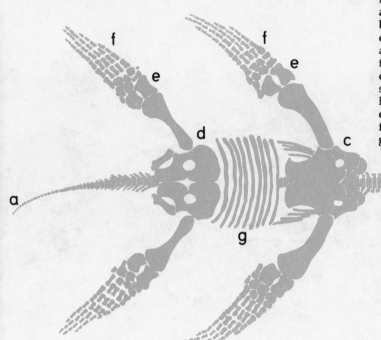

Ichthyosaurs

Ichthyosaurs ("fish lizards") were aquatic reptiles that flourished in shallow seas about 220 million to 90 million years ago. They had fins, flippers, long narrow jaws, and a superbly streamlined body. The largest individuals measured about 7.6m (25ft) but some species were no longer than a man. Ichthyosaurs ate cephalopods and fish, and gave birth to their young in water.

These reptiles spread all around the world, but their fossils are most plentiful in Lower Jurassic rocks about 180 million years old. Fossil ichthyosaur bones or faeces occur in, for instance, Britain's Lias shales and limestones, and in rocks in Germany, North and South America, Australia, and Indonesia.

Skin and bones
Fine-grained Lower Jurassic rock preserved this fine *Stenopterygius* ichthyosaur fossil from Boll Holzmaden in southern Germany. Even skin remains survive. The fossil's biconcave vertebrae, bony eye ring, and slim, toothy jaws are typical of ichthyosaurs. Experts familiar with such features can often identify an ichthyosaur from just one bone.

These three ichthyosaurs represent evolutionary trends in the order Ichthyosauria.

1 **Cymbospondylus,** a Triassic ichthyosaur, had relatively long arm and thigh bones, short skull, and small tail, recalling the body build of the ichthyosaurs' land-based ancestors.

2 **Ichthyosaurus,** from Lower Jurassic rocks, was much more streamlined than *Cymbospondylus*. It had a large dorsal fin and tail, forelimbs broadened into paddles, and a long, tapered skull.

3 **Ophthalmosaurus,** from Upper Jurassic rocks, reveals further evolutionary changes, including enlarged eyes and propulsive tail, tiny hind limbs, and teeth in a groove instead of separately socketed as in early ichthyosaurs.

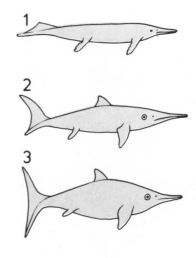

Dolphin's lookalike
Ichthyosaurs could have sheared swiftly through the waves. Their bodies were designed for speed, much like those of dolphins – living sea mammals evolved from ancestors that dwelt on land. But ichthyosaurs had vertical not horizontal tails, longer jaws, and simpler brains that made them less intelligent than dolphins.

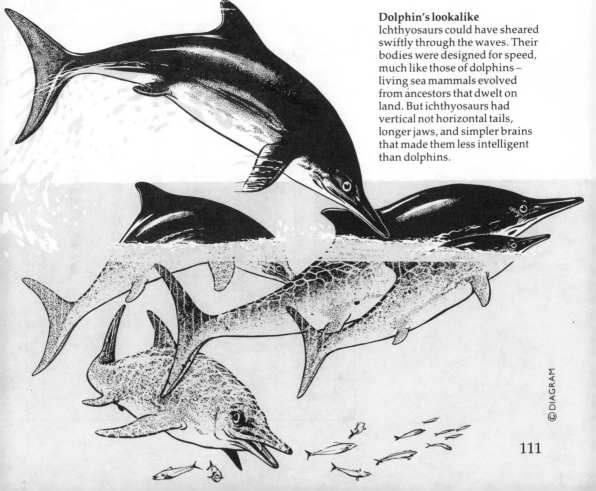

© DIAGRAM

111

Eosuchians

Eosuchians ("dawn crocodiles") were mostly small, lizard-like creatures, although some of these gave rise to larger, crocodile-like reptiles. The order Eosuchia lasted from 290 to 50 million years ago, but their heyday was 280–200 million years ago. Fossils crop up mostly in Upper Permian rocks of South Africa, though finds occur elsewhere, especially in North America and Europe.

Eosuchians evolved from cotylosaurs ("stem reptiles") and were the first diapsids (reptiles of the subclass with two openings in the skull behind each eye). Their infraclass, Lepidosauria, includes the snakes and lizards.

1 **Youngina** had slim limbs, a long, slim tail and body, and a pointed skull of "old-fashioned" design, with teeth inside the mouth as well as set in sockets

Diapsid family tree
This family tree shows likely relationships of orders of the reptile subclass Diapsida.
1 Cotylosauria (stem reptile ancestors)
2 Eosuchia (eosuchians)
3 Rhynchocephalia (beak-heads or rhynchocephalians)
4 Squamata (lizards and their offshoot, snakes)
5 Thecodontia (thecodonts)
6 Crocodilia (crocodilians)
7 Saurischia (saurischian dinosaurs)
8 Ornithischia (ornithischian dinosaurs), derived from either thecodonts or saurischian dinosaurs
9 Pterosauria (pterosaurs)

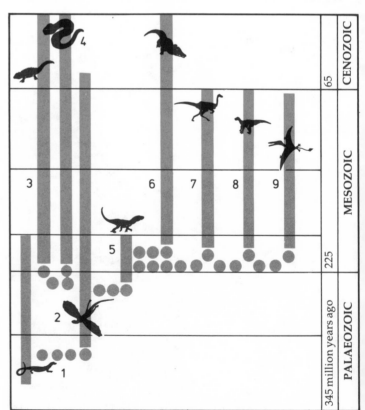

on the jaw rim. Length: 45cm (18 in). Time: Late
Permian. Place: South Africa. Suborder:
Younginiformes (lizard-like).
2 **Weigeltisaurus** glided from tree to tree on skin
wings stretched between enormously long ribs.
Length: 50cm (20in). Time: Late Permian. Place:
England and Germany.
3 **Askeptosaurus** had a long, slim body, long
sharp-toothed skull, and small, paddle-like limbs.
Length: 1.5m (5ft). Time: Mid Triassic. Place:
Europe. Suborder: Thalattosauria (aquatic hunters).
4 **Champsosaurus** resembled a slim-snouted
crocodilian. Length: 1.5m (5ft). Time: Late
Cretaceous-Eocene. Place: North America and
Europe. Suborder: Choristodera (fish hunters).

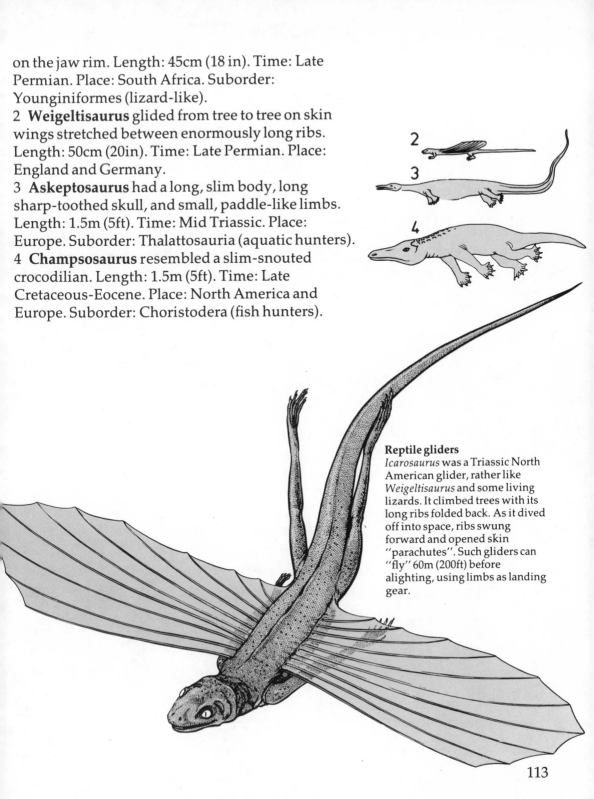

Reptile gliders
Icarosaurus was a Triassic North
American glider, rather like
Weigeltisaurus and some living
lizards. It climbed trees with its
long ribs folded back. As it dived
off into space, ribs swung
forward and opened skin
"parachutes". Such gliders can
"fly" 60m (200ft) before
alighting, using limbs as landing
gear.

113

Lizards and snakes

With their eosuchian ancestors, lizards, snakes, and rhynchocephalians ("beak-heads") form an infraclass, the lepidosaurs.

Lizards appeared about 230 million years ago and gave rise to some strange aquatic forms. Mosasaurs ("Meuse lizards") were a family of huge Late Cretaceous sea lizards that seized ammonites and fish in sharp-toothed jaws. Mosasaur fossils are especially plentiful in Niobrara chalk from Kansas.

By 70 million years ago, lizards had given rise to snakes. Between them, snakes and lizards make up the Squamata – the most numerous, varied, and widespread reptile order in the world today. Here are four prehistoric examples.

1 **Paliguana** was a small, early lizard, or maybe a lizard ancestor, from Early Triassic South Africa.

2 **Acteosaurus** had a long, skinny body, slim tail, and short limbs. It was in the dolichosaurid family of semi-aquatic Cretaceous European lizards. Length: about 40cm (16in).

3 **Tylosaurus** was a mosasaur up to 8m (26ft) long. Place: North America and New Zealand.

4 **Dinilysia** might have been an early relative of modern boas and pythons. Length: 1.8m (6ft). Time: Late Cretaceous. Place: Patagonia, South America.

Mosasaur body plan
a Huge size
b Long head
c Nostrils high on skull
d Long jaws, with a joint in the lower jaw
e Sharp teeth set in sockets
f Short neck
g Long body
h Flattened, paddle-shaped limbs to steer and balance
i Long tail flattened from side to side for swimming

Rhynchocephalians

Rhynchocephalians ("beak-heads") had an upper jaw ending in a down-curved beak. Some were lizard-like, some bulkier. They evolved and spread in the Triassic Period, then largely fizzled out. Only one species is alive today.

1 **Homoeosaurus** was a lizard-like Late Jurassic beast possibly related to the living tuatara found only in New Zealand. Length: 19cm (7.5in). Place: south-west Germany.

2 **Scaphonyx** belonged to the rhynchosaurid family of heavy-bodied beak-heads with a deep skull, toothless, tong-like jaws, and rows of crushing toothplates in the mouth. Probably it chopped and crushed up husked fruits. Length: 1.8m (6ft). Weight: about 90kg (200lb). Time: Mid Triassic. Place: Brazil. (Others lived in Africa and India.)

1

2

Clues to diet

Rhynchosaurs probably ate plants. Clues include features of skull and teeth, shown here from below (**A**) and side (**B**):
a rows of teeth on tooth plates;
b groove to take the lower jaw as jaws shut like a penknife:
c tong-like food-gathering beak, manipulated by a large tongue. Body restorations also give clues to diet. Strong hind limbs (**C**) could have scratched up roots and tubers. The "barrel body" (**D**) left room for a large gut, digesting bulky food.

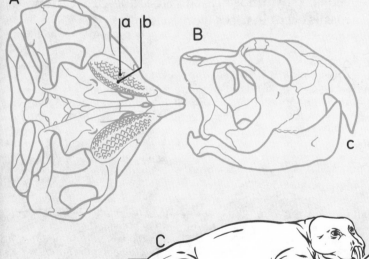

Thecodonts

Thecodonts ("socket toothed" reptiles) were a hugely important reptile order, for they gave rise to the crocodiles, dinosaurs, and pterosaurs. All four made up one infraclass: the archosaurs or "ruling reptiles". Archosaurs dominated life on land in Mesozoic times.

Thecodonts probably evolved from eosuchians about 226 million years ago. Most were large, four-legged flesh-eaters with distinctive skeletons. They came, perhaps, in four suborders. First were sprawling, heavy-bodied proterosuchians. From these sprang pseudosuchians, including small, light creatures with bodies held well off the ground. Pseudosuchians gave rise to four-legged, armoured herbivores called aëtosaurs, and to the flesh-eating, crocodile-like phytosaurs.

Thecodonts died out about 193 million years ago, replaced by dinosaurs and crocodiles. Thecodont fossils occur in rocks worldwide.

Two ways of standing (above)
A Many thecodonts sprawled, knees and elbows stuck out and feet flat on the ground.
B Pseudosuchians tucked the knees and elbows down and in to lift the body and gain speed. Some ran on their toes.

Thecodont body plan (right)
A typical thecodont had these features.
1 Skull lightened on each side by two holes behind the eye (**a**), one in front of the eye (**b**), and one in the lower jaw (**c**).
2 Teeth set in sockets
3 About two dozen vertebrae between head and hip region
4 Two sacral vertebrae (linked to the hips)
5 Hip socket shaped as a solid bony basin
6 Fairly straight thighbone, not sharply inturned at the top
7 Hind limbs (**a**) longer than front limbs (**b**)
8 Shin no longer than thigh
9 Five digits per "hand" and foot

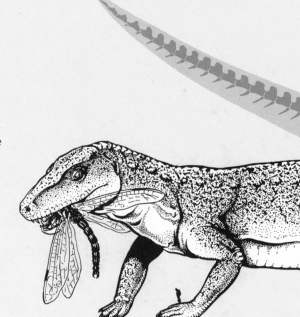

These species represent the thecodont suborders.

1 **Erythrosuchus** was an early proterosuchian, with a stout, squat body, thick limbs, large head, and rather short tail. Length: 4.5m (14ft 9in). Time: Early Triassic. Place: South Africa.

2 **Euparkeria** was a small, light pseudosuchian hunter. It rose on long hind legs to sprint, the tail balancing the head and neck. Rows of armour plates ran down the back. Length: 60cm (2ft). Time: Early Triassic. Place: South Africa.

3 **Stagonolepis** was a heavy-bodied aëtosaur encased in bony armour plates. It had a pig-like snout with weak teeth. Perhaps it grubbed for roots. Length: 3m (10ft). Time: Late Triassic. Place: Europe.

4 **Rutiodon** was a phytosaur – a crocodile-like reptile but with nostrils in a bump almost between the eyes, not at the snout tip. Such fish-eaters dominated pools and rivers until replaced by crocodiles. Length: 3m (10ft). Time: Late Triassic. Place: North America and Europe.

Crocodilians

Crocodilians are living fossils, the last surviving archosaurs. Their bulky, armoured bodies, long, deep, flattened swimmers' tails, short, sturdy limbs, and long, strong, toothy, flesh-eaters' jaws resemble those of crocodiles alive 100 million years ago.

Like dinosaurs, crocodilians evolved from thecodonts 200 million years or more ago. Fossils show they dominated pools and rivers worldwide when climates almost everywhere were warm. Small, early protosuchian crocodiles gave rise to the larger mesosuchians, some designed for life at sea. From mesosuchians came the strange, land-based sebecosuchians, and the eusuchians – the one suborder that survives today.

Where they lived
This map shows finds of the four fossil crocodiles described on these pages.
1 *Protosuchus*
2 *Metriorhynchus* (which lived in seas worldwide)
3 *Baurusuchus*
4 *Deinosuchus*

Nostrils fore or aft
Nostril position helps us to distinguish crocodilian fossils from those of phytosaurs.
a Crocodilian nostrils (an enclosed nasal passage permits breathing while eating)
b Phytosaur nostrils

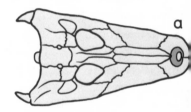

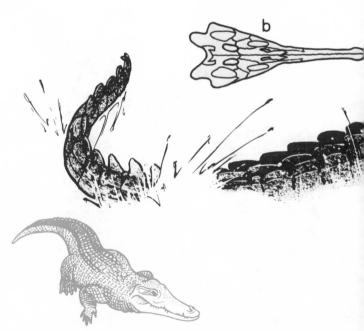

Here is one example from each suborder.

1 Protosuchus from Arizona was a protosuchian crocodile. It had a short, sharp-toothed skull and rather long legs. Maybe it lived mainly on land. Time: Late Triassic or Early Jurassic.

2 Metriorhynchus was a marine mesosuchian with limbs evolved as flippers, a tail fin, very long jaws with sharp fish-eater's teeth, and no bony armour. Length: 3m (10ft). Its Mid-Late Jurassic fossils occur in Europe and South America.

3 Baurusuchus was a sebecosuchian from Brazil. It had a short, deep, flattened skull, with few teeth (the front ones very large), and sideways-facing eyes. Probably it lived on land. Length: maybe 1.5m (5ft). Time: Late Cretaceous.

4 Deinosuchus, an eusuchian, was the largest-ever crocodile, with immense jaws. It lived in Late Cretaceous Texas and must have eaten small and medium sized dinosaurs. Length: 16m (52ft 6in).

Terror of the dinosaurs
The huge crocodile *Deinosuchus* lurked in rivers and ambushed dinosaurs that came to drink. Contrast this monster's size with that of a modern crocodile, shown to the same scale (bottom of facing page).

Pterosaurs

Pterosaurs ("winged lizards") included the first and largest flying backboned animals. They evolved from Triassic pseudosuchian gliding reptiles, lasted 130 million years, and died out at the end of the Cretaceous Period, 65 million years ago. They had skin wings stretched between the limbs and body; light, but strong, skeletons; and well-developed powers of sight and wing control. Some walked like birds, others shuffled awkwardly and roosted bat-like. Most caught fish or other prey. Two main groups evolved: first rhamphorhynchoids, with teeth and tails; then pterodactyloids, which lacked these and so shed needless weight. Large pterodactyloids were poor fliers and mostly soared or glided, where air rose over heated land or winds blew up sea cliffs.

How pterosaurs began (above)
The gliding pseudosuchian *Sharovipteryx* might have given rise to pterosaurs (though these flew with their fore limbs). It lived in Early Triassic times in what is now Soviet Central Asia.

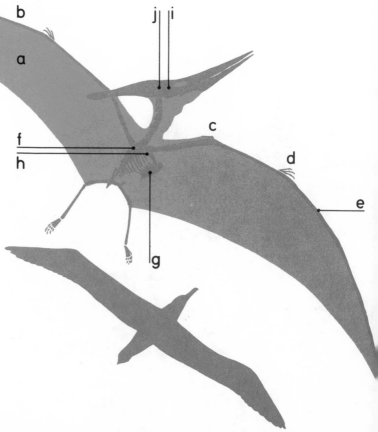

Pterosaur body plan
A pterosaur skeleton is shown here to scale with an albatross. These features adapted a pterosaur for flight.
a Skin wing membrane
b Arm bones and long fourth-finger bone support wing membrane
c Pteroid bone supports a front wing flap that helps to prevent stalling
d Clawed fingers 1–3 serve as hooks for roosting
e Light, strong skeleton with hollow, air-filled bones
f Fused vertebrae reinforce the shoulder area
g Enlarged breastbone anchors wing muscles
h Strong joint linking shoulder blade to spine and breastbone
i Large eyes with keen vision
j Brain well developed for sight and co-ordinated flight

These examples show special features and trends.

1 **Dimorphodon,** an Early Jurassic pterosaur, from southern England, had a big head, biting teeth, and a long tail. Wingspan: 1.5m (5ft).

2 **Sordes** had fur to trap body heat (maybe all pterosaurs were warm-blooded). This small rhamphorhynchoid comes from Late Jurassic rocks in the USSR (Kazakhstan).

3 **Pteranodon** was a huge, tailless, toothless pterodactyloid. Its "weather vane" head crest might have kept it heading into wind. It zoomed off sea cliffs, caught fish in its beak, and stored them in a throat pouch for its young. Wingspan: 8m (26ft). Time: Late Cretaceous. Place: USA (Delaware, Kansas, and Texas) and Japan.

4 **Quetzalcoatlus,** or "feathered serpent", was the largest known pterosaur: a long-necked beast that soared on hot air and gobbled carrion. Wingspan: 11–12m (36–39ft). Weight: 86kg (190lb). Time: Late Cretaceous. Place: Texas and Alberta.

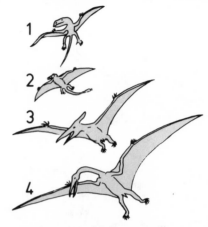

"Vultures" and "albatrosses"
Weak fliers, giant pterosaurs relied on moving air to keep them borne aloft.
A *Quetzalcoatlus* soared on thermal currents rising from land heated by the Sun.
B *Pteranodon* launched into the winds that blew up sea cliffs.

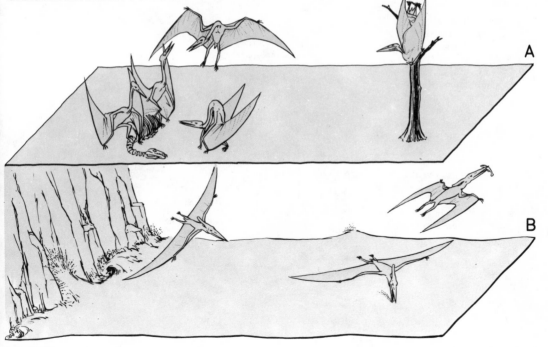

Saurischian dinosaurs 1

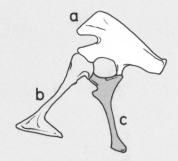

Dinosaurs ("terrible lizards") were probably the most successful-ever backboned animals to live on land. They evolved from pseudosuchian thecodonts some 205 million years ago and dominated lands worldwide for 140 million years – all through the rest of Mesozoic time. Their more than 350 genera included bipeds and quadrupeds: flesh-eaters and plant-eaters. Some were the largest-ever land animals. A few grew no bigger than a chicken. Many were probably warm-blooded. Almost all walked and ran on the toes, with legs straight down below the body, like a horse or an ostrich.

Two orders of these archosaurs evolved. Saurischians (the "lizard-hipped") comprised two suborders. One, the theropods or "beast feet", contained the flesh-eating dinosaurs. These bipeds ranged from small lizard-catchers to monsters as heavy as an elephant. A few small theropods were brainier than any reptile now alive.

Saurischian hip bones (above)
A saurischian's hip girdle had a forward-pointing pubis, as in most reptiles.
a Ilium **b** Pubis **c** Ischium

Theropod skulls (right)
Shown here are four theropod skulls of contrasting shapes and sizes. A human skull has also been included for scale.
A *Tyrannosaurus*, one of the largest of all flesh-eating dinosaurs, had big, sharp teeth shaped like serrated "steak-knife" blades.
B *Allosaurus*, a large Jurassic scavenger or big-game hunter, had sabre-like teeth designed for piercing flesh.
C *Ornithomimus*, a toothless "ostrich dinosaur", probably ate insects, lizards, leaves, fruit, and seeds.
D *Compsognathus*, no bigger than a chicken, had small, sharp teeth and ate lizards.
E Human skull, to scale.

© DIAGRAM

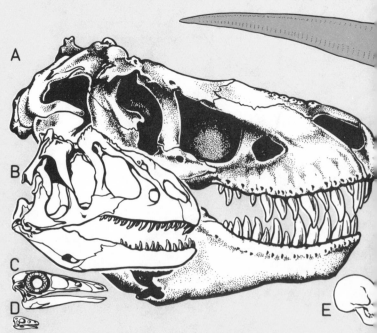

Some experts divide theropods into three infraorders. Here is one example of each.

1 **Coelurus** belonged to the coelurosaurs – small, light, sprinters with long legs, a long tail, and sharp claws and teeth (but some had toothless beaks). Length: 2m (6ft 6in). Time: Late Jurassic. Place: Wyoming, USA.

2 **Deinonychus** was a deinonychosaur – a small, fierce hunter that could stand on one leg, balanced by its stiffened tail, and slash out with a deadly "switchblade" toe claw. Length: 2.4–4m (8–13ft). Time: Early Cretaceous. Place: western USA.

3 **Tyrannosaurus** was one of the largest carnosaurs (great flesh-eating dinosaurs). It had a mighty head and body, huge legs and toe claws, savage fangs, and massive jaws. Arms were small but muscular. It killed big plant-eating dinosaurs or ate ones already dead. Length: 12m (39ft). Height: 5.6m (18ft 4in). Weight: 6.4 tonnes. Time: Late Cretaceous. Place: North America and China.

Theropods compared
Differences in body build reflect different modes of life in theropods.
a *Daspletosaurus*, a bulky carnosaur, perhaps waddled slowly like a duck. If so, such monsters might have been too slow for active hunting and perhaps lived on carrion instead.
b *Dromiceiomimus* was an ostrich dinosaur, capable of running faster than a horse.
c *Stenonychosaurus* was an agile sprinter with a brain larger than an emu's.
d Man, shown to scale.

123

Saurischian dinosaurs 2

The second saurischian suborder, the sauropodomorphs ("lizard-feet forms"), included the largest dinosaurs of all. Pioneers were the prosauropods: two-legged and four-legged dinosaurs, some lighter than a man. Their mostly Triassic infraorder included perhaps the first plant-eating dinosaurs. In the Jurassic Period these gave way to the colossal, four-legged sauropods – the largest-ever land animals. These giants raised long necks to browse on trees fringing warm, sluggish, lowland rivers. Their vast bulk made most of them safe from all enemies except carnosaurs and giant crocodiles. Sauropods endured as long as any dinosaurs, but their heyday was in Late Jurassic times. Rich fossil finds have come from China, the United States, and Tanzania.

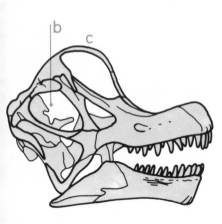

Sauropod features
The *Brachiosaurus* skull (above) and skeleton with outlined body (right) include features typical of many sauropods.
a Relatively small head
b Very small brain (bulk for bulk the smallest brain of any vertebrate)
c Some skull bones reduced to struts to save weight
d Long, flexible neck
e Spinal bones hollowed out to reduce weight
f Vast, heavy body

g Massive limbs, with thick, heavy bones
h Short, strong hind feet; toes 1–3 had long claws
i Each thumb had one long claw
j Long tail

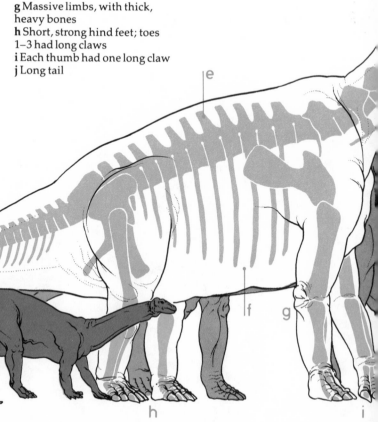

1 **Plateosaurus,** a big prosauropod, had a bulky body, rather long neck, small head, long tail, big, strong hind limbs, and shorter forelimbs. The five-fingered "hand" had a great thumb claw. This beast ate plants and maybe meat, and lived in Late Triassic Western Europe. Length: 8m (26ft).

2 **Brachiosaurus,** a brachiosaurid sauropod, was one of the most massive dinosaurs of all. Length: about 23m (75ft) or more. Height: 12m (39ft). Weight: 78 tonnes or more. Time: Late Jurassic. Place: USA, Portugal, Algeria, and Tanzania.

3 **Diplodocus** belonged to the diplodocids – a family of lightweight sauropods (some perhaps the longest dinosaurs of all) with snaky necks and whiplash tails. Length: 26.6m (87ft). Weight: 10.6 tonnes. Time: Late Jurassic. Place: western USA.

Sauropodomorph sizes (left)
Three sauropods are here shown to the same scale as a giraffe and a man.
1 *Plateosaurus*
2 *Brachiosaurus*
3 *Diplodocus*

Where they lived (below)
This map shows sites of fossil finds of the dinosaurs shown on these two pages.
1 *Plateosaurus*
2 *Brachiosaurus*
3 *Diplodocus*

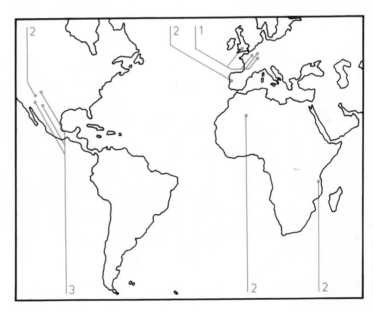

125

Ornithischian dinosaurs 1

Ornithischian ("bird hipped") dinosaurs, the second dinosaur order, get their name from hip bones designed like those of birds. Almost all were herbivores, with jaws superbly engineered for cropping and chewing leaves. Maybe this is why ornithischian species in time outnumbered those of sauropods, whose teeth were less effective. Saurischians or perhaps thecodonts gave rise to ornithischians in Late Triassic or Early Jurassic times. We shall look briefly at their four suborders, first the ornithopods.

Ornithopods ("bird feet") comprised dozens of species that walked or ran on long hind limbs. Small, early, agile sprinters no larger than a big dog gave rise to beasts as heavy as elephants. Here are examples from four families.

1 **Hypsilophodon** was a dinosaur "gazelle": a small, lightweight sprinter with long shins and feet, short arms, and a long stiffened tail to balance the head and body as it ran. It had ridged, self-sharpening teeth. Length: 1.8m (6ft). Time: Early Cretaceous. Place: England, Spain, Portugal, and USA.

Ornithischian hip bones
In ornithischian dinosaurs the pubis pointed backward.
a Ilium b Pubis c Ischium

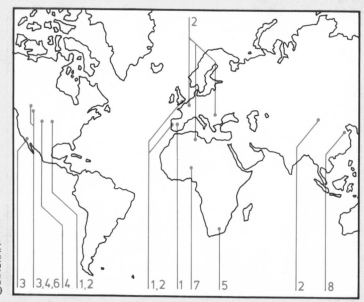

Where they lived
This map shows fossil finds of ornithopods described or pictured on these two pages. Items 1–4 tally with those so numbered in the text.
1 *Hypsilophodon*
2 *Iguanodon*
3 *Lambeosaurus*
4 *Pachycephalosaurus*
5 *Fabrosaurus*
6 *Stegoceras*
7 *Ouranosaurus*
8 *Shantungosaurus*

©DIAGRAM

2 **Iguanodon** was big and bulky with relatively large arms, a toothless beak, and hoofed nails. It roamed swampy lowlands, mostly on all fours. Length: up to 9m (29ft 6in). Weight: up to 4.5 tonnes. Time: Early Cretaceous. Place: northern continents.

3 **Lambeosaurus** belonged to the hadrosaurids or duckbilled ornithopods – Late Cretaceous beasts with wide, toothless beaks but up to 2000 cheek teeth: more than any other dinosaurs. Some hadrosaurids had head crests or "nose flaps" which they could blow up like balloons. Some grew far larger than their iguanodontid ancestors. Duckbills roamed all northern continents. *Lambeosaurus* from western North America grew up to 15m (49ft).

4 **Pachycephalosaurus,** a pachycephalosaurid ("bone-headed lizard") had a thick crash-helmet skull. Maybe rival males banged heads and winners ruled herds of females. Length: 4.6m (15ft). Time: Late Cretaceous. Place: North America. (Most other boneheads came from China or Mongolia.)

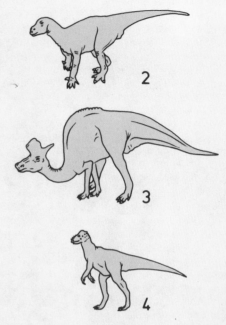

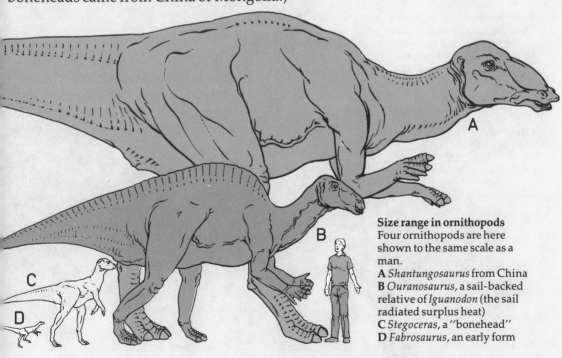

Size range in ornithopods
Four ornithopods are here shown to the same scale as a man.
A *Shantungosaurus* from China
B *Ouranosaurus,* a sail-backed relative of *Iguanodon* (the sail radiated surplus heat)
C *Stegoceras,* a "bonehead"
D *Fabrosaurus,* an early form

Ornithischian dinosaurs 2

Our brief look at the dinosaurs ends with three suborders of mostly four-legged ornithischians: the plated, armoured, and horned dinosaurs. Like their ornithopod relatives, all three had horny beaks for cropping plants. But unlike ornithopods most were not designed to run fast from danger. Instead they relied seemingly on body armour to protect them from the carnosaurs.

Plated dinosaurs (stegosaurs) had thick skin with bony plates or spikes that jutted from the back and tail. Armoured dinosaurs (ankylosaurs) had back and flanks encased in flexible armour made up of bony chunks and spikes covered with horny sheaths. Horned dinosaurs (ceratopsians) had a backswept bony crest protecting neck and shoulders; many sprouted rhinoceros-like horns. Stegosaurs evolved first, ankylosaurs second, and ceratopsians third. But ankylosaurs might have sprung from *Scelidosaurus*, an Early Jurassic dinosaur pre-dating even stegosaurs.

Bulky bodies
Three ornithischian dinosaurs, each representing a different suborder, are here shown to scale with a light battle tank.
A *Stegosaurus*, a plated dinosaur (stegosaur)
B *Ankylosaurus*, an armoured dinosaur (ankylosaur)
C *Triceratops*, a horned dinosaur (ceratopsian)

A

B

1 **Scelidosaurus,** perhaps a proto-ankylosaur, had seven rows of bony studs and spikes set in its back. Length: 3.5m (11ft 6in). Time: Early Jurassic. Place: England and Tibet.

2 **Stegosaurus,** the largest stegosaur, had two rows of plates (some huge) along the neck and back, and four spikes to guard the tail. Length: up to 9m (30ft). Time: Late Jurassic. Place: USA.

3 **Ankylosaurus,** the largest ankylosaur, had a thickened skull and crosswise bands of bony plates and studs protecting back and tail. The tail ended in a bony club. Length: up to 10.7m (35ft). Time: Late Cretaceous. Place: western North America.

4 **Triceratops,** among the last and largest of the ceratopsians, reached 9m (30ft) and 5.4 tonnes. It had a short neck frill, short nose horn, and two long brow horns. Time: Late Cretaceous. Place: western North America.

C

©DIAGRAM

129

Mammal-like reptiles 1

Mammal-like reptiles form the subclass Synapsida – reptiles with one hole low in each side of the skull behind the eye. All were quadrupeds. Early kinds were sprawlers like their cotylosaur ancestors. But evolution produced species that stood more erect, had body hair and several kinds of teeth, and were warm-blooded – features found in their better-known descendants, mammals.

Pelycosaurs or "basin-shaped (pelvis) lizards" were the earlier, more primitive, of two synapsid orders to appear. They arose by 280 million years ago. In Early Permian times, big, sprawling pelycosaur flesh-eaters and herbivores dominated life on land, at least in North America and Europe, where almost all their fossils have been found. They died out about 250 million years ago.

Synapsid family tree
This shows orders (1–3) and suborders (a–f).
1 Cotylosaur ancestors
2 Pelycosauria (primitive mammal-like reptiles)
a Edaphosauria (herbivores)
b Sphenacodontia (carnivores)
c Ophiacodontia (early forms)
3 Therapsida (advanced mammal-like reptiles)
d Phthinosuchia (ancestral therapsids)
e Theriodontia (advanced flesh-eaters)
f Anomodontia (herbivores)

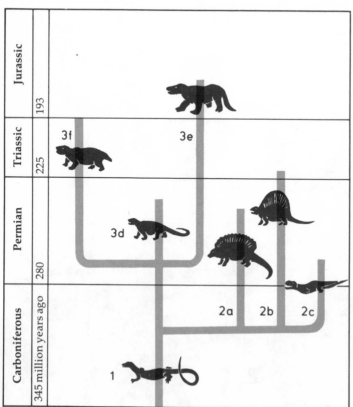

Two types of jaws (right)
Here we compare the skulls of flesh-eating *Dimetrodon* and plant-eating *Edaphosaurus*.
A *Dimetrodon*
a Long, deep, narrow skull
b Strong jaws with wide gape
c Small chewing teeth
d Two pairs of upper canine teeth: saw-edged blades
e "Step" in upper jaw
f Biting and grasping "incisors"
B *Edaphosaurus*
a Short, fairly shallow skull
b Straight-edged jaws (no step)
c Blunt teeth, all more or less alike
d Toothplates in mouth roof – *Edaphosaurus* could have crushed tough seed-fern leaves and maybe even mollusc shells
Sails as radiators (far right)
Long spines jutting from the backbone supported *Dimetrodon's* skin sail. If the pelycosaur stood sideways to the Sun this heated the sail's blood supply, so warming the whole body. If *Dimetrodon* faced away from the Sun, its sail shed heat, and its body cooled.

These examples show features of the pelycosaurs' three suborders: ophiacodonts, sphenacodonts, and edaphosaurs.

1 **Ophiacodon,** an ophiacodont, had a low-slung, lizard-like body, long hind legs, narrow, deep, long head, and jaws equipped with many sharp teeth. Probably it hunted fish in rivers. Length: 3.7m (12ft). Time: Early Permian. Place: Texas, USA.

2 **Dimetrodon** was a sphenacodont – a big flesh-eater with long, sharp "steak-knife" teeth and powerful jaws. It was one of the first backboned land animals able to kill beasts its own size. A huge skin "sail" rose from its back. Length: 3.5m (11ft 6in). Time: Early–Mid Permian. Place: Texas and Oklahoma, USA.

3 **Edaphosaurus,** an edaphosaur, was a large herbivore with blunt teeth, some in the mouth roof. A long, high skin "sail" ran down its back. Length: 3.3m (11ft). Time: Late Pennsylvanian–Early Permian. Place: USA and Europe.

© DIAGRAM

131

Mammal-like reptiles 2

In time an advanced, varied order of mammal-like reptiles, the therapsids (literally "mammal arch"), took over from their ancestors the pelycosaurs. Therapsids flourished from Mid Permian to Early Jurassic times. In the Late Permian they formed the chief flesh-eaters and plant-eaters living on dry land. Scientists have found Permian therapsid fossils worldwide, especially in South Africa and the USSR.

The therapsids comprised three suborders. Here we give examples of two: the primitive, ancestral phthinosuchians, and their "dead-end" offshoot the anomodonts. The latter were small to large plant-eaters and flesh-eaters in four infraorders: dromasaurs, dinocephalians ("terrible heads"), venyukoviamorphs, and dicynodonts ("two dog-like teeth").

1 **Phthinosuchus** belonged to the phthinosuchians – "old-fashioned" therapsids but with a large skull opening behind each eye, one pair of canine teeth per jaw, and a more upright stance than the pelycosaurs. Length: 1.5m (5ft). Time: Mid Permian. Place: USSR.

2 **Galepus** belonged to the dromasaurs – little, lightweight insect-eaters with jaws hinged well below the tooth row. Length: maybe 30cm (1ft). Time: Mid Permian. Place: South Africa.

3 **Moschops** was a plant-eating dinocephalian. It had a thick, short, dome-shaped skull, peg-like cropping teeth, squat, heavy body, sloping back, and short tail. Stocky limbs held its body well off the ground. Big flesh-eating dinocephalians might have attacked it. Length: 2.4m (8ft). Time: Mid Permian. Place: South Africa.

4 **Venyukovia** belonged to the venyukoviamorphs – big, partly beaked herbivores with a deep lower jaw, mostly short teeth, and a few big, stubby front teeth. Time: Mid Permian. Place: USSR.

5 **Lystrosaurus** belonged to the dicynodonts – abundant, worldwide plant-eaters. They had a short, broad body, short tail, strong legs, big holes in the skull behind the eyes, a horny beak, and a toothless mouth or just two upper tusks. *Lystrosaurus* was an Early Triassic reptilian "hippo", at home by lakes and rivers. Length: 1m (3ft 3in). Place: Antarctica, South Africa, India, and China.

Hunters and hunted
South Africa's Karroo rock beds hold fossil bones of big dinocephalians like *Moschops* (**A**) and *Titanosuchus* (**B**). *Moschops* peaceably ate plants, but *Titanosuchus*'s long, heavy jaws had sharp incisors and long, stabbing canine teeth, for tackling big game.

©DIAGRAM

Mammal-like reptiles 3

The most mammal-like of all mammal-like reptiles belonged to the therapsids' third suborder: the theriodonts ("mammal toothed"). These flesh-eaters were mostly small to medium-sized, with teeth and many bones designed astonishingly like a mammal's. Some were certainly warm-blooded and had a covering of body hair. A few perhaps even suckled young. Only the jaw and hearing mechanism marks these off from mammals, their direct descendants.

Theriodonts flourished about 250–170 million years ago. They spread worldwide, but South Africa's Permian rocks are the richest source of fossils. Here are examples from the theriodonts' six infraorders (some known well only from skulls). All but example 5 are from South Africa.

1 **Lycaenops** belonged to the gorgonopsians – plentiful Permian flesh-eaters derived from pelycosaur forebears. They had rather low-slung bodies, heavy skeletons and "sabre" teeth. Length 1m (3ft 3in). Time: Late Permian.

2 **Lycosuchus** represents the therocephalians. These had as few toe and finger bones as mammals, a skull crest, and a large hole in the skull behind each eye. Some were powerfully-built carnivores. Length: 1.8m (6ft). Time: Mid Permian.

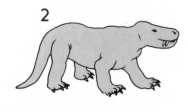

3 **Bauria** was one of the bauriamorphs, with mammal-like teeth and skull but old-fashioned lower jaw. *Bauria* seemingly had broad, grinding back teeth inset from the jaw rim, so perhaps space for cheeks storing half-chewed food. Length: maybe 1m (3ft 3in). Time: Early Triassic.

4 **Cynognathus** was a wolf-sized cynodont predator or scavenger, with dog-like skull and teeth, and limbs held fairly well below the body, a help in running fast. Length: 1.5m (5ft). Time: Early–Mid Triassic.

5 **Oligokyphus** belonged to the tritylodonts: small, rodent-like last survivors of all mammal-like reptiles. Length: 50cm (20in). Time: Late Triassic–Early Jurassic. Place: England and Portugal.

6 **Diarthrognathus** was an ictidosaur ("weasel lizard"), in the group arguably ancestral to the mammals. Its jaw hinged almost like a mammal's. Length: maybe 40cm (16in). Time: Late Triassic.

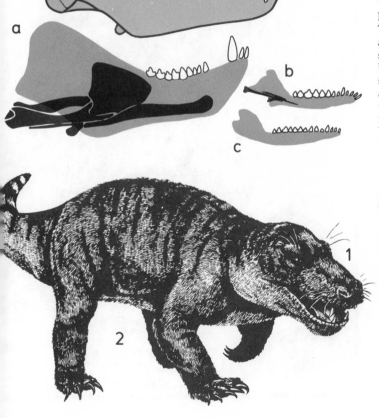

Cynognathus skull
This reveals a tendency towards mammalian features.
A Mouth and nasal passage separated, allowing breathing while eating
B Differentiated teeth: incisors, canines and ridged cheek teeth for fast chewing (speeding digestion to provide a high energy output)

Inner jaws
The inner jaws of *Cynognathus* and two early mammals (not to scale) illustrate a trend to fewer bones. The dentary is here shown tinted.
a *Cynognathus* jaw: seven bones, the dentary relatively larger than in early synapsids
b *Morganucodon* jaw: four bones, the dentary by far the largest
c *Spalacotherium* jaw: one bone, the dentary

Mammal-like features
This whole-body restoration of *Cynognathus* is based on key features found in skeletal remains.
1 Body covering of hair inferred from whisker pits in the snout
2 Mammal-like posture (knees and elbows held beneath the body) inferred from bones of limbs, hips and shoulders

Chapter 7

FOSSIL BIRDS

Aves (birds) is the most recently evolved of all nine classes of backboned animals. We can divide birds into three subclasses: Archaeornithes, comprising only *Archaeopteryx*, the first known bird; Odontoholcae, the toothed Cretaceous birds; and Neornithes, the rest. This short chapter gives prehistoric examples from the birds' more than 30 orders, most of which survive today.

A nineteenth-century book illustration shows a flock of moa skeletons in New Zealand's Canterbury Museum. Unlike most prehistoric birds, these recently extinct flightless giants left behind a wealth of bones. They enabled scientists to build up detailed reconstructions. (The Mansell Collection.)

ELEPHANTIPUS

ROBUSTUS

DINORNIS CASUARINUS

137

About birds

Birds are warm-blooded, backboned animals with a covering of feathers. Most have front limbs evolved as wings capable of flapping, soaring, or gliding flight. Early birds had teeth and a long bony tail core. Later birds evolved a more lightweight structure: with a toothless beak, a tail of lightweight feathers only, and hollow, air-filled bones for buoyancy. Flying birds gained a powerful breastbone to help anchor big, strong wing muscles. The first birds evolved about 140 million years ago. Some experts think birds' ancestors were small carnivorous coelurosaurs: dinosaurs like little *Compsognathus*. Other experts believe birds evolved directly from the dinosaurs' own ancestors, pseudosuchian thecodonts like *Euparkeria*. At first birds shared the air with pterosaurs. But birds have long outlived these creatures. This might be partly because birds'

Euparkeria
Euparkeria walked on all fours but could rise on its long hind legs to chase prey or escape from enemies. Early small pseudosuchian thecodonts like this gave rise to crocodiles, pterosaurs, dinosaurs, and maybe birds. But a long time gap separates the last known pseudosuchian from the first known bird.

Compsognathus
Tiny coelurosaur dinosaurs similar to *Compsognathus* might have given rise to the first known birds. Both shared over 20 anatomical similarities. Five are shown here.
a Short body
b Slim, flexible neck
c Very long legs
d Stiff ankle joints
e Long fingers

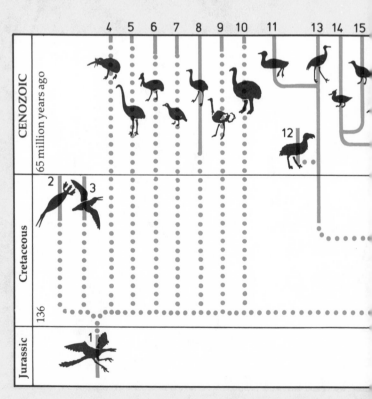

138

feathered wings survive injuries better than the pterosaurs' more fragile skin wings.

The class Aves (birds) contains some 30 orders, most with living representatives. No one knows exactly how and when each order began, or how all orders are related. This is largely because birds' fragile bones are seldom fossilized. Indeed, some species are known from little more than the fossil impression of a feather, or from a fossil footprint. But certain rocks have produced key fossil bird remains: for instance fine-grained limestone from Solnhofen in southern West Germany, the Niobrara Chalk of Kansas, and mudstone rocks of Utah and Wyoming. Finds in places such as these tell us that a number of orders of modern birds – and maybe even a few modern families – have flourished 50 million years or more.

Family tree of birds
Zoologists divide the class Aves (birds) into three subclasses, containing three superorders, subdivided into the 33 orders numbered below.
A Subclass Archaeornithes
1 Archaeopterygiformes (*Archaeopteryx*)
B Subclass Odontoholcae (Odontognathous or toothed Cretaceous birds)
2 Hesperornithiformes
3 Ichthyornithiformes
C Subclass Neornithes: Palaeognathous birds
4 Apterygiformes (kiwis)
5 Dinornithiformes (moas)
6 Casuariformes (emus etc)
7 Tinamiformes (tinamous)
8 Rheiformes (rheas)
9 Struthioniformes (ostriches)
10 Aepyornithiformes (elephant birds)
C Subclass Neornithes: Neognathous birds
11 Podicipediformes (grebes)
12 Diatrymiformes (*Diatryma*)
13 Gruiformes (cranes, rails)
14 Anseriformes (ducks)
15 Galliformes (grouse etc)
16 Charadriiformes (shorebirds)
17 Procellariiformes (albatrosses etc)
18 Sphenisciformes (penguins)
19 Pelecaniformes (pelicans)
20 Gaviiformes (divers)
21 Ciconiiformes (storks)
22 Columbiformes (pigeons)
23 Psittaciformes (parrots)
24 Strigiformes (owls)
25 Falconiformes (hawks etc)
26 Caprimulgiformes (nightjars)
27 Apodiformes (swifts)
28 Cuculiformes (cuckoos)
29 Coliiformes (mousebirds)
30 Passeriformes (perching birds)
31 Coraciiformes (rollers)
32 Piciformes (woodpeckers)
33 Alcediniformes (kingfishers)

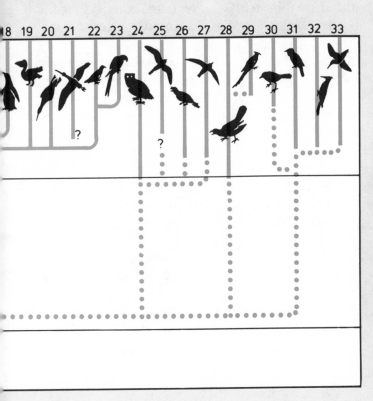

18 19 20 21 22 23 24 25 26 27 28 29 30 31 32 33

© DIAGRAM

Early birds

Birds shown here represent three early groups with teeth. *Archaeopteryx*, the first known bird, resembled a small coelurosaurian dinosaur but had feathers and long "arms". Wings, wishbone, and the angled bones of its shoulder girdle adapted *Archaeopteryx* for flight. Some experts think it climbed trees and fluttered down. Others believe it took off by sprinting into a headwind after insects. *Hesperornis* and its kin probably hunted fish just below the sea surface, and nested on lonely coasts or offshore islands. *Ichthyornis* may have flown above the sea and plunged to seize small fish, as terns do now. It might have given rise to modern shore birds.

1 **Archaeopteryx** ("ancient wing") had feathered wings but also unbird-like features: small teeth in the jaws, three-clawed fingers jutting from each wing, and a tail with a long, thin, bony core. Length: about 1m (about 3ft). Time: Late Jurassic. Place: Bavaria, southern West Germany. Order: Archaeopterygiformes.

2 **Hesperornis** ("western bird") resembled a large diver. It had a long, slim, pointed beak rimmed with teeth, and vestigial wings. It swam by thrusting water back with big lobed feet. Legs joined the body far back, so *Hesperornis* shuffled clumsily on land. Length: up to 1.5m (5ft). Time: Late Cretaceous. Place: North America. Order: Hesperornithiformes (toothed divers).

3 **Ichthyornis** ("fish bird") was a small, stout, tern-like bird with long, pointed wings, small feet and a long, slim beak armed with small curved teeth. *Ichthyornis* is the earliest known bird with a keeled breastbone to help support flight muscles. Height: 20cm (8in). Time: Late Cretaceous. Place: North America (eg Kansas, Texas, Alabama). Order: Ichthyornithiformes (toothed tern-like birds).

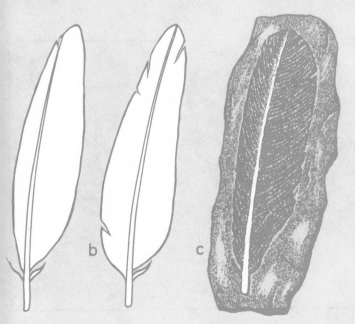

The first known bird (above)
Archaeopteryx might have used
its clawed limbs for climbing
trees, then flapped and fluttered
down. Ability to climb and fly
would have helped it to escape
enemies and capture agile prey.

Feathers and flight (left)
Feathers help to prove that
Archaeopteryx flew.
a Flying bird's wing feather: its
asymmetrical design helps it act
as an aerofoil.
b Flightless bird's wing feather:
its symmetrical design is useless
for flight.
c *Archaeopteryx* wing feather: its
design is asymmetrical.

141

Flightless birds

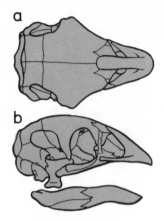

Most dead land birds make poor fossils: their fragile skeletons are soon eaten or just rot away. But big, flightless running birds have left their remains in Cenozoic rocks. Many of these birds are collectively called "ratites" from the Latin *ratis* ("raft"), for the breastbone was flat and raft-like, not keeled to anchor powerful flight muscles like the breastbone of a flying bird. Some experts think all prehistoric ratites and their living descendants the emu and ostrich shared one ancestor; others argue that different ratites had separate beginnings.

The flightless birds shown here were huge. Some filled the roles of cattle, others behaved like big cats. Such monsters tended to appear in lands with no large mammal herbivores or carnivores. Early ratites left only fossils, but *Dinornis* and *Aepyornis* survive as actual bones, found preserved in swamps. Archaeologists have even identified some moas' stomach contents. There are also *Aepyornis maximus* eggs – the largest known eggs to have been laid by any bird.

A moa's skull (above)
These views show a moa skull seen from above (**a**) and from the right (**b**). The beak was shaped for cropping plants, not rending flesh.

Where they lived (below)
This map depicts homes of the four groups of extinct flightless birds represented by those named on the facing page.
1 Elephant birds
2 Terror cranes
3 Moas
4 Phorusrhacids

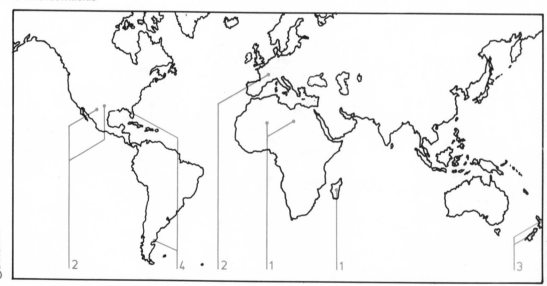

©DIAGRAM

142

1 **Aepyornis maximus** ("greatest of the high birds") was possibly the heaviest-ever bird at 440kg (970lb). A Madagascan elephant bird, it was the "roc" of legend, extinct by AD1700.

2 **Diatryma steini,** a "terror crane", stood 2m (6ft 6in) high and probably killed prey with its huge clawed feet and massive "parrot's" beak. It lived in North America 50 million years ago.

3 **Dinornis giganteus** (" giant terrible bird") was the tallest known bird, 3.5m (11ft 6in) high. This New Zealand giant moa was a browser that may have died out only about 400 years ago.

4 **Phorusrhacos longissimus** stood 2m (6ft 6in) tall and probably devoured goat-sized creatures with its huge "eagle's" beak. This savage hunter prowled Patagonia about 20 million years ago.

Grounded giants
Four giant flightless birds appear here on the same scale as a chicken.
1 *Aepyornis maximus*
2 *Diatryma steini*
3 *Dinornis giganteus*
4 *Phorusrhacos longissimus*

143

Water birds

Birds shown here are fossil or living examples of the eight orders of water birds and marsh birds. All except Ciconiiformes were probably derived from shorebirds.

1 **Stictonetta,** Australia's freckled duck is a primitive living member of the order Anseriformes (ducks, geese, swans), an order dating from the Eocene.

2 **Presbyornis** was a long-legged wader with a mainly duck-like skull. It might have given rise to ducks, geese, and swans. Vast flocks bred and fed on algae in salty, shallow lakes. Time: Early Eocene. Place: North America, and maybe worldwide. Order: Charadriiformes – shorebirds including gulls, terns, auks, and many waders.

3 **Osteodontornis** was among the largest-ever flying birds. Maybe it seized squid from the sea surface in its long bill rimmed with tooth-like bony spikes. Wingspan: up to 5.2m (17ft). Time: Miocene. Place: California. Order: Pelecaniformes – web-footed, fish-eating seabirds including cormorants, gannets, and pelicans.

4 **Puffinus,** an early shearwater, lived in Eocene Europe and Miocene North America. Order: Procellariiformes (albatrosses, petrels, shearwaters) – oceanic birds with long, slim, strong wings, webbed feet, and tube-shaped nostrils.

5 **Pachydyptes,** was a giant penguin about 1.6m (5ft 3in) high. Time: Eocene. Place: New Zealand. Order: Sphenisciformes (penguins) – flightless birds with wings evolved as swimming flippers. Penguins possibly evolved from petrels.

6 **Colymboides** was a diver (loon) no bigger than a small duck. Time: Eocene. Place: England. Order: Gaviiformes (divers) – swimming birds with webbed feet on legs set far back; strong fliers.

7 **Podiceps** was an early grebe. Time: Oligocene. Place: Oregon, USA. Order: Podicipediformes (grebes) – swimming birds resembling divers but with lobed not webbed feet. They probably evolved from gruiformes (cranes, rails, etc).

8 **Proardea** was an early heron. Time: Eocene. Place: England. Order: Ciconiiformes (herons, ibises, and other long-legged, long-necked wading birds) – ancestry uncertain.

8

Miocene seabirds
Strange birds in the order Pelecaniformes once fished the North Pacific Ocean.
a *Osteodontornis* seized prey at the surface.
b Plotopterids hunted under water, swimming with their wings. They were the north's equivalent of penguins.

a

b

Land birds

These pages give early examples from 13 orders of land birds, including birds of prey.

1 **Raphus,** the dodo, was a flightless pigeon as big as a large turkey. It died out a mere three centuries ago. Order: Columbiformes (doves and pigeons) – birds derived from shorebirds as early as the Eocene epoch.

2 **Archaeopsittacus** was an early parrot. Time: Late Oligocene. Place: France. Order: Psittaciformes (parrots) – tropical birds with a strong, hooked bill and two toes per foot turned back for perching. They might have evolved from pigeons.

3 **Argentavis,** a gigantic vulture-like bird of prey, was the largest-known bird able to fly. Wingspan: up to 7.6m (25ft). Weight: 120kg (265lb). Time: Early Pliocene. Place: Argentina. Order: Falconiformes – birds of prey that mostly hunt by day. These date from Late Eocene times.

4 **Ogygoptynx,** the first known owl, dates from Palaeocene times. Order: Strigiformes (owls) – nocturnal birds of prey unrelated to falconiformes.

5 **Gallinuloides** was an early member of the Galliformes (chickens and their kin). These possibly derived from ducks and geese. Time: Eocene. Place: Wyoming, USA.

6 **Dynamopterus** was an early cuckoo. Time: Oligocene. Place: France. Order: Cuculiformes (cuckoos, hoatzins, touracos) – primitive land birds, most with an outer toe turned back.

7 **Colius** has reversible hind and outer toes, and swings acrobatically from twigs. Time: Recent. Place: Africa. Order: Coliiformes (mousebirds), dating from Miocene times.

8 **Caprimulgus** has long, slim wings and a short, broad bill. It hunts at twilight. Time: Pleistocene on. Place: worldwide. Order: Caprimulgiformes (nightjars, oilbirds, frogmouth, etc).

9 **Aegialornis,** an early swift-like bird, had scimitar-shaped wings and flew fast to catch insects. Time:

Oligocene. Place: France. Order: Apodiformes
(swifts and maybe hummingbirds).

10 **Geranopterus** was an early roller from Oligocene
France. Order: Coraciiformes (rollers, hornbills,
hoopoes) – colourful hole-nesters with the three
front toes partly joined. These were the main land
birds of the Oligocene epoch.

11 **Archaeotrogon** was an early trogon from Eocene
France. Order: Alcediniformes (bee-eaters,
kingfishers, motmots, todies, trogons) – colourful
hole-nesters resembling coraciiforms but with a
unique kind of middle ear bone.

12 **Neanis,** a primobucconid, lived in Eocene
Wyoming. Order: Piciformes (barbets, toucans,
woodpeckers, etc) – perching birds with a reversed
outer toe.

13 **Lanius,** the shrike genus, goes back to Miocene
France. Order: Passeriformes ("true" perching
birds) – birds with the first toe turned back. They
include the songbirds and account for three-fifths of
all living bird species.

©DIAGRAM

Teratornid tar trap
Teratornis, a huge condor-like
vulture, here plucks flesh from a
mammoth fallen in a tar pool.
Feasting teratornids also
tumbled in, became stuck and
died. In fact the tar pools of Los
Angeles are rich in Late
Pleistocene bird bones.
Palaeontologists have recovered
those of condors, eagles, hawks,
owls, ducks, geese, herons,
storks, cranes, pigeons, ravens,
turkeys and many perching birds.

Chapter 8

FOSSIL MAMMALS

Members of the class Mammalia have dominated life on land for the last 65 million years. But mammal origins go back almost three times as far as that. Starting with the first primitive mammals, this chapter covers the mammals' four subclasses: the extinct Eotheria and Allotheria; the egg-laying Prototheria with few surviving species; and the pouched and placental mammals, collectively called Theria.

Most pages deal with the placental mammals, by far the largest mammal group. We look at prehistoric members of their nearly 30 orders, 13 of which are now extinct.

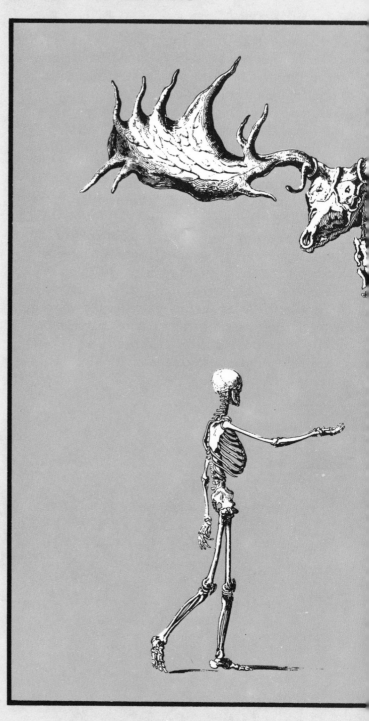

An old engraving contrasts skeletons of modern man and *Megaloceros*, the giant "Irish elk" from Pleistocene Eurasia. Dissimilar in form and size, man and deer remind us of the great diversity of mammals that evolved from tiny, shrew-like pioneers. (The Mansell Collection.)

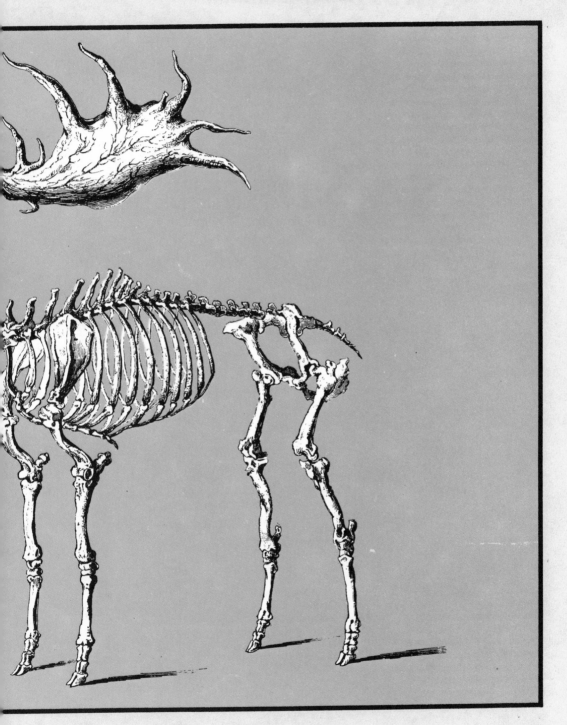

About mammals

By 190 million years or so ago, mammal-like reptiles had given rise to mammals: warm-blooded backboned animals with hair, an efficient four-chambered heart, and a muscular sheet (the diaphragm) that helps to work the lungs. Most give birth instead of laying eggs, and all feed babies milk from special glands. Such soft parts do not survive as fossils, and experts distinguish fossil mammals from reptiles by differences between their bones – especially the jaws.

People disagree about how to group all mammals. Arguably the class Mammalia holds four main groups or subclasses, two extinct. Eotheria ("dawn mammals") were small, primitive, early mammals, known from rather scanty fossil finds. Prototheria

Mammal family tree
This family tree shows all 39 orders of the class Mammalia (mammals). Relationships of some are largely guesswork.
A Therapsid ancestors
B Subclass Eotheria
1 Docodonta (docodonts)
2 Triconodonta (triconodonts)
C Subclass Prototheria
3 Monotremata (monotremes)
D Subclass Allotheria
4 Multituberculata
E Subclass Theria, infraclass Pantotheria
5 Eupantotheria (eupantotheres)
6 Symmetrodonta (symmetrodonts)
F Subclass Theria, infraclass Metatheria, superorder Marsupialia
7 Marsupicarnivora
8 Paucituberculata
9 Peramelina (bandicoots)
10 Diprotodonta (koalas etc)
G Subclass Theria, infraclass Eutheria (placental mammals)
11 Taeniodontia (taeniodonts)
12 Edentata (edentates)
13 Pholidota (pangolins)
14 Lagomorpha (rabbits etc)

15 Rodentia (rodents)
16 Tillodontia (tillodonts)
17 Primates
18 Chiroptera (bats)
19 Dermoptera (gliders)
20 Insectivora (insectivores)
21 Creodonta (creodonts)
22 Carnivora (carnivores)
23 Cetacea (whales)
24 Amblypoda (amblypods)
25 Condylarthra (condylarths)
26 Tubulidentata (aardvarks)
27 Perissodactyla (horses etc)
28 Litopterna (litopterns)
29 Astrapotheria (astrapotheres)
30 Notoungulata (notoungulates)
31 Trigonostylopoidea
32 Xenungulata (xenungulates)
33 Pyrotheria (pyrotheres)
34 Artiodactyla (cows etc)
35 Hyracoidea (hyraxes)
36 Embrithopoda (embrithopods)
37 Proboscidea (elephants etc)
38 Sirenia (sea cows)
39 Desmostylia (desmostylans)

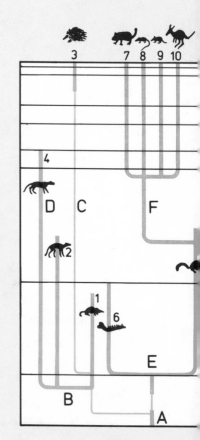

are primitive egg-layers such as the living platypus.
Allotheria were early, mostly small mammals, with
teeth much like a rodent's. Theria include marsupials
(pouched mammals) and placentals (mammals
whose babies develop in the mother well nourished
by food from a special structure, the placenta).

Only small, unobtrusive mammals co-existed with
the dinosaurs. When these died out, therians more
than took their place, evolving shapes and sizes that
suited them for life in almost every habitat. Small,
early forms gave rise to bulky herbivores and
carnivores. Others took to air or water. Mammals
have ruled most lands for the last 65 million years,
the Cenozoic Era.

Early mammals

Of most known early mammals little more than teeth or jaws survive. Eotheria, the earliest of all, were mostly small and shrew-like Triassic and/or Jurassic creatures. Some had traces of the old, reptilian type of jaw joint. Their two main subdivisions, docodonts and triconodonts, are named from differences in their types of teeth.

A 175-million-year gap separates the Eotheria from their possible descendants the Prototheria. These egg-laying mammals are known only from the monotremes, an order that includes the spiny anteater.

The Allotheria formed a long line of early mammals with multicuspid molars like rodents' (cusps are points on a tooth's grinding surface). Seemingly these were the first plant-eating mammals. Multituberculates, their only order, persisted 90 million years, into the Eocene, when rodents took their place.

The Theria ("true" mammals) include not only modern mammals but their fossil ancestors the pantotheres, with two orders: eupantotheres and symmetrodonts. Eupantotheres had complex molars shaped to shear and crush. Symmetrodonts had simpler teeth. Both died out by Mid Cretaceous times (some 100 million years ago).

Mammal features
Unlike most reptiles, fossil mammals show these and other features.
a Distinctive jaw hinge
b Only one bone on each side of the lower jaw
c Three auditory bones in each ear, inside the skull
d Teeth only on jaw rims
e Complex cheek teeth, with two roots or more
f A single bony nasal opening
g Growing bones show epiphyses – ends separated by cartilage from the main parts of the bones
h Enlarged braincase
i No "third eye" in the skull
j No ribs in neck or lower back area
k Limbs hinged to move fore and aft below the body
l Distinctive shoulder girdle
m Distinctive hip bones

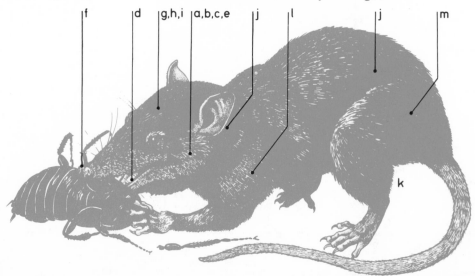

1 **Megazostrodon,** one of the first mammals, was a tiny shrew-like triconodont with a slim lower jaw. Length: 10cm (4in). Time: Late Triassic or Early Jurassic. Place: South Africa.

2 **Echidna,** a living spiny anteater, here stands for the monotremes of which the only early fossils known are 15-million-year-old teeth from Australia. Some bones are primitive and reptilian. Like reptiles, monotremes expel all body wastes and give birth through one hole. They lack ears and have poor temperature control.

3 **Taeniolabis,** a beaver-sized multituberculate, had a heavy skull, strong jaw muscles, and big, chisel-like teeth. It probably ate nuts. Time: Early Palaeocene. Place: North America.

4 Unnamed eupantothere. This agile, squirrel-like climber lived in Late Jurassic Portugal. Its skeleton was the first found (in 1977) for any mammal dating from that time.

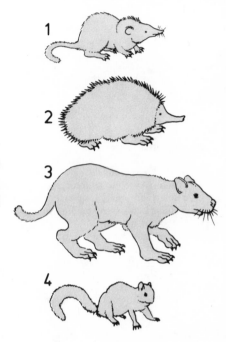

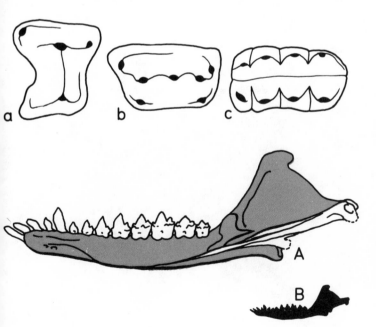

Teeth of early mammals
Almost microscopically tiny teeth are the only known remains of many early fossil mammals. Shown here are crown views of three distinctive types of upper cheek teeth (mostly much enlarged).
a Docodont tooth shaped like a dumb-bell
b Triconodont tooth, with three cusps in a row
c Multituberculate tooth, with many cusps

Distinctive jaw bones
Shown here much enlarged (**A**) and actual size (**B**) is the jaw of the early mammal *Morganucodon.* Like reptile jaws this still comprises several bones, but the dentary (shaded) is by far the largest. In later mammals only this bone persists.

153

Pouched mammals

Marsupials (pouched mammals) give birth to tiny undeveloped young. Many grow up in a pouch located on the mother's belly and supported by so-called marsupial bones. Such bones and the teeth help experts to identify fossil marsupials.

Marsupials probably evolved from eupantotheres in Mid Cretaceous times (100 million years ago). They reached all continents except Africa and Asia, but only in the island continent Australia did they escape competition from advanced placental mammals. Placental counterparts abound among the four marsupial orders. Marsupicarnivores, perhaps the parent stock, include extinct cat-like and wolf-like flesh-eaters. Paucituberculates are the insect-eating opossum rats and marsupial moles and mice. Peramelines comprise bandicoots – Australia's long-snouted rabbits. Diprotodonts, the main surviving group, include herbivorous kangaroos, koalas, wombats, and their extinct relatives – some of "giant" size, notably numbers 2–4 of the four extinct marsupials shown here (right).

Marsupial features
The American opossum shows primitive marsupial features not found in placental mammals.
a Relatively small braincase
b Holes in the bony palate (the roof of the mouth)
c Four molar teeth on each side of each jaw
d Shelf on the outer rim of the cheek teeth
e No full set of milk teeth
f Inturned lower jaw below the jaw hinge
g Marsupial (hip) bones
h Clawed second and third toes on the hind foot form a comb for grooming fur

1 **Thylacosmilus** was a leopard-sized South American marsupicarnivore equivalent of the placental sabre-toothed cat: with long, curved, stabbing upper canines sheathed in the lower jaw. One of the last in its family (the borhyaenids), *Thylacosmilus* dates from Pliocene times.

2 **Thylacoleo,** an almost lion-sized marsupial "lion" from Pliocene and Pleistocene Australia, had unique tusk-like incisors and huge premolar cheek teeth that sheared like scissors. It might have eaten flesh or fibrous fruits and tubers.

3 **Sthenurus** was a giant kangaroo with short jaws, short tail, and huge fourth toe. It stood about 3m (10ft) high and browsed on trees. Time: Pliocene and Pleistocene. Place: Australia.

4 **Diprotodon,** the largest known marsupial, was hippo-like, with big incisor tusks, but cheek teeth like kangaroos'. Length: 3.4m (11ft). Time: Pleistocene. Place: Australia.

Australia's "hippo"
Diprotodon munched plants on salt flats at Lake Callabonna in south-east Australia. Fossil finds show that individuals fell through a dried salt crust and drowned in mud beneath.

Placental pioneers

Placental mammals (the infraclass Eutheria) are so named from the pregnant females' placenta – the organ nourishing young in the womb. Placentals evolved from eupantotheres in Mid Cretaceous times. Advanced brains, and babies already well developed at birth gave placentals major advantages. In the last 65 million years they have produced 95 per cent of all known mammal genera, past and present. Here we look briefly at five early orders of placentals, two extinct.

The firstcomers were insectivores – small, nocturnal insect-eaters, some ancestral to those living insectivores shrews and hedgehogs. Closely related orders include bats (chiropterans); gliding mammals (dermopterans); the large, rat-like taeniodonts; and the bear-like tillodonts. The last two died out in Eocene times. We give one prehistoric example from each group.

Three different designs
Here we show basic differences in body structure between three groups of living mammals: monotremes (**A**), marsupials (**B**), and placentals (**C**). All release body waste from gut and bladder, and all produce egg cells in ovaries. But wastes and eggs or babies exit differently.

a Ovaries   e Anal canal
b Fallopian tubes   f Cloaca
c Uterus(es)   g Vagina
d Bladder   h Urethra

One exit (A)
In monotremes, such as the platypus, uteruses, bladder, and gut lead to a cloaca – a common exit by which eggs, liquid waste, and solid waste all leave the body. Hence the name monotreme, meaning "one opening".

Two exits (B)
In marsupials, such as the wallaby, uteruses and bladder lead to a cloaca but the gut's anal canal ends separately. Babies and liquid waste leave the body through the cloaca. Solid waste leaves from the gut's anal canal.

Three exits (C)
In placental mammals, such as the shrew, uterus, bladder, and gut each have a separate body exit. Babies leave via a vagina, liquid waste via a urethra, solid waste via the anal canal.

1 **Zalambdalestes** was an agile insectivore, with a long face, large eyes, small brain, long, sharp incisor teeth, "old-fashioned" molars, clawed feet, and a long tail. It belonged to the proteutherians: a suborder close to the ancestry of all main placental groups. Length: 20cm (8in). Time: Late Cretaceous. Place: Mongolia.

2 **Planetetherium,** a squirrel-sized dermopteran, glided between trees on skin webs joining legs and tail. Time: Palaeocene. Place: Montana, USA.

3 **Icaronycteris,** the first known bat, lived 50 million years ago in (Eocene) Wyoming, USA. Order: Chiroptera (bats).

4 **Stylinodon,** a taeniodont leaf-eater, resembled a huge rat with short, strong limbs, short toes, and powerful claws. Teeth were high-crowned, rootless pegs. Time: Mid-Eocene. Place: North America.

5 **Trogosus,** a tillodont, resembled a big bear, with flat, clawed feet but chisel-shaped incisors like a rodent's, and low-crowned molars. It was a herbivore. Time: Mid Eocene. Place: North America.

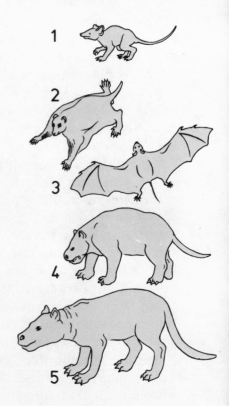

©DIAGRAM

The first fossil bat
Shown is a fossil of the first known bat, *Icaronycteris.* Fingers held up skin-membrane wings as in modern bats, and like them it slept upside down. Even such specialized placentals shared the following common features.
a Relatively big brain case
b Solid bony palate
c Lower jaws lacking inturned angle
d No marsupial hip bones
e Distinctive teeth

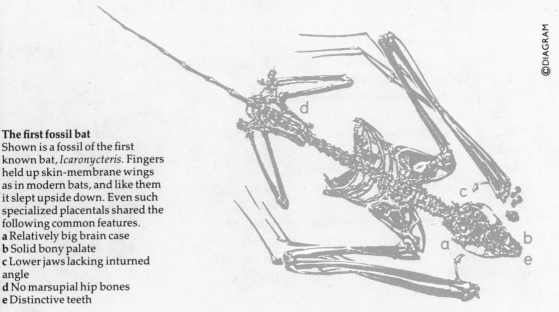

Primates 1

Primates – man, apes, monkeys, and their kin – include the most intelligent of all animals. Yet their direct ancestors were insectivores, the earliest, most primitive placental mammals. Primates primitively kept five toes and fingers, but gained grasping hands, forward-facing eyes, and a big "thinking" region of the brain. Eyes and hands co-ordinated by the brain adapted early primates for an agile life among the trees. Meanwhile changes in the teeth adapted them for eating fruits or most other foods.

Primates resembling today's tree shrews probably appeared by 65 million years ago, giving rise to several groups. First came the relatively small and small-brained prosimians, including ancestors of today's lemurs and tarsiers. These were plentiful 60 to 40 million years ago in North America, Eurasia, and Africa. Prosimians led to the brainer and often bigger anthropoids (monkeys, apes, and men) that largely took their place. Here are examples from the primates' five suborders.

1 **Plesiadapis**, a squirrel-sized, squirrel-like prosimian, had a long snout, side-facing eyes, big chisel-like teeth, long bushy tail, and claws not nails. It could not grasp with its hands. It ate leaves, largely on the ground. Time: Mid Palaeocene. Place: North America and Europe. Suborder: Plesiadapoidea (extinct by Mid Eocene times).

2 **Adapis** had a shorter snout than *Plesiadapis*, forward-facing eyes, relatively bigger brain, and grasping hands and feet with nails not claws. It climbed forest trees, eating shoots, fruits, eggs, and insects. Length: 40cm (16in). Time: Mid–Late Eocene. Place: Europe. Suborder: Lemuroidea (lemurs), probably ancestral to apes.

3 **Necrolemur** was small with huge eyes, big ears, a small "pinched" nose, long tail, and long tree-climber's limbs with gripping toe pads. It was an Eocene European ancestor of modern tarsiers. Suborder: Tarsioidea.

4 **Dolichocebus** was small and squirrel-like with a bushy tail, and claws not nails. It lived in Oligocene South America. Suborder: Platyrrhini ("flat noses") – New World monkeys. These have widely spaced, outward facing nostrils and long tails, and lack good thumb-and-finger grip.

5 **Mesopithecus** was probably ancestral to the slim, long-legged, long-tailed monkeys known as langurs. It lived in Late Miocene Greece and Asia Minor. Suborder: Catarrhini ("down-facing noses") – Old World monkeys, apes, and men. These have nostrils close together, good thumb-and-finger grip, and relatively big brains; tails become short in many later monkeys and are absent in apes and man.

Primate features
These clues help experts recognize a primate fossil.
a Large brain case
b Jaws often short
c Low-crowned cheek teeth with low, blunt cusps (designed to cope with almost any food)
d Usually no third molar or first premolar tooth
e Orbits (eye sockets) large and facing forward
f Bony bar behind the orbits
g Small nasal opening
h Long, flexible limbs
i Five toes and fingers, usually with nails not claws
j Gap between thumb and big toe and other digits – an aid in grasping

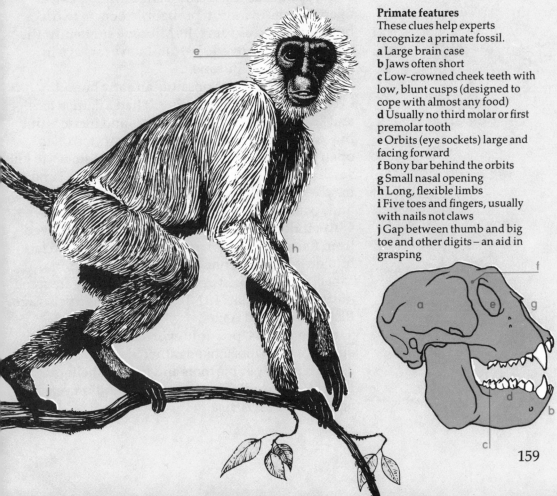

159

Primates 2

Hominoids, the catarrhine superfamily containing apes and men, arose from prosimian ancestors some 35 million years ago, probably in Africa. Bigger and brainier than monkeys, apes spread through Europe and Asia 25–10 million years ago.

Four hominoid families evolved: oreopithecids, a dead end; pongids (great apes); hylobatids (gibbons); and hominids (men and ape men). Our examples stress those that led to modern man.

1 **Aegyptopithecus,** from Oligocene Egypt (26–28 million years ago) was a small ape with a short tail. It had low brows and jutting snout but bony eye sockets not just bars like early primates. It might have led to modern apes.

2 **Dryopithecus** was a chimpanzee-like ape with relatively short arms. It probably stood on two legs but climbed on all fours. Its Miocene subfamily, the dryopithecines, lived 25–10 million years ago in Africa, Asia, and Europe.

3 **Ramapithecus** was possibly an early hominid evolved from a dryopithecine. It had a flattish face, and "human" teeth: small canines and incisors but big, crushing, grinding molars. It ventured on to open grassland, maybe using sticks and stones to kill small prey, or in defence. Height: 1.2m (4ft). Time: maybe 15–7 million years ago. Place: Africa, Europe, and Asia.

4 **Australopithecus,** an ape man, possibly evolved from *Ramapithecus*. It had an ape-like skull and face, human-type teeth, one-third modern man's brain capacity, but walked upright and probably ate some meat. Height: 1.2m (4ft). Time: 5–1 million years ago. Place: Africa and Asia.

5 **Homo habilis,** possibly evolved from a slender australopithecine, looked rather like a small-brained child. It made pebble tools and maybe shelters. Height: 1.2m (4ft). Time: about 4–2 million years ago. Place: East Africa.

6 **Homo erectus** had thick eyebrow ridges, heavy teeth, and receding chin and forehead, but a bigger brain and larger body than *Homo habilis*, its likely ancestor. It pioneered the use of fire, standard (hand axe) tools, and big-game hunting. Height: 1.7m (5ft 6in). Time: 1,750,000–200,000 years ago. Place: Africa, Asia, and Europe.

7 **Homo sapiens,** our own species, evolved about 200,000 years ago, probably from *Homo erectus*. Key innovations included larger brain, higher, more rounded forehead, and increased body size. We show the Neanderthal subspecies of 100,000–35,000 years ago. Our own subspecies (*Homo sapiens sapiens*) appeared perhaps only 40,000 years ago.

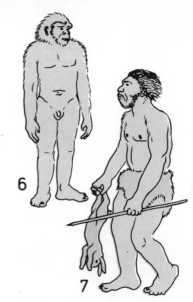

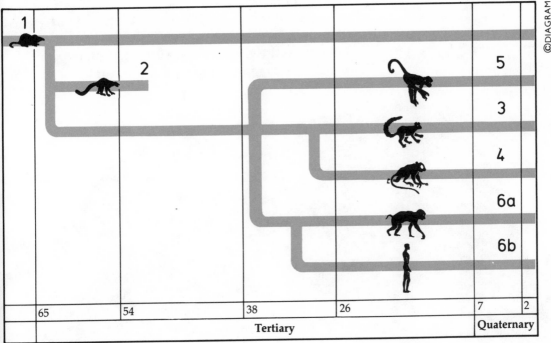

				5
				3
				4
				6a
				6b
65	54	38	26	7	2
		Tertiary			Quaternary

©DIAGRAM

Primates' family tree
This shows: insectivores ancestral to the primates (**1**); the primates' five suborders (**2–6**); and the catarrhines' superfamilies including apes and men (**a–b**).

1 Insectivores
2 Plesiadapoidea (early primates)
3 Lemuroidea (lemurs and lorises)
4 Tarsioidea (tarsiers)
5 Platyrrhini (New World monkeys
6 Catarrhini
a Cercopithecoidea (Old World monkeys)
b Hominoidea (apes and men)

161

Creodonts

Creodonts were the first successful flesh-eating placental mammals. Many walked "flat footed" on short, heavy limbs tipped with claws. The tail was long, the brain was small, and the teeth were less efficient than a cat's for stabbing flesh or shearing through it. Creodonts ranged from weasel size to beasts bigger than a bear. Small species might have eaten insects. Larger kinds probably included big-game hunters, carrion eaters, and omnivores. Creodonts roamed northern continents about 65–5 million years ago – as "top dogs" in the Eocene (54–38 million years ago). Then there were still hoofed animals slow enough for them to catch, and they lacked competition from brainier, more lethal killers than themselves.

There were two creodont suborders: the early deltatheridians and the hyaenodonts. The hyaenodonts were the main group, with two families: oxyaenids and hyaenodontids.

A flesh-eating monster
Megistotherium might have been the largest-ever flesh-eating mammal. This huge creodont weighed about 900kg (1980lb). The head was twice as large as any bear's and was armed with mighty canine teeth. *Megistotherium* quite likely killed elephant-like mastodonts. Miocene rocks in Libya contain this monster's fossils.

1 **Deltatheridium,** a deltatheridian, had creodont-like teeth (but just might have been a marsupial). Length: maybe 15cm (6in). Time: Late Cretaceous. Place: Mongolia.

2 **Patriofelis** was a bear-size oxyaenid. Oxyaenids had a short, broad head, deep strong jaws, and cheek teeth that crushed rather than sheared. Time: Mid Eocene. Place: North America.

3 **Hyaenodon** included wolf-sized species that tackled big, hoofed animals. It belonged to the hyaenodontids – a far bigger, longer-lasting group than oxyaenids, and usually with a longer head, more slender jaws, a slimmer body, longer legs, and a tendency to walk on the toes. Teeth sheared rather than crushed. *Hyaenodon* lived in northern continents from Late Eocene to Miocene times.

1

2

3

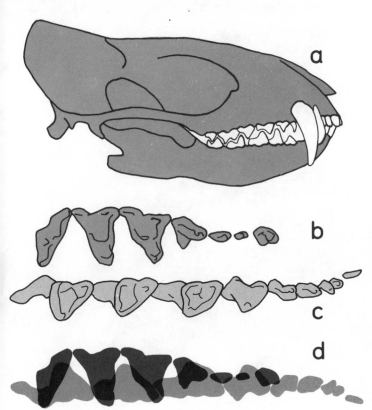

a

b

c

d

Teeth to shear and chop
Early mammals had cheek teeth with triangular crowns. Those in early creodonts worked to shear and chop, as shown here for *Deltatheridium*.
a Skull with teeth, shown enlarged.
b Crown view of upper cheek teeth, shown much enlarged.
c Crown view of lower cheek teeth, shown much enlarged.
d Upper and lower cheek teeth here shown meshed together. The front edges of the lower teeth slid along the back edges of the upper teeth ahead of them. The low "heel" behind the main part of each lower tooth fitted into the surface of the upper tooth behind it, stopping further sliding. Partial meeting of cusps ("peaks") on the tooth crowns helped chop up food.

Modern carnivores 1

Modern carnivores form one order, Carnivora, with two suborders: fissipeds or "split feet" (dogs, cats, and their kin) and pinnipeds (the "fin footed" eared seals, earless seals, and walruses).

Fissipeds had bigger brains, keener ears, more deadly piercing canines and shearing cheek teeth, and longer limbs than creodonts. Cunning, agile fissipeds replaced creodonts as the major carnivores, for only fissipeds could catch and kill new kinds of speedy herbivores that replaced slower types.

Fissipeds evolved from insectivores by 60 million years ago. By 35 million years ago they dominated life on most continents. Three superfamilies arose: first miacids, then their descendants aeluroids or feloids (civets, hyaenas, and cats) and arctoids or canoids (dogs, bears, and their kin). Beasts from the first two groups are pictured on this spread.

Victor and victim
Big stabbing cats like *Smilodon* probably preyed on big, slow-moving herbivores like the large, prehistoric ground sloth *Mylodon*. Remains of both occur in Los Angeles' famous La Brea tar pits. Tar trapped big herbivores that drank rainwater concealing tar beneath. The creatures' struggles attracted predators.

1 **Miacis** was a small, weasel-like, tree-climbing miacid. Length: 60cm (2ft). Time: Eocene. Place: North America, Europe, Asia.

2 **Genetta,** the genet, resembles Eocene members of its Old World civet family (viverridae). It is a cat-like forest dweller with short limbs, long tail, and retractile claws. Length: 96cm (3ft 2in). Time: Pleistocene on. Place: Africa, Eurasia.

3 **Percrocuta** had a big head, strong jaws, bone-cracking teeth, and longish legs. Species of this early hyaena were wolf size to lion size. Time: Miocene. Place: Africa and Asia.

4 **Dinictis** was a puma-sized ancestor of biting and stabbing cats (all in the felid family). Time: Early Oligocene–Early Miocene. Place: North America.

5 **Felis leo spelaea,** the great "cave lion", was a biting cat one-third larger than the largest lion. It lived in Mid Pleistocene Europe.

6 **Smilodon,** the "sabre-tooth tiger", was a heavy, lion-sized stabbing cat. Its dagger-like upper canines stabbed large prey or slashed blood vessels in their necks. Time: Late Pliocene–Pleistocene. Place: North and South America.

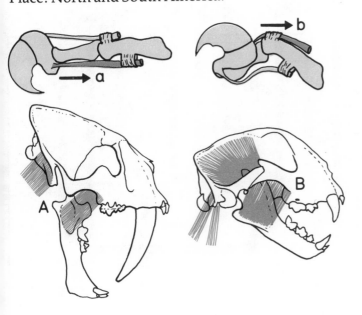

Cats' claws
Ability to unsheathe claws and spread toes provides most cats with formidable weapons.
a Unsheathing: pulling a toe's lower tendon makes a claw curve down and out.
b Sheathing: pulling a toe's upper tendon makes a claw curve up and back.

Stabbing and biting
Stabbing cats and biting cats evolved different techniques for using teeth as weapons.
A Stabbing cats like *Smilodon* used strong neck muscles to strike downward with the head.
B Biting cats like the cheetah have strong jaw muscles and exert a formidable bite.

©DIAGRAM

165

Modern carnivores 2

Arctoids (or canoids) evolved from miacids over 40 million years ago. These carnivores mostly share a skull peculiarity involving the middle ear. They have less specialized teeth than cats and cannot sheathe their claws. Some eat almost any food. Four families evolved. Canids include dogs, wolves, and extinct bear-dogs. The primitive mustelids include weasels, otters, and badgers. Ursids are bears and the giant panda. Procyonids are raccoons, coatis, kinkajous, and lesser pandas.

By 25 million years ago arctoids gave rise to pinnipeds – seals and walruses. This suborder of aquatic carnivores developed sleek, streamlined bodies, webbed limbs, and teeth designed for catching fish or (in walruses) for crushing clams. But unlike whales or ichthyosaurs, seals kept a flexible neck and the tail is reduced to a stump.

1 **Cynodictis,** a fox-sized early dog, had a long, low body, long neck, long tail, sharp shearing cheek teeth, and longer legs and larger brain than miacids. Time: Late Eocene–Early Oligocene. Place: Europe and East Asia.

Carnivore family tree
This shows likely links between families of the order Carnivora.
1 Miacid fissipeds
a Miacidae
2 Aeluroid fissipeds
b Ursidae (bears)
c Canidae (dogs)
d Procyonidae (raccoons)
e Mustelidae (weasels)
3 Pinnipeds
f Odobenidae (walruses)
g Otariidae (eared seals)
h Phocidae (earless seals)
4 Aeluroid fissipeds
i Hyaenidae (hyaenas)
j Viverridae (civets)
k Felidae (cats)

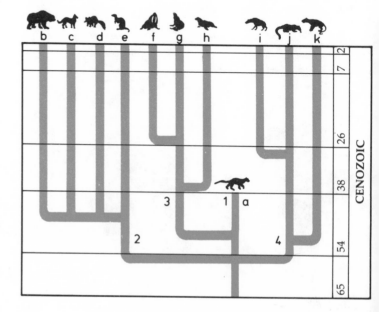

2 **Daphoenodon,** the largest canid of its day, was a wolf-like "bear-dog" from Early Miocene North America. It had rather short limbs, spreading toes, a long, strong tail, and was perhaps omnivorous.

3 **Osteoborus,** a hyaena-like canid scavenger, had a bulging forehead and strong jaws. It lived in Miocene North America.

4 **Megalictis** from Early Miocene North America resembled a wolverine but was as big as a black bear. It was the largest-ever mustelid.

5 **Ursus spelaeus** the great "cave bear" of Pleistocene Europe, was descended from the dog family. Bears walk flat footed, not on toes like dogs, and most are omnivores. Length: 1.6m (5ft 3in).

6 **Phlaocyon** was a small, raccoon-like procyonid (some say canid) from Miocene North America. It climbed trees, had grinding but not shearing cheek teeth, and was omnivorous.

7 **Allodesmus,** an early relative of sea-lions, probably resembled a sea-elephant, the largest earless seal alive today. Time: Early Miocene–Mid Miocene. Place: North Pacific Ocean.

Formidable jaws
Here we compare skulls of a cave bear (**A**) and a fox (**B**), drawn to scale. Cave bears were omnivores, despite their massive jaws.

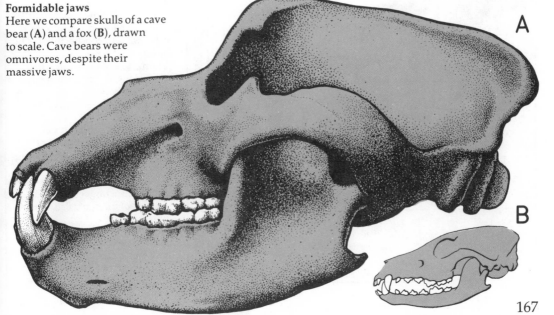

Condylarths

These were the first ungulates (hoofed mammals), abundant 65 to 40 million years ago. Early kinds had claws, not hooves or nails, and some ate flesh, not plants. Later kinds developed chopping, grinding teeth for pulping leaves, and long limbs tipped with nails or hooves, for running fast away from carnivores. The pioneers were rabbit-sized; some later forms were longer than a large bear.

Condylarths evolved from small insectivores, and probably gave rise to all the more advanced hoofed mammals (maybe also to the strange, long-snouted aardvark). Condylarths spread through northern continents and into South America and Africa. Most fossils come from Palaeocene rocks in North America, and from Eocene rocks there and in South America, Europe, and Asia. We give examples from contrasting families in the order Condylarthra.

Early ungulates
This family tree shows likely relationships between both orders of primitive hoofed mammals described on pages 168–171. The modern aardvark may be descended from the condylarths.
1 Condylarthra (condylarths)
2 Amblypoda (amblypods)
a Pantodonta (pantodonts)
b Dinocerata (uintatheres)
3 Tubulidentata (aardvarks)

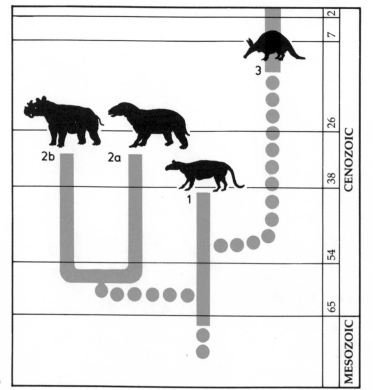

Early ungulate features (right)
The skull and feet bones of *Phenacodus* show some features found in early carnivores, but others stressed in herbivores like sheep and horses.
a Long, low skull (shown half actual length), resembling that of early carnivores
b Large canine teeth, as in early carnivores
c Ridged, square-crowned molars forming a battery of cheek teeth capable of crushing vegetable food
d Bones of hind foot, tipped with small, blunt hooves not nails
e Bones of forefoot, also tipped with hooves

© DIAGRAM

1 **Protungulatum,** the first-known ungulate, was rabbit-sized. It ate plants and maybe other foods. Time: Late Cretaceous–Early Palaeocene. Place: North America. Its arctyonid family featured early, mostly small condylarths with a supple back, short limbs tipped with claws, a long tail, long, low head, and "old-fashioned" molars with triangular crowns.

2 **Andrewsarchus** was a huge, heavy, bear-like condylarth, probably omnivorous. Length: 4m (13ft). Time: Late Eocene. Place: East Asia. Family: Mesonychidae, largely dog-like condylarths. Many had blunt-cusped cheek teeth capable of crushing bones, and ran dog-like on toes with flattened nails not claws. Packs maybe hunted big herbivores.

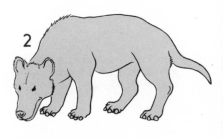

3 **Phenacodus,** a phenacodontid condylarth, was a sheep-sized plant-eater, with big canine teeth and square-crowned molars. It was the earliest known mammal with hooves not nails or claws. Fossil skeletons show that it roamed woods and shrublands in North America and Europe. Time: Late Palaeocene–Early Eocene.

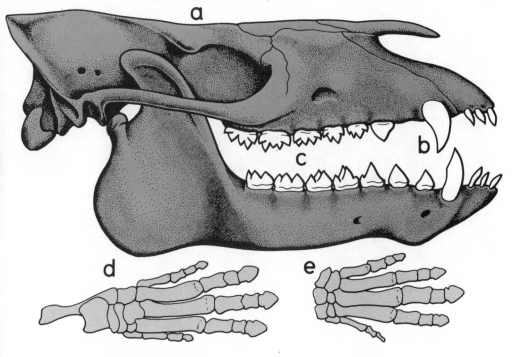

Amblypods

Amblypods ("slow footed") were early, ponderous hoofed herbivores with broad, low, ridged cheek teeth. Some had tusks or horns. Many might have lived in swamps. Size, weapons, or habitat helped save them from attack by creodonts.

Amblypods flourished 60 to 30 million years ago, mostly in Palaeocene and Eocene North America, but in Europe and East Asia too.

There were two suborders. Pantodonts ranged from dog size to pony size. They had short legs and short, broad feet. Canine teeth were long and cheek teeth had a simple pattern.

Uintatheres included rhino-sized mammals – among the largest of their age. Many had strange bony horns and long, wicked-looking upper canines. Upper molars bore a broad, v-shaped crest.

Here we show three pantodonts from different families, and the best known uintathere.

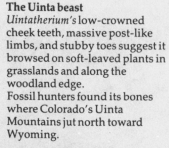

The Uinta beast
Uintatherium's low-crowned cheek teeth, massive post-like limbs, and stubby toes suggest it browsed on soft-leaved plants in grasslands and along the woodland edge.
Fossil hunters found its bones where Colorado's Uinta Mountains jut north toward Wyoming.

1 **Pantolambda,** was one of the first hoofed mammals as big as a sheep. It had heavy legs, short feet, and a long, low head with large canine teeth. It might have wallowed, hippo-like, and browsed on land. Time: Mid Palaeocene. Place: North America and Asia.

2 **Barylambda** was pony-sized, with heavy body, small head, and primitive teeth. It might have sat on its haunches to browse high up. It lived in Late Palaeocene–Early Eocene North America.

3 **Coryphodon** had a long, heavy body, large head, wide muzzle, and knife-like upper canines. Length: 3m (9ft 10in). Some think it lived largely in water. Time: Late Palaeocene–Early Eocene. Place: northern continents.

4 **Uintatherium** was a massive uintathere the size of a large rhinoceros. Thick limbs with spreading feet held up its heavy body. Three pairs of bony knobs sprouted from the head, and males had wickedly long, strong, upper canines. Time: Eocene. Place: North America.

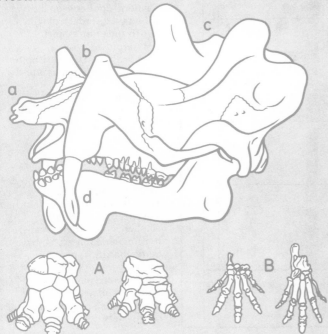

A "six-horned" head
Uintatherium's great grotesque skull 76cm (30in) long shows weapons used by jousting males or in defence.
a Two small knobs on the nose
b Two larger knobs between the nose and eyes
c Two big broad knobs above and behind the eyes
d Huge upper canines, protected by a flange in the lower jaw

Two kinds of feet
Big heavy herbivores tended to develop much more massive foot and toe bones than their smaller, lighter relatives.
A The short, thick, strong bones that bore the weight of mighty *Uintatherium*
B The relatively slim, thin bones of *Pantolambda's* fore and hind feet

171

Subungulates

These "not quite ungulates" comprise seemingly quite unrelated hyraxes, sirenians (sea cows), proboscideans (elephants and their kin), and the extinct embrithopods and desmostylans.

Close study of their bones and teeth in fact reveal relationships. Subungulates tend to lack a collar bone but keep five-toed feet, with nails rather than hooves. Many have few front teeth, though a pair often become tusk-like, also enlarged premolars and crosswise ridges on the grinding cheek teeth. Their yet unknown ancestors might have lived in swamps in Africa some 60 million years ago. We show examples from four of the five orders of subungulates.

1 **Saghatherium** was an agile, rodent-like hyrax with chisel-like incisors but toes with hoof-like nails. Time: Early Oligocene. Place: Egypt. Length: 40cm (16in). Living hyraxes are also small but one prehistoric kind was rhinoceros size.

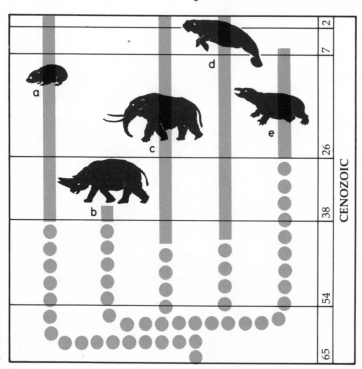

Subungulate family tree (left)
This family tree shows likely relationships between the five subungulate orders, of which two (**b** and **e**) are extinct.
a Hyracoidea (hyraxes)
b Embrithopoda (embrithopods)
c Proboscidea (elephants and their ancestors)
d Sirenia (sea cows)
e Desmostylia (desmostylans)

Mystery monster (right)
Twin-horned *Arsinoitherium* had a big head, and limbs like elephants'. The cross-crested, high-crowned cheek teeth were better than those of its contemporaries for crushing tough-leaved plants. Lack of known fossil relatives makes this creature's origin something of a mystery.

CENOZOIC

©DIAGRAM

172

2 **Arsinoitherium,** from Early Oligocene Egypt, is the only known embrithopod. This rhinoceros-sized mammal had a huge pair of horns side by side on its head. It probably munched coarse-leaved plants in swamps. Length: 3.4m (11ft).

3 **Desmostylus,** a heavy, walrus-like desmostylan, swam or waded on North Pacific coasts from Early Miocene to maybe Early Pliocene times. Short tusks grew forward from both jaws. Rear teeth were close-packed, heavily enamelled bundles, perhaps used for crunching shellfish or munching seaweed.

4 **Protosiren** was an early sea cow about 2.4m (8ft) long, from Middle Eocene North Africa and Europe. Sea cows developed broad, beak-like snouts, flipper-like front limbs, and a broad, flat tail; they lost hind limbs. Living species browse on aquatic plants in tropical river mouths.

2

3

4

Proboscideans 1

Elephants, their ancestors, and other kin make up the subungulate order Proboscidea ("long snouted"). The first pig-sized, trunkless proboscideans lived in Africa 40 million years ago. From Africa, their descendants invaded all continents except Australia and Antarctica. Meanwhile they grew in size. Many were immense slab-sided beasts with "tree-trunk" limbs, and a huge head armed with tusks and brandishing a flexible muscular trunk – a "hand" bringing leaves and water to the mouth. In four pages we show something of the astonishing variety of prehistoric proboscideans.

Our examples come from three of their four suborders: moeritheres, deinotheres, and euelephantoids (mastodonts and elephants). The barytheres from Eocene Egypt are little known.

1 **Moeritherium,** a pig-sized, heavy footed moerithere, had a short, flexible snout and forward-jutting incisor teeth. It lived in swampy lands. Time: Late Eocene–Early Oligocene. Place: northern Africa (Egypt, Mali, Senegal).

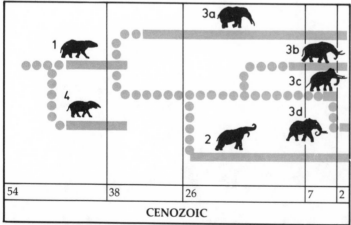

54	38	26	7	2
		CENOZOIC		

Spade blades (below)
Platybelodon's lower jaw ended in two flat, broad tusks (here shaded) like spade blades.

Proboscidean family tree
1 Moeritherioidea (moeritheres)
2 Deinotherioidea (deinotheres)
3 Euelephantoidea (mastodonts and elephants)
a Gomphotherioidea (long-jawed mastodonts)
b Mastodontoidea (crested toothed mastodonts)
c Stegodontoidea (stegodonts)
d Elephantoidea (mammoths and elephants)
4 Barytherioidea (barytheres)

2 **Deinotherium** belonged to the deinotheres: beasts up to 4m (13ft) high, with down-curved tusks in the lower jaw – forks for digging roots perhaps. Time: Miocene–Pleistocene. Place: Africa and Asia.

3 **Phiomia,** an early mastodont ("nipple toothed" proboscidean), had pig-like cheek teeth, long lower jaw, four short tusks, and short trunk. Height: 2.4m (8ft). Time: Early Oligocene. Place: Egypt.

4 **Platybelodon,** a mastodont "shovel tusker", dug up plants with broad blade-like teeth jutting from its long lower jaw. Time: Miocene. Place: Asia and North America.

5 **Mastodon** had long, curved upper tusks, strong crests across its cheek teeth, and a coat of reddish hair. It browsed in forests of Pleistocene North America, dying off 8000 years ago.

6 **Stegodon** was a long-tusked possible ancestor of mammoths (see next page). Time: Pliocene–Pleistocene. Place: Asia and Africa.

A living bulldozer
Platybelodon might have used its tusks to scoop up land- or water-plants as shown here.

175

Proboscideans 2

Mammoths ("giants") included all extinct members of the Elephantinae, the elephant subfamily. They evolved from mastodons (see page 175) but were mostly taller, with higher skulls, shorter jaws, and more complex molar teeth, designed for grazing. Some were the largest-ever elephants; some, dwarfs. Cold-adapted types were hairy.

Mammoths appeared in Africa five million years ago, colonized the Northern Hemisphere, and died out about 10,000 years ago, after giving rise to modern elephants. There were several genera and many species. Our examples give an idea of the range of types and sizes.

1 **Elephas falconeri** from the island of Sicily was one of several dwarfed island elephants – an agile beast only one-quarter the size of its mainland ancestors.

Mammoth skull and teeth
A woolly mammoth skull appears here to scale with a human skull.
a Huge, curved tusks served in defence and maybe helped clear snow from pasture.
b As it grew, each huge molar pushed out its predecessor(s).
c A crown view shows the many crosswise ridges that helped this molar grind like a mill.

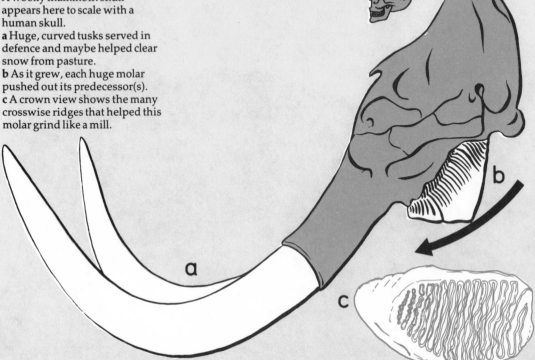

© DIAGRAM

176

2 **Elephas trogontherii,** one of the earliest mammoths, was the largest-ever elephant, about 4.3m (14ft) at the shoulder.

3 **Mammuthus imperator,** the second-largest elephant, stood over 4m (13ft) high and its tusks measured up to 4.3m (14ft). It flourished in North America in Pleistocene times.

4 **Mammuthus primigenius,** known as the woolly mammoth, had long, dark, brown hair and thick, woolly underfur. The vast tusks curved forward, up, and back. Height: 2.9m (9ft 6in). This mammoth ranged north of the Arctic Circle.

5 **Palaeoloxodon** was a massive straight-tusked forest dweller more than 4m (13ft) high. It lived in Europe, including England.

Giant and dwarf
This illustration compares a woolly mammoth with a dwarf island species. The woolly mammoth's great bulk and shaggy coat helped to conserve body heat, enabling this elephant to survive in cold northern climates. The dwarf elephant's head and feet differ from those of larger elephants in ways that reflect an overall design for more agility.

South American ungulates

A whole "zoo" of hoofed mammals evolved in South America while sea cut it off from other lands. Many amazingly resembled rodents, horses, and other beasts that evolved elsewhere. There were six orders. Notoungulates, litopterns, astrapotheres, and trigonostylopoids probably evolved from early condylarths. Xenungulates and pyrotheres might have come from amblypods. Some endured for many million years. But none survived long once land linked North and South America about two million years ago, and carnivores and ungulates invaded from the north.

1 **Notostylops** was a small primitive hoofed mammal from Early Eocene Patagonia. It belonged to the Notoprogonia suborder of the notoungulates, the largest order of South American ungulates.

2 **Toxodon** was a rhinoceros-sized toxodont ("bow-toothed") – one of the largest and last notoungulates. It had short legs, broad, three-toed feet, big cropping incisor teeth and tall cheek teeth. The molars curved in toward each other. Time: Pliocene–Pleistocene.

3 **Protypotherium** resembled a large rodent. It was a typothere notoungulate with claws not hooves. Length: 51cm (20in). Time: Miocene.

4 **Pachyrukhos** had long hind limbs and a stumpy tail. Probably it ran and even leaped like a hare. This Miocene creature belonged to the hegetotheres, the smallest notoungulates.

5 **Thoatherium** looked astonishingly like a small horse. It belonged to the proterotheriids, a family of litopterns. Time: Early Miocene.

6 **Macrauchenia** – camel-sized, and camel-like – might have had a trunk. It was a macraucheniid litoptern. Time: Pleistocene.

7 **Astrapotherium** belonged to the big astrapotheres. It had a "sawn-off" face, dagger-like canine teeth, and possibly a trunk. Hind limbs were weak. Perhaps it lived in water. Length 2.7m (9ft). Time: Oligocene–Miocene.

8 **Trigonostylops** was a small beast once grouped with astrapotheres, but now put in its own order: Trigonostylopoidea. Time: Palaeocene–Eocene.

9 **Carodnia**, the only known xenungulate, might have resembled a uintathere. Time: Late Palaeocene.

10 **Pyrotherium**, a pyrothere, amazingly resembled a large early elephant in its trunk, chisel-like tusks, and type of teeth. Time: Oligocene.

Life by a pool
This restoration shows a likely scene in South America about three million years ago. A *Macrauchenia* (**a**) drinks from a pool inhabited by two *Toxodon* (**b,c**). Alignment of the nose, eyes, and ears suggests that *Toxodon* could swim all but submerged. Perhaps this bulky mammal lived like a hippopotamus.

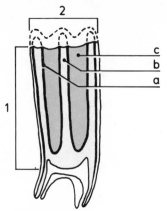

Teeth for grazing
The cheek tooth of a modern horse is designed to chew hard, abrasive grass.
1 High crown
2 Uneven grinding surface produced by wear on enamel (**a**), dentine (**b**), cement (**c**).

Evolving hooves
These illustrations show the hooves of *Hyracotherium* (**A**), *Miohippus* (**B**), *Merychippus* (**C**), and *Equus* (**D**). Notice that side toes shrink and disappear, until each foot has only an enlarged middle toe ending in a big, broad hoof. Changes in foot design – like changes in skull design – adapted horses for life as grazers on open plains.

Horses

Horses, rhinoceroses, tapirs, and others make up the perissodactyls, or "odd-toed" ungulates – one of the two great surviving orders of hoofed mammals. Evolving over 55 million years ago, perissodactyls became the most abundant ungulates, but then declined.

Few creatures left a richer fossil record than the horses – a superfamily in the hippomorph suborder. Dog-sized forest browsers living more than 50 million years ago gave rise to big, speedy grazers, with teeth for chewing tough-leaved grasses that began replacing forests. From their home in North America, horses reached Eurasia, Africa, and South America. Only members of one genus still survive.
1 **Hyracotherium,** the first known horse, was fox-sized. It had a short neck, curved back, long tail, slim limbs, and long four-toed forefeet and three-toed hind feet. Its low-crowned cheek teeth munched soft leaves in swampy North American and European forests. Time: Late Palaeocene–Early Eocene.
2 **Mesohippus** was bigger than *Hyracotherium*, with a straighter back, longer legs, three-toed forefeet, and enlarged premolar teeth. Height: 60cm (2ft) Time: Early–Mid Oligocene.

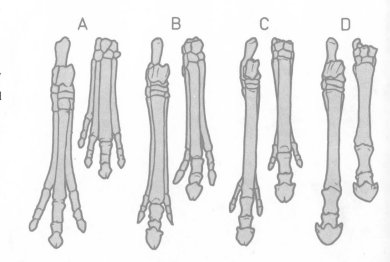

3 **Miohippus** was a little larger still, with big middle toe, big, molar-type premolars, and ridged cheek teeth. Time: Mid Oligocene–Early Miocene.

4 **Merychippus** was pony-sized, walked on each middle toe, and had a long neck and tall ridged cheek teeth for chewing grasses. Time: Miocene.

5 **Pliohippus**, 1.2m (4ft) tall, was the first one-toed horse. Time: Pliocene.

6 **Equus,** the modern horse, 1.5m (5ft) tall, evolved two million years ago, and reached most continents. Surviving wild species are Przewalski's horse, the wild ass, and zebras.

7 **Palaeotherium** belonged to the palaeotheriids, a family evolved from early horses. It had three hooves per foot and seemed half horse, half tapir. Height: 75cm (30in). Time: Late Eocene–Early Oligocene. Place: Europe.

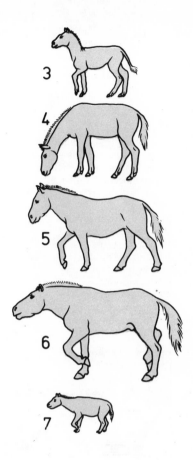

Horse trends

These two early horses show important evolutionary trends.
a *Orohippus*, a middle Eocene descendant of *Hyracotherium*, was still only whippet-sized, but its teeth could cope with tougher leaves, and it lived in drier woodlands.
b *Merychippus*, a larger horse, could eat the hard, abrasive grasses of Miocene North America. A big gap separated its cropping front teeth from grinding cheek teeth, and cement filled "pools" between the enamelled ridges of each cheek tooth's crown. When *Merychippus* ran fast, its side toes did not reach the ground.

©DIAGRAM

Brontotheres and chalicotheres

These were the strangest of all odd-toed ungulates. Brontotheres included massive beasts with elephantine limbs and a blunt, bony prong jutting from the nose. They ate only soft-leaved plants. The brontothere (titanothere) superfamily was in the same suborder as horses and, like them, came from condylarths. The group flourished in all northern continents and lived about 50–30 million years ago.

Chalicotheres ranged from sheep size to cart-horse size. They looked a bit like horses, yet their three-toed feet had big, curved claws that could be sheathed. Some think they walked on knuckles, using claws to dig up edible roots or pull down leafy branches. The suborder was never plentiful but persisted more than 50 million years until less than two million years ago. Fossils come from Europe, Asia, Africa, and North America.

Perissodactyls
This family tree shows links between suborders (**b–d**) and superfamilies (**i–iv**) in the order Perissodactyla ("odd-toed" ungulates) described on pages 180–185.
a Condylarth ancestors
b Hippomorpha
i Equoidea (horses)
ii Brontotherioidea (titanotheres or brontotheres)
c Ancylopoda (chalicotheres)
d Ceratomorpha
iii Tapiroidea (tapirs)
iv Rhinocerotoidea (rhinoceroses)

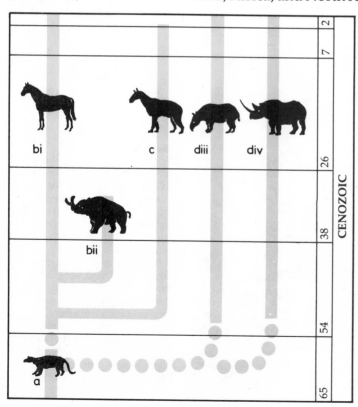

©DIAGRAM

1 **Brontotherium** was between a rhinoceros and elephant in size. Thick legs and short, broad feet with hoofed toes (four per forefoot, three per hindfoot) supported its massive body. From its blunt snout jutted a thick Y-shaped horn, perhaps brandished at attacking creodonts. (Males might have used horns in jousting contests.) It lived in Oligocene North America, probably on open plains, crushing soft leaves between the big, square, low-crowned but enamel-hardened molars. Shoulder height: 2.5m (8ft).

2 **Moropus** resembled a big horse, but like all chalicotheres, had claws, low-crowned teeth unsuited for eating grass, and longer front limbs than hind limbs so that its back sloped down from shoulders to hips. Length: 3m (10ft). Time: Early–Mid Miocene. Place: North America.

Clash of titans (below)
As shown here, rival male titanotheres (brontotheres) might have locked horns like stags or banged heads like bighorn rams. Winning stags and rams rule herds of females; perhaps successful male titanotheres had harems, too.

Tapirs and rhinoceroses

Tapirs and rhinoceroses appeared some 50 million years ago. Early kinds were small, agile beasts like early horses. But each of the two superfamilies had distinctive teeth for browsing. Rhinoceroses produced three families with scores of genera – some huge, many sprouting nasal horns. Tapirs remained small and primitive, with a heavy body, rounded back, short, stubby legs, and a flexible trunk-like snout for grasping forest plants. Several families evolved.

Rhinoceroses and tapirs reached all northern lands and Africa, and tapirs entered South America. But now rhinos only live in Africa and Asia; tapirs in Asia and South America.

1 **Hyracodon** was a small agile member of the hyracodontids, early "running" rhinos with long, slim legs, long three-toed feet, and a browser's molars. Length: 1.5m (5ft). Time: Oligocene. Place: North America.

2 **Metamynodon** was a hippo-like amynodontid rhino with short, thick limbs and short, broad feet. Time: Oligocene. Place: North America.

3 **Paraceratherium (Baluchitherium),** the largest-ever land mammal, was a rhinocerotid rhino. Long legs and longish neck helped its small hornless head browse giraffe-like among high branches. Shoulder height: 5.5m (18ft). Time: Oligocene–Miocene. Place: Asia.

4 **Elasmotherium,** an elephant-sized rhinocerotid, had a huge horn measuring almost 2m (6ft 6in). Time: Pleistocene. Place: Eurasia.

5 **Coelodonta,** the (rhinocerotid) woolly rhinoceros, survived Ice Age cold helped by its shaggy coat. Its head bore two horns – fore and aft. Time: Pleistocene. Place: Eurasia.

6 **Helaletes** was a small, early, agile helaletid tapir from Eocene North America and East Asia.

7 **Miotapirus** was a tapirid ancestor of modern tapirs, from Early Miocene North America.

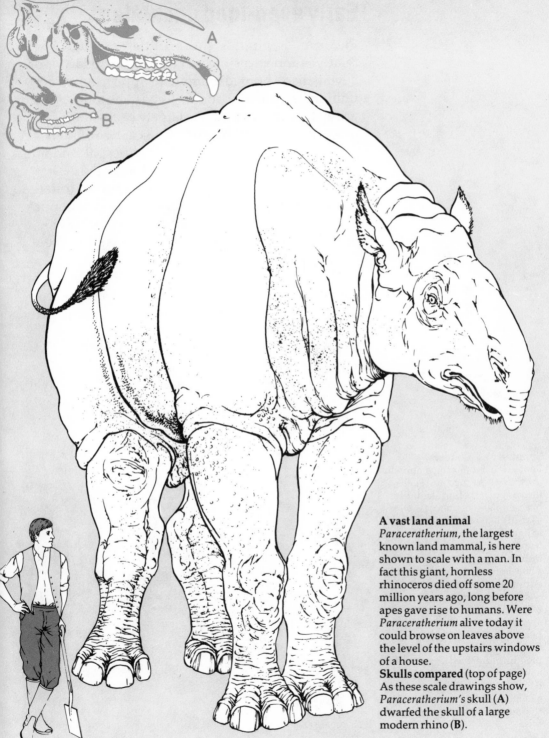

A vast land animal
Paraceratherium, the largest
known land mammal, is here
shown to scale with a man. In
fact this giant, hornless
rhinoceros died off some 20
million years ago, long before
apes gave rise to humans. Were
Paraceratherium alive today it
could browse on leaves above
the level of the upstairs windows
of a house.
Skulls compared (top of page)
As these scale drawings show,
Paraceratherium's skull (**A**)
dwarfed the skull of a large
modern rhino (**B**).

©DIAGRAM

Early even-toed ungulates

Pigs, camels, giraffes, sheep, cattle, and their relatives and ancestors make up the artiodactyls or "even-toed" hoofed mammals. Appearing over 50 million years ago, artiodactyls became today's major group of ungulates thanks largely to advances in digestion. We start by a brief look at two of their more primitive suborders. The extinct palaeodonts comprise two superfamilies: dichobunoids (ancestral artiodactyls) and entelodontoids (often called "giant pigs"). The suborder Suina contains three superfamilies: suoids (pigs and peccaries), anthracotheres (extinct pig-like beasts), and the hippopotamoids (hippopotamuses).

1 **Dichobune**, a small dichobunoid, had short limbs, four-toed feet, and low skull with long canine teeth and low-crowned molars. Time: Mid Eocene–Early Oligocene. Place: Europe.

2 **Archaeotherium** was a huge warthog-like entelodont. It had humped shoulders, thin legs, and a long, heavy head with lumpy cheeks. Probably it grubbed up roots. Height: 1m (3ft 3in). Time: Oligocene. Place: North America and East Asia.

Sprinters and plodders
Bones from four living mammals show that odd-toed and even-toed ungulates of similar build have similar foot bones (shown shaded). These bones are longer in lightweight sprinters than in heavy plodders, and sprinters bear their weight on fewer toes.
a Foot bone of a horse, a speedy odd-toed ungulate that bears its weight on one toe.
b Foot bones of a chevrotain, a speedy even-toed ungulate that bears its weight on two toes.
c Foot bones of a rhinoceros, a plodding odd-toed ungulate that spreads its weight on three toes.
d Foot bones of a hippopotamus, a plodding even-toed ungulate that spreads its weight on four toes.

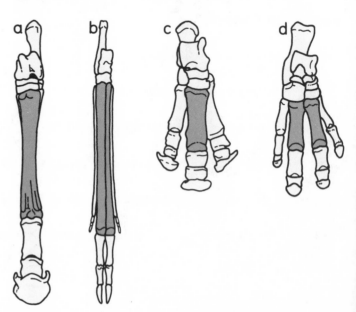

3 **Bothriodon,** an anthracothere, had a long body, short limbs with four-toed feet, and a long skull with 44 teeth including low-crowned molars. Length: 1.5m (5ft). Time: Late Eocene–Early Miocene. Place: Northern continents and Africa.

4 **Platygonus,** a peccary, had a short deep skull, big, shearing canine teeth, and long legs with reduced side toes. Length: 1m (3ft 3in). Time: Pliocene–Pleistocene. Place: North and South America.

5 **Hippopotamus,** the sole known genus in its superfamily, is a huge, heavy, aquatic mammal with short, thick, four-toed legs, tiny ears and eyes, and a vast mouth armed with tusk-like canine teeth. Length: up to 4.3m (14ft). Time: Mid Miocene to today. Place: Africa (once Eurasia too).

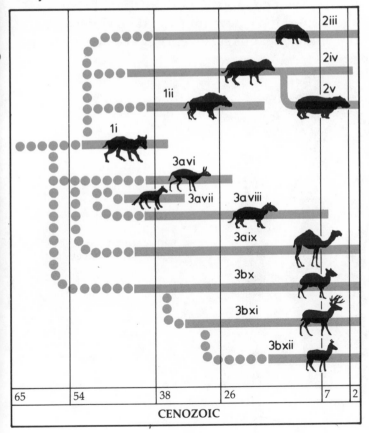

Artiodactyl family tree
This family tree shows the Artiodactyla ("even toed" ungulates), described on pages 186–195. We give suborders (1–3), infraorders (a–b), and superfamilies (i–xii).
1 Palaeodonta (palaeodonts)
i Dichobunoidea (ancestral artiodactyls)
ii Entelodontoidea (entelodonts)
2 Suina (pig-like artiodactyls)
iii Suoidea (pigs and peccaries)
iv Anthracotherioidea (anthracotheres)
v Hippopotamoidea (hippopotamuses)
3 Ruminantia (ruminants)
a Tylopoda (camels etc)
vi Cainotherioidea (cainotheres)
vii Anoplotherioidea (anoplotheres)
viii Merycoidodontoidea (oreodonts)
ix Cameloidea (camels and llamas)
b Pecora (pecorans)
x Traguloidea (tragulids etc)
xi Cervoidea (deer, giraffes, and kin)
xii Bovoidea (cattle, antelopes, prongbucks, etc)

65	54	38	26	7	2
		CENOZOIC			

Camels and their kin

The most successful of all hoofed animals alive are those that ruminate, or "chew the cud". These so-called ruminants can snatch a meal of leaves, run away if set upon by carnivores, and then digest their meal at leisure. The suborder's more primitive members form the tylopods, an infraorder containing camels and their prehistoric kin. Their front teeth and cheek teeth differ less strikingly than those of pigs, and cheek teeth bear crescent-crested crowns. Tylopods mostly browse on trees and shrubs.

There were four superfamilies: small, rabbit-like cainotheres; primitive anoplotheres; heavily built, rather pig-like merycoidodonts (also often known as oreodonts); and cameloids, including camels and llamas. Tylopods appeared about 45 million years ago. Some kinds grew numerous and widespread. But only cameloids survive. They evolved in North

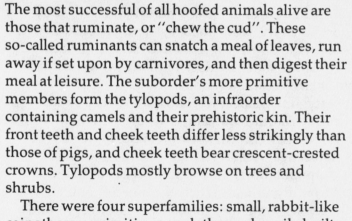

Giant and dwarf
Two prehistoric camels shown here to scale illustrate the range of sizes in their superfamily 10 million years ago.
A *Alticamelus* had a long neck and stilt-like limbs. This "giraffe camel" held its head 3m (10ft) high. It could have browsed on leafy twigs high off the ground.
B *Stenomylus* was tiny, and browsed on low vegetation. But it had long, slender limbs designed for sprinting. *Stenomylus* and its close kin were the camel counterparts of gazelles.

America, reached South America, Eurasia, and Africa, but then died out in North America, Europe, and Africa.

1 **Cainotherium** was a rabbit-sized cainothere. Big eyes and keen ears warned of danger, and it escaped by bounding off on long hind limbs. It lived in Mid Oligocene–Mid Miocene Europe.

2 **Anoplotherium** belonged to the anoplotheres: heavily built, tapir-sized beasts with clawed toes and long tails. Shoulder height: 1m (3ft 3in). Time: Late Eocene–Early Oligocene. Place: Europe.

3 **Merycoidodon** was a sheep-sized oreodont, with rather pig-like proportions: large head, short neck, long body, and short limbs. Time: Early–Mid Oligocene. Place: North America.

4 **Agrichoerus** belonged to the agrichoeres – a family of slim oreodonts with long head, body, and tail, and clawed feet used perhaps in climbing trees or digging up edible roots. This sheep-sized beast lived in Oligocene–Early Miocene North America.

5 **Poëbrotherium** was a sheep-sized early camel with short legs, two-toed feet, and a full set of teeth. It looked like a tiny llama. Time: Oligocene. Place: North America.

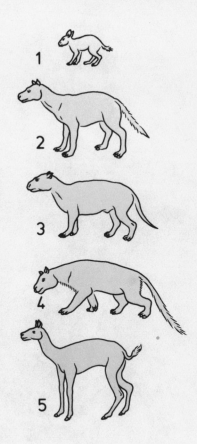

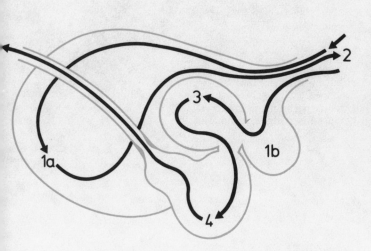

Ruminant stomach
This section diagram shows what happens to swallowed food inside a ruminant's four-chambered stomach.
1a Most swallowed food enters the rumen, the largest chamber; digestion starts here.
1b Tough plant fibres and stones go to the reticulum.
2 Pulped food returns to the mouth for extra crushing.
3 Re-swallowed food has water squeezed out in the omasum.
4 Enzymes extract proteins from food in the abomasum.

189

Primitive pecorans

Cattle, sheep, deer, and close relatives make up the pecorans – a more advanced cud-chewing infraorder than the tylopods (camel and their kin). Pecorans have very complex stomachs. Most are graceful, with long, slim limbs. Their main defence is speed, though most have horns or antlers, or long, sharp canine teeth. They crop leaves by pressing them between the lower front teeth and an upper horny pad; the top front teeth are missing.

Pecorans first appeared over 40 million years ago, seemingly in Asia, and spread to other continents. Three superfamilies evolved. First we look at the traguloids: primitive mostly prehistoric pecorans including small hornless deer-like beasts with less complicated stomachs than true deer. Our examples come from different families.

Horned traguloid
Syndyoceras from Miocene North America belonged to the protoceratids, traguloids of which the males developed spectacularly strange horns. It included these features.
a Small deer size
b A pair of curved horns growing above the eyes
c A diverging pair of horns forming a prong above the nose
d Long, slim limbs
e Only the third and fourth toes functional
f Metacarpal bones on front feet not fused as in higher ruminants

1 **Archaeomeryx** was no bigger than a large rabbit. It had a curved back and longer hind limbs than front limbs. It looked like living mouse deer (chevrotains), but belonged to the primitive hypertragulids which had a long, canine-like premolar tooth in each lower jaw. Time: Late Eocene. Place: East Asia.

2 **Synthetoceras** belonged to the protoceratids, deer-sized traguloids whose males grew horns. *Synthetoceras* males had a Y-shaped horn jutting up and forward from the nose and shorter horns curving back and up behind the eyes. Time: Early–Late Miocene. Place: North America.

3 **Tragulus,** the Asian chevrotain, is one of only two survivors of the traguloids. Like others in its family (the tragulids), this looks like a rodent about 30cm (1ft) high. It lacks horns but has long, tusk-like upper canines, and four toes per foot (the outer two are short and useless). Time: Late Pliocene to today. Place: South-East Asian forests.

Leptomeryx skull
This life size skull of an early pecoran has the following features.
A No horns
B Sharp upper canine teeth
C Toothless cropping pad
D Cropping front teeth
E Gap developing between front and back teeth

Deer and giraffes

The cervoids (deer, giraffes, and relatives) and bovoids (cattle and their kin) are more "progressive" ruminants than their ancestors the traguloids. Mostly their upper canine teeth are short or missing, but their heads sprout horns or antlers. Certain bones inside their long, two-toed limbs have shrunk or almost vanished. And living members of these groups have a four-chambered stomach, better able to digest tough plant foods than the three-chambered stomach of the chevrotains.

Members of the cervoid superfamily live mostly in wooded countryside. Their low-crowned cheek teeth are designed to browse on trees or shrubs. Male deer grow and shed antlers each year; giraffes grow permanent skin-covered horns. We show five prehistoric examples from the three cervoid families, all established 25 million years ago.

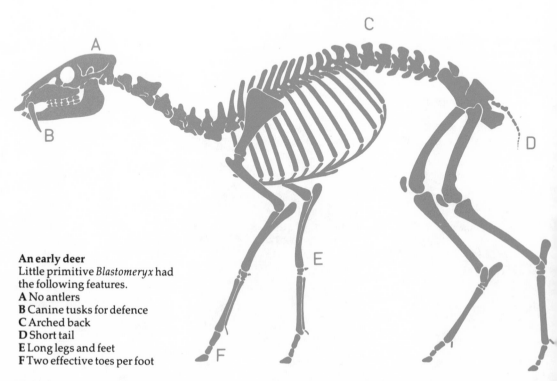

An early deer
Little primitive *Blastomeryx* had the following features.
A No antlers
B Canine tusks for defence
C Arched back
D Short tail
E Long legs and feet
F Two effective toes per foot

1 **Blastomeryx** had "advanced" limbs but old-fashioned tusk-like upper canines and no horns. It belonged to the early, primitive palaeomerycid family. Length: 76cm (30in). Time: Miocene. Place: North America.

2 **Cranioceras,** a North American palaeomerycid, had tall, pronged horns and a backswept third horn. Length: 1.5m (5ft). Time: Miocene–Pliocene.

3 **Megaloceros,** misnamed the "Irish elk", was a giant relative of the fallow deer, with antlers 3m (10ft) across. It lived in Pleistocene Eurasia and died off about 2500 years ago.

4 **Palaeotragus,** a giraffid ancestor of okapis and giraffes, resembled the okapi. Time: Miocene. Place: Eurasia, Africa.

5 **Sivatherium** was a large giraffid with a big head, but shorter legs and neck than a giraffe. Males grew two pairs of horns, one long and branched. Shoulder height: 2.2m (7ft). Time: Pleistocene. Place: South Asia and Africa.

The antler cycle
Like living deer, *Megaloceros* would have grown and shed its antlers every year. The annual cycle went like this.
A New antlers sprout in summer, nourished and protected by a covering called velvet.

B Bony antlers reach their full extent in autumn, when rival stags perhaps locked them in combat.
C Antlers shed in spring reveal the pedicle or base from which next season's antlers grow.

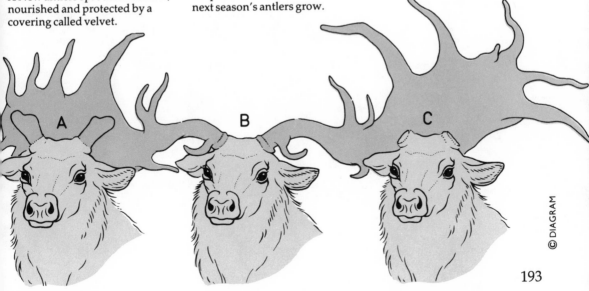

© DIAGRAM

193

Cattle and their kin

The top grass eaters in the world today are bovoids – members of the cattle superfamily. Cattle, sheep, goats, antelopes, musk-oxen, and prongbucks between them share out almost all the grasslands of the world, with horses as their only large competitors. The bovoids' teeth, stomachs, legs, and feet superbly suit them for a life as grazers on the plains. Both sexes grow strong horns they never shed – living spears on which they can impale attacking carnivores.

Bovoids seemingly evolved from traguloids some 15 million years ago, later than giraffes or deer. Five million years ago great herds were spreading through the north. In time they swarmed on grassy slopes and plains across Eurasia, Africa, and North America. Experts disagree about how to subdivide bovoids. Our examples show five contrasting types.
1 **Gazella** is a small, delicately built antelope, a desert and savanna sprinter. Shoulder height: 65–100cm (26–40in). Time: Late Miocene onward. Place: Africa, Asia, and once also Europe.

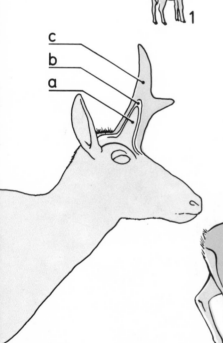

Unusual horns (left)
Unlike other bovoids, prongbucks grow forked horns and annually shed the outer sheath, seen here in section.
a Permanent bony horn core
b Hair covering the core
c Horn formed by outgrowth of the hair – growth takes four months from when the old horn has been shed

Prongbuck duellists (below)
Rival *Merycodus* males could have locked their long forked horns in ritual combat, as shown here. As in most bovoid duels, such fights were probably trials of strength, not combats to the death. The weaker creature would have given up and run away.

2 **Mesembriportax** was a large, rather heavy-bodied antelope related to the living nilgai. It had long legs and uniquely forked horns. Time: Early Pliocene. Place: South Africa.

3 **Bison** The long-horned species shown lived in Late Pleistocene North America. Shoulder height: up to 1.8m (6ft).

4 **Myotragus,** the Balearic cave goat, lived on West Mediterranean islands in Pleistocene and Recent times. Its large lower middle incisor teeth resembled rodents'. Shoulder height: 50cm (20in).

5 **Merycodus** was a small, early, deer-like prongbuck "antelope" with long forked horns. Prongbucks keep the bony horn cores but annually shed the horny sheaths. Time: Mid Miocene–Late Miocene. Other prongbucks (some bizarrely horned) evolved, all in North America. Only one survives.

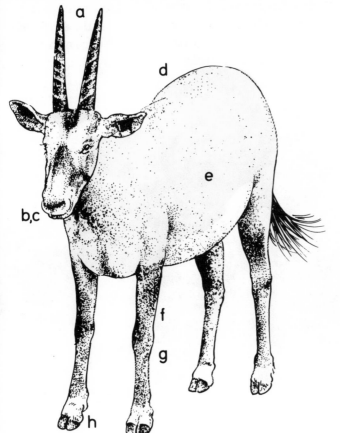

Bovoid features (left)
Palaeoreas was a small Miocene ruminant of Europe, Asia, and North Africa. Its spiral horns foreshadowed those of the big modern antelopes called oryxes. *Palaeoreas* and its bovoid kin owed their success to features such as the following.
a Both sexes permanently horned
b Wear-resistant teeth, with high crowns strengthened by folded ridges of enamel
c Lower incisors cropping against a hard toothless pad
d Sturdy body
e Four-chambered stomach for digesting tough leaves
f Long limbs for running
g Fused cannon bone reducing risk of sprains
h Two-toed "cloven" hooves

"Toothless" mammals

Strange mammals with few teeth make up the order Edentata. Edentates evolved in the Americas while these were isolated. They include the living anteaters, sloths, and armadillos, and their extinct relatives the huge, astonishingly armoured glyptodonts and unwieldy ground sloths. Most edentates lack front teeth and have a few, simple, rootless cheek teeth. The brain is small. The limbs are short, strong, and tipped with long, curved claws. All edentates were and are slow-moving eaters of plants, carrion, or insects.

Perhaps the first-known kinds were the so-called palaeanodonts from the North America of 60 million years ago. But maybe those gave rise to another order (Pholidota): the scaly Old World pangolins. Most fossil edentates arose in what was then the island continent of South America, though a land-bridge later let some into North America. Local tales and finds of hairy hides hint that ground sloths survived in southern Argentina until four centuries ago. Quite likely man killed off these last big edentates.

Glyptodon's body plan
a Deep, short skull with a small brain case
b Teeth only in sides of jaw
c Each tooth like three short pillars stuck together in a row, and without enamel
d Many vertebrae fused to help support the shell
e Short, massive limb bones
f Broad shoulder girdle
g Massive hip bones
h Heavy, bony tail
i Five-toed feet with hoof-like claws
j Solid armour (evolved from bony studs and plates set in the skin)

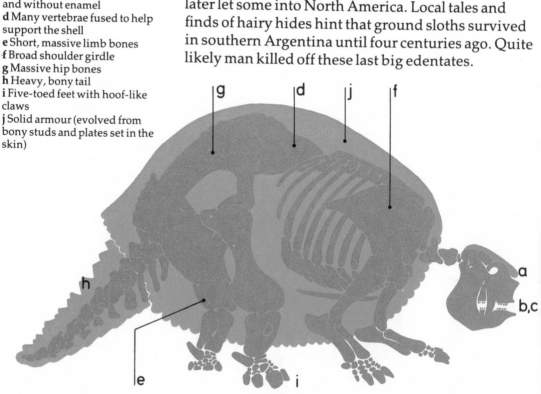

These three examples show a palaeanodont and two enormous fossil edentates.

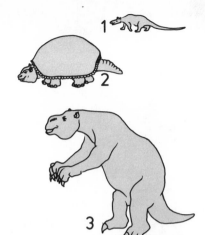

1 **Metacheiromys** was a palaeanodont with a long, low head, sharp canine teeth, but maybe horny pads instead of cheek teeth. It had short legs, sharp claws, and a long, heavy tail. Length: 45cm (18in). Time: Middle Eocene (about 48 million years ago). Place: North America.

2 **Glyptodon** was a huge "mammal tortoise"–a glyptodont descended from early armadillos. Horny sheaths covered the many bony plates that formed a great dome-shaped shell around its body. A bony cap crowned its skull, and bony rings armoured a thick, heavy tail, swung to fend off enemies. Length: about 3m (10ft). Time: Pliocene–Pleistocene. Place: South America.

3 **Megatherium,** the largest ground sloth, was an elephantine tree-top browser. Length: 6m (20ft). Time: Late Pliocene–Pleistocene. Place: South America to south-east USA.

Megatherium's life style
Megatherium is here shown to scale with a man. The monster walked on its knuckles and the sides of its feet. Females might have carried a baby on their back. A strong tail propped up *Megatherium* as it reared to claw leaves into its mouth and chewed them with its peg-like teeth.

Whales

Whales are mammals so well adapted for water life that they swim as easily as fishes and cannot walk on land. Fossil clues suggest that their order (Cetacea) evolved from mesonychids – a flesh-eating group of condylarths, the first hoofed mammals. By 52 million years ago these probably gave rise to *Pakicetus*, known from a skull unearthed in Pakistan and described in 1983. Sharing features found in whales and tapirs, *Pakicetus* was about 1.8m (6ft) long, lived near water, but could not dive deep or hear well when submerged. By 40 million years ago such beasts gave rise to three suborders of true whales worldwide. Buoyed up by water, some evolved bigger bodies than the largest-ever land mammal.

A snaky whale
Here the big early archaeocete *Basilosaurus*, also known as *Zeuglodon*, is shown in pursuit of herring. At over 20m (66ft), this monster rivalled many a large modern whale for length, but it had a slender rather snake-like body. The saw-edged teeth were suitable for seizing fishes. *Basilosaurus* fossils crop up in rocks of what were ancient sea beds as far apart as Africa and North America.

1 **Prozeuglodon,** found in Mid-Late Eocene North Africa, grew maybe 3m (10ft) long. It had a long snout, peg-like front teeth, and three-ridged saw-like cheek teeth. It belonged to the archaeocetes, a primitive Eocene–Late Oligocene suborder.

2 **Prosqualodon,** a more advanced small whale, had its blowhole above and behind the eyes. But its triangular shark-like cheek teeth were "old fashioned". It resembled dolphins and, like those, belonged to the odontocetes (toothed whales). Length: 2.3m (7ft 6in). Time: Oligocene–Early Miocene. Place: southern oceans.

3 **Cetotherium** probably resembled a small version of the living grey whale – a mysticete (baleen whale). These toothless whales trap swarms of shrimp-like creatures on the fringed baleen plates that hang from their upper jaws. Length: perhaps 4m (13ft). Time: Mid–Late Miocene. Place: Europe.

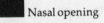
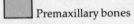

Back to the blowhole
Skulls of prehistoric whales show the nasal opening moving back to become a blowhole.
a *Prozeuglodon,* an Eocene whale from North Africa
b *Aulophyseter,* a Miocene whale from North America

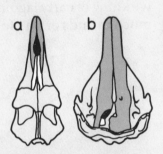

■ Nasal opening

▨ Premaxillary bones

□ Nasal bones

Rodents and rabbits

In numbers, variety, and distribution, the most successful mammals ever have been members of the rodent order: squirrels, rats, mole rats, cavies, beavers, porcupines, and many more. Between them rodents live in trees, on mountains, underground, in streams and swamps – everywhere from polar wastes to steamy forests in the tropics.

Small size, fast breeding, and ability to gnaw and digest foods as hard as wood contributed to their success. This started when an insectivore-type ancestor gave rise to the first squirrel-like rodent more than 60 million years ago. From northern continents, that pioneer's descendants reached every continent except Antarctica.

Soon after rodents came the rodent-like lagomorphs: rabbits, hares, and pikas. Lagomorphs have eight long chisel-like incisor teeth (rodents have just four) and there are other differences. What we know of early lagomorphs and rodents owes much to finds of tiny fossil teeth.

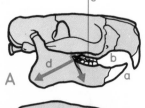

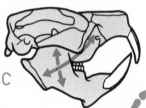

Rodent skeleton
This *Paramys* skeleton shows the following rodent features.
e Long, low skull
f Flexible fore limbs
g Strong hind limbs
h Five clawed toes per limb

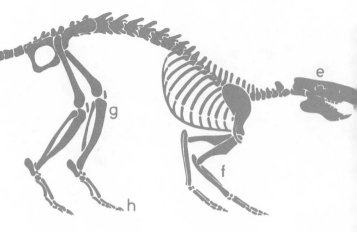

Rodent skulls (above)
Shown here are the skulls of extinct creatures from three different rodent suborders:
A *Paramys*, a protrogomorph;
B *Cricetops*, a myomorph;
C *Palaeocastor*, a sciuromorph.
All have the following features:
a four chisel-like incisors;
b gap behind the incisors;
c tall, complex molars;
d masseter muscles allowing lower jaw to move up and down, sideways, and to and fro.

1 **Paramys,** the first-known rodent, was a squirrel-like climber in the primitive suborder protrogomorphs. Length 60cm (2ft). Time: Late Palaeocene–Mid Eocene. Place: North America and Europe.

2 **Epigaulus,** a two-horned gopher, was a burrowing protrogomorph rodent. Its horns served in defence or maybe as a pair of shovels. Length: 26cm (10in). Time: Miocene. Place: North America.

3 **Sciurus,** the squirrel genus found today in North America and Europe, is a "living fossil" in the rodent suborder Sciuromorpha; its history dates back 38 million years.

4 **Castoroides** was a land-based beaver almost as big as a black bear. Length: 2.3m (7ft 6in). Time: Pleistocene. Place: North America.

5 **Eocardia** was a Miocene guinea-pig-like member of the cavioids, a South-American rodent suborder. (*Eumegamys,* an almost hippo-sized Pliocene cavioid, was the largest ever rodent.)

6 **Lepus,** the modern hare, evolved five million years or so ago. Place: northern continents and Africa. Order: Lagomorpha.

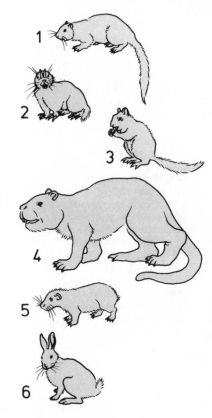

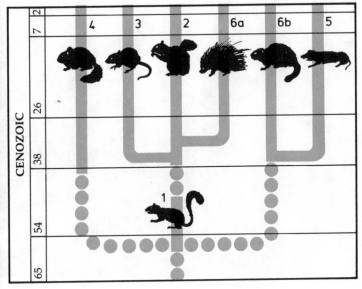

Rodent family tree
Zoologists group most of the 45 or so rodent families into suborders based on cheek teeth and jaw muscle design. We show the five suborders (**1–5**) and two well-known families unassigned to any suborder (**6**).

1 Protrogomorpha (primitive rodents)
2 Sciuromorpha (squirrels)
3 Myomorpha (rats, mice, etc – the largest rodent group)
4 Caviomorpha (South American rodents such as guinea pigs and chinchillas)
5 Phiomorpha (mole rats etc)
6 Unassigned to a suborder:
a Hystricidae (Old World porcupines)
b Castoridae (beavers)

©DIAGRAM

201

Chapter 9

RECORDS IN THE ROCKS

Immense changes have transformed the surface of the Earth since its creation. Prolonged rains filled the ocean basins. Later, continents collided and broke up. Their shifting created mountains and altered climates: deserts grew and shrank; hot lands cooled down and vice versa. Rarely, huge rocks hurtled in from space and punched holes in the Earth's crust, sometimes perhaps explosively enough to alter climates suddenly.

Living things responded to such changes with waves of extinction followed by bursts of evolution. All this helped to give the phases of Earth's history their special characters.

The next pages draw a thumbnail sketch of life through the periods and epochs of our ever-changing planet.

Eighteenth-century antiquaries examine rock layers crammed with ancient sea shells. Fossils in rocks laid down at different times allow palaeontologists to reconstruct past life period by period, epoch by epoch. (Illustration from *The History and Antiquities of Harwich and Dovercourt* by Samuel Dale.)

Precambrian time

Precambrian time – time before the Cambrian Period began – occupied Earth's first 4000 million years – 87 per cent of our planet's past. Rocks laid down then still largely form the cores of continents. Some hold the fossil clues that show when life began and early living things evolved. Microscopic one-celled life forms appeared in seas at least 3500 million years ago. By 3200 million years ago, blue-green algae had begun enriching the atmosphere with oxygen. In time they made life possible for more complex living things. Soft-bodied, many-celled water animals were burrowing through underwater mud 1000 million years ago. By 680 million years ago, soft corals, jellyfish, and worms all flourished off a sandy shore in South Australia. Similar creatures also thrived in what today are England and Newfoundland.

Precambrian world
This world map shows land masses in Late Precambrian time. Lines represent Equator, Tropics, and Polar Circles.

Precambrian life
These fossil organisms lived in Precambrian time. They are not shown to scale.

MONERA
a *Kakabekia*

PLANTS
b Green alga

INVERTEBRATES
c Sponge
d Annelid worms
e Coelenterates

Cambrian Period

Fossils suddenly appear more numerous and varied in Cambrian times (600–500 million years ago), the dawn of the Palaeozoic Era ("age of ancient life"). Scientists first studied them in Wales (called Cambria in Latin). But Cambrian rocks occur in North America and other continents as well as Europe. The fine-grained Burgess Shales of south-west Canada preserved a rich sample of life below the waves 550 million years ago. Down here lived jellyfishes, sponges, starfishes, worms, and velvet worms – all relatives of animals alive today. Most striking, though, were creatures that had gained the knack of building hard protective shells from chemicals dissolved in water. The commonest were trilobites, many of them sea-bed scavengers. There was even a little lancelet-like creature – forerunner of the jawless fishes which appeared just before the Cambrian Period closed.

Cambrian world
This world map shows land masses in Cambrian time. Lines represent Equator, Tropics, and Polar Circles.

Cambrian life
These fossil organisms lived in Cambrian time. They are not shown to scale.

INVERTEBRATES
a Annelid worm
b Molluscs
c Coelenterate
d Arthropods
e Lancelet

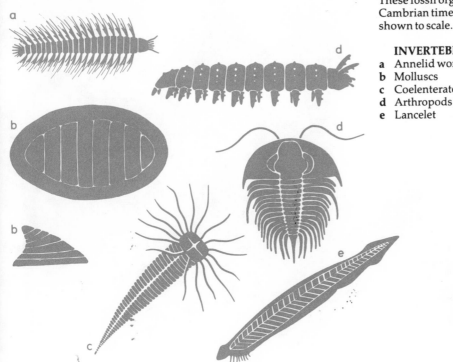

Ordovician world
This world map shows land masses in Ordovician time. Lines represent Equator, Tropics, and Polar Circles.

Ordovician life
These fossil organisms lived in Ordovician time. They are not shown to scale.

INVERTEBRATES
a Coelenterate
b Molluscs
c Arthropod
d Tentaculate
e Echinoderms
f Branchiotreme

Ordovician Period

The Ordovician Period (500–440 million years ago) is named after the Ordovices, an ancient Celtic tribe of western Wales, where scientists first studied Ordovician fossils. In Ordovician times, northern landmasses were coming together, and southern continents already formed a single mass of land. The South Pole lay over North Africa, much of which was under ice. Shallow seas repeatedly invaded North America.

Ordovician fossils show that animals and plants still lived only in the sea. Many resembled Cambrian ancestors. Trilobites were now most numerous. Graptolites and lampshells teemed below the waves. Molluscs evolved apace – bivalves resembling modern clams and oysters, and gastropods (one-shelled molluscs such as whelks and limpets). Now, too, those early vertebrates the jawless fishes increased in number.

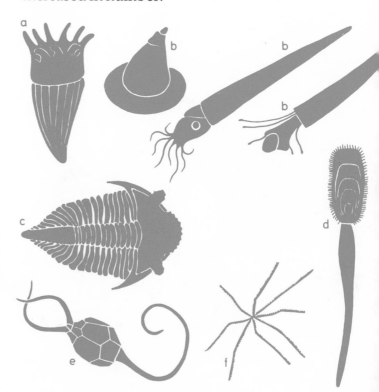

Silurian Period

The Silurian Period (440–395 million years ago) is named after the Silures, an ancient tribe astride the Welsh–English border. Silurian fossils occur in all landmasses but Antarctica.

By the end of Silurian time, colliding continents had raised up mountains and forged two supercontinents: Laurasia in the north, and Gondwanaland in the south. Life thrived largely in warm shallow seas that invaded much of Laurasia. The first jawed fishes (acanthodians and placoderms) appeared, some of them doubtless hunted by the giant "sea scorpions". Solitary corals built great reefs, and shallow sea floors supported a rich array of sea lilies, lampshells, corals, trilobites, graptolites, and molluscs. Jawless fishes invaded lakes and rivers. The first land plants appeared, followed soon by creatures such as scorpions and millipedes.

Silurian world
This world map shows land masses in Silurian time. Lines represent Equator, Tropics, and Polar Circles.

Silurian life
These fossil organisms lived in Silurian time. A re-dating of this period may put some items later. They are not shown to scale.

PLANTS
a Rhyniophyte

INVERTEBRATES
b Mollusc
c Arthropods
d Annelid worm
e Echinoderm

FISHES
f Agnathans

Devonian Period

The Devonian Period (395–345 million years ago) is named from shales, slates, and Old Red Sandstone laid down in Devon, England. But such rocks occur in every continent. During Devonian times Laurasia and Gondwanaland drew closer together until they met at what are now North and South America to form a single supercontinent: Pangaea ("all Earth"). Shallow sea invaded Laurasia, and mountains rose where North America and Europe had collided. Climates everywhere were warm.

The Devonian is aptly called the Age of Fishes, for now these backboned animals diversified and multiplied enormously. Jawless fishes shared the waters with the more progressive jawed fishes destined to replace them. Among these were huge placoderms, early sharks, and early bony fishes,

Devonian world
This world map shows land masses in Devonian time. Lines represent Equator, Tropics, and Polar Circles.

Devonian life
These fossil organisms represent some of those that flourished in Devonian time. They are not shown to scale.

PLANTS
a *Asteroxylon*
b Fern ancestor

INVERTEBRATES
c Mollusc
d Arthropod

FISHES
e Agnathan
f Placoderms
g Acanthodians
h Chondrichthyan
i Sarcopterygian

AMPHIBIANS
j Labyrinthodont

including the fleshy-finned ancestors of
amphibians. In seas, vast coral reefs produced
immense tracts of limy rock, and the ancestors of
ammonites appeared. But trilobites and graptolites
grew scarcer.

By Early Devonian times, there were many land
plants equipped with tubes for sucking moisture
from the soil. Scale trees, giant horsetails, and
feathery tree ferns produced the world's first forests.
These mostly thrived on swampy lands, but the first
seed-bearing plants foreshadowed kinds that later
colonized dry land. Greenland's swampy tropical
forests became home to the first, low-slung
amphibians. Here, too, crawled spider-like
arachnids and the first wingless insects – bristletails
and springtails.

g

h

i

j

Carboniferous Period

The Carboniferous Period (345–280 million years ago) gets its name from thick bands of carbon in the form of coal laid down when shallow seas drowned tropical forests in what are now North America, Europe, and elsewhere.

The period had two distinct halves, with different names in the Americas. Mississippian, or Early Carboniferous, rocks (345–300 million years old) include limestones formed when limy muds were laid down in a shallow sea in what is now the Mississippi Valley. Arthropods, bryozoans, crinoids, corals, and molluscs flourished in these waters.

Pennsylvanian, or Late Carboniferous, rocks formed 300–280 million years ago. They include Pennsylvania's coal measures, produced when shallow sea drowned tropical forests that covered much of lowland North America. The rich variety of

Carboniferous world
This world map shows land masses in Carboniferous time. Lines represent Equator, Tropics, and Polar Circles.

Carboniferous life
These fossil organisms represent some of those that flourished in Carboniferous time. They are not shown to scale.

PLANTS
a Horsetail
b Gymnosperm
c Club moss

INVERTEBRATES
d Arthropods

FISHES
e Chondrichthyan
f Acanthodian

AMPHIBIANS
g Labyrinthodont

REPTILES
h Cotylosaur

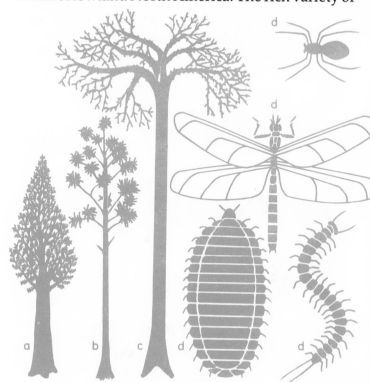

life in Pennsylvanian coal swamps helps to earn the
Carboniferous Period its nickname "Age of
Amphibians". Sprawling amphibians – some as big
as crocodiles – lurked on mudbanks below the tall
scale trees, tree ferns, and giant horsetails. Others
hunted fishes in the pools they shared with
lungfishes. Amphibians with sturdy limbs walked
easily on land. Those with tiny limbs or none swam
like eels or burrowed through leaf litter on the forest
floor. Here snails slithered through the humid
undergrowth. Rotting logs sheltered centipedes,
scorpions, and countless kinds of cockroach. Old tree
stumps were the dens of small, early, insect-eating
reptiles. Above flew or hovered giant "dragonflies"
and lesser insects. Meanwhile, vast ice sheets
smothered much of the world's great southern
landmass.

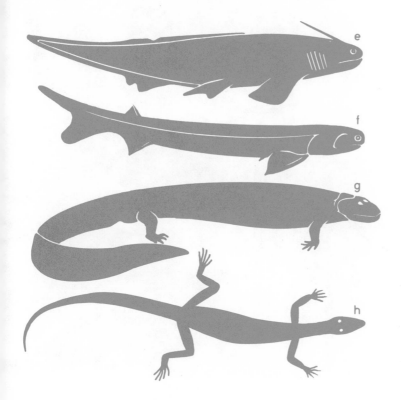

Permian Period

Fossil-rich rocks near Perm in Russia explain the name given to the Permian Period (280–225 million years ago). But much of what we know of Permian life on land comes from the Red Beds of Texas and South Africa. As the supercontinent Pangaea drifted north, glaciers retreated southwards. Great Earth movements threw up mountain ranges ancestral to the Appalachians, Rockies, Alps, and Urals. Fossil plants betray distinct climatic zones, and a trend to seasonally cooler, drier climates. Conifers began replacing scale trees and giant horsetails. The new forests largely featured firs and pines.

Now, reptiles took over from amphibians to dominate life on land. First came cotylosaurs, then mammal-like reptiles – primitive pelycosaurs and the more advanced therapsids. Now, too, the lizards' ancestors were scampering around, and sturdy

Permian world
This world map shows land masses in Permian time. Lines represent Equator, Tropics, and Polar Circles.

Permian life
These fossil organisms represent some of those that flourished in Permian time. They are not shown to scale.

PLANTS
a Tree fern
b Conifer

INVERTEBRATES
c Arthropods

FISHES
d Chondrichthyan
e Bony fish

AMPHIBIANS
f Labyrinthodonts
g Lepospondyls

REPTILES
h Cotylosaurs
i Mesosaur
j Therapsid
k Pelycosaur
l Eosuchian

amphibians like *Cacops* and *Diadectes* lumbered overland. But the last anthracosaur and lepospondyl amphibians died out. New major insect groups – beetles, bugs, and cicadas – were emerging.

Changes also happened in the waters. The last acanthodians died out, along with rhipidistians, the fleshy-finned fishes that had given rise to amphibians. But bony fishes teemed in lakes and rivers, and some found their way into the sea. Here, ammonoids were plentiful, but trilobites and sea scorpions became extinct at last. In fact, dozens of major groups of creatures (most marine invertebrates) vanished from the fossil record as the Permian Period ended. This mass death marks the ending of the Palaeozoic Era – the "age of ancient life".

Triassic Period

Triassic world
This world map shows land masses in Triassic time. Lines represent Equator, Tropics, and Polar Circles.

Triassic life
These fossil organisms represent some of those that flourished in Triassic time. They are not shown to scale.

PLANTS
a Gymnosperm

INVERTEBRATES
b Arthropod
c Molluscs

AMPHIBIANS
d Anuran
e Labyrinthodonts

REPTILES
f Euryapsid
g Euryapsid or eosuchian
h Chelonian
i Ichthyosaur
j Cotylosaur
k Crocodilian
l Therapsids
m Rhynchocephalian
n Thecodont
o Saurischian dinosaur
p Ornithischian dinosaur

MAMMALS
q Eotherian

The Triassic Period (225–193 million years ago) marks the start of the Mesozoic Era or "age of middle life", also called the Age of Dinosaurs. The Triassic gets its name from the Latin *trias* ("three") after three rock layers dating from this period and found in Germany.

In Triassic times, the vast supercontinent Pangaea was splitting up to reform the northern and southern units, Laurasia and Gondwanaland. Land animals could still colonize the world by simply walking overland, but deserts occupied vast inland tracts. Conifers and other plants designed for cool or dry conditions thrived, at the expense of moisture-loving kinds, though ferns and horsetails flourished by the waterside. Now, too, there were palm-like cycads and bennettitaleans, also yews and ginkgoes. This was indeed an age of reptiles. Early on there was

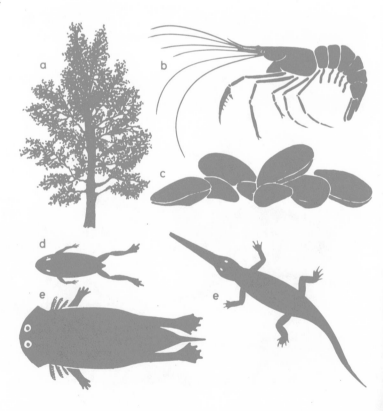

a wealth of rhynchosaurs, thecodonts, and mammal-like reptiles. But all these faded out, to be replaced by others. By Mid-Late Triassic times, early saurischian dinosaurs shared the lands with tortoises and lizards, and maybe with the earliest ornithischian dinosaurs and shrew-like mammals. Above their heads flapped skin-winged pterosaurs. Lakes and rivers formed the homes of crocodiles, while early frogs hopped amid damp herbage at the water's rim. Shallow seas afforded hunting grounds for strange marine reptiles: placodonts, nothosaurs, and those uncannily dolphin-like creatures, the flippered ichthyosaurs. Dinosaurs, crocodiles, and pterosaurs – the masters on land, in fresh water, and in air – were all members of one group: the archosaurs. These ruling reptiles dominated life on land all through the rest of Mesozoic time.

Jurassic Period

The Jurassic Period (193–136 million years ago) is named from rocks formed at this time in the Jura Mountains of France and Switzerland. Gondwanaland began breaking up; the North Atlantic Ocean opened and shallow seas invaded parts of Central North America and Europe. But land animals could still walk freely everywhere. Climates now were largely warm and moist, and rains watered what had been Triassic deserts.

Palm-like bennettitaleans and cycads, along with ferns and tree ferns, flourished thickly on moist river banks. A zoo of plant-eating dinosaurs browsed on this vegetation. Early armoured dinosaurs probably munched low-growing ferns and fungi. Knee-high leaves provided food for agile ornithopods. Stegosaurs might have reared on hind limbs to browse on low tree branches. But the "giraffes" of the Jurassic would have been huge sauropods

Jurassic world
This world map shows land masses in Jurassic time. Lines represent Equator, Tropics, and Polar Circles.

Jurassic life
These fossil organisms represent some of those that flourished in Jurassic time. They are not shown to scale.

PLANTS
a Conifer
b Bennettitalean

INVERTEBRATES
c Tentaculate
d Arthropod
e Molluscs

FISHES
f Bony fish
g Chondrichthyan

AMPHIBIANS
h Salamander

REPTILES
i Plesiosaur
j Ichthyosaur
k Crocodilian
l Pterosaur
m Ornithischian dinosaur
n Saurischian dinosaurs

BIRDS
o *Archaeopteryx*

216

capable of cropping treetop leaves beyond the reach of other dinosaurs. These monsters beat paths through forests, making gaps where smaller dinosaurs could follow and find food. Sauropod droppings fed the soil, and nourished seedlings that in time became large trees. But there were hunting dinosaurs as well. Small sprinters like *Coelurus* chased lizards through the undergrowth or scavenged at the kills of beasts like *Allosaurus* – a great flesh-eating carnosaur that could have preyed upon the sauropods. Now early birds took to the air, joining strange furry-bodied pterosaurs. There was also a crawling, buzzing menagerie of brand new insects, ancestral to the living earwigs, flies, caddis flies, bees, wasps, and ants. Meanwhile long-necked and short-necked plesiosaurs, and ichthyosaurs swam through the shallow seas in search of fish or ammonites.

Cretaceous world
This world map shows land masses in Cretaceous time. Lines represent Equator, Tropics, and Polar Circles.

Cretaceous life
These fossil organisms represent some of those that flourished in Cretaceous time. They are not shown to scale.

PLANTS
a Flowering plant

INVERTEBRATES
b Molluscs
c Arthropod
d Echinoderm

FISHES
e Chondrichthyan
f Sarcopterygian

REPTILES
g Chelonians
h Plesiosaurs
i Eosuchian
j Squamata
k Crocodilian
l Pterosaur
m Saurischian dinosaurs
n Ornithischian dinosaurs

BIRDS
o *Ichthyornis*
p *Hesperornis*

MAMMALS
q Insectivore

Cretaceous Period

The Cretaceous Period (136–65 million years ago) owes its name to vast thicknesses of chalk laid down in shallow seas (*creta* is the Latin name for chalk). By Late Cretaceous times, the break-up of the supercontinents Gondwanaland and Laurasia was well advanced, and continents were taking on their modern outlines and positions. Southern continents became vast islands. Seas split the northern landmass Laurasia in two: Asiamerica (East Asia with Western North America) and Euramerica (Europe with eastern North America). Mountain building began pushing up the present Rockies, Andes, and other mighty groups of mountains.

These changes isolated some groups of dinosaurs, so kinds that evolved in Asiamerica could no longer easily reach other continents, and vice versa. Meanwhile climates tended to grow cooler. Flowering plants evolved and spread explosively during the Cretaceous Period, along with the pollinating bees and butterflies. By 70 million years ago hickories, magnolias, and oaks formed stream-side forests in what is now Alberta, with china firs, giant sequoias, and swamp cypresses covering more

swampy tracts. Duckbilled dinosaurs evolved powerful batteries of grinding teeth to chew new tough-leaved kinds of vegetation. Ferns and cycad cones probably provided food for the horned dinosaur *Anchiceratops*. Higher, drier land away from swamps was the likely home of boneheaded dinosaurs like *Pachycephalosaurus*. Toothless ostrich dinosaurs roamed lowland clearings. *Tyrannosaurus* and its kin probably killed off enough duckbilled dinosaurs to stop these damaging the forests by overbrowsing. Other creatures living in or near this swampy delta region included such "modern" forms as frogs, salamanders, softshell turtles, snakes, gulls, waders, and opossums. Meanwhile *Quetzalcoatlus*, the biggest ever pterosaur, soared overhead. Mosasaurs and long-necked plesiosaurs were the great sea-going reptiles of their age, and *Pteranodon* swooped to pluck fishes from the waves.

About 65 million years ago, dinosaurs, pterosaurs, big marine reptiles, ammonites, and many other groups became extinct. Experts still dispute the cause of this catastrophe that brought the Mesozoic Era to a close.

© DIAGRAM

219

Palaeocene world
This world map shows land masses in Palaeocene time. Lines represent Equator, Tropics, and Polar Circles.

Palaeocene life
These fossil animals represent some of those that flourished in Palaeocene time. They are not shown to scale.

INVERTEBRATES
a Tentaculate

AMPHIBIANS
b Caecilian

MAMMALS
c Amblypod
d Dermopteran
e Primate
f Condylarth
g Multituberculate
h Perissodactyl
i Rodent

Palaeocene Epoch

The Palaeocene ("old recent life") of 65–54 million years ago marks the first epoch of the Tertiary Period occupying most of the Cenozoic Era ("age of recent life"). Retreating seas exposed dry land in much of inland North America, Africa, and Australia. But South America was cut adrift with its own unique evolving "ark" of mammals. Everywhere, new kinds of mammal were appearing. Primitive early species waned as more advanced placentals took their place: condylarths (the first hoofed herbivores), rodents, and squirrel-like primates shared their world with bulky amblypods and primitive, early flesh-eating creodonts. Carnivorous mammals met some competition from big flightless birds of prey like *Diatryma*. Most fossil mammals come from North America, Europe, and Central Asia; in other places no extensive land-based sediments were laid down at this time.

At sea, gastropods and bivalves replaced ammonites as the leading molluscs. New kinds of sea urchin and foraminiferan replaced old ones. Among fishes, sharks seem to have been particularly plentiful.

Eocene Epoch

Mountains rose and fissures leaked great lava flows in India and Scotland in the Eocene or "dawn of recent life" (54–38 million years ago). The rifting North Atlantic cut off North America from Europe, and South America lost links with Antarctica. Seas invaded much of Africa, Australia, and Siberia. Climates were generally warm or mild. Tropical palms even flourished in the London Basin.

Mammals continued to diversify. The first whales and sea cows swam in seas. Rodents ousted multituberculates as the main small mammals. Insectivores gave rise to bats. Primates included forest-dwelling ancestors of today's lemurs and tarsiers. Ungainly uintatheres stomped around North America and Asia. But condylarths were giving way to more modern ungulates – early horses, tapirs, and rhinoceroses, and pig-like anthracotheres in Asia and Europe. Ancestors of elephants roamed Africa. Meanwhile, the isolated condylarths of South America produced a unique zoo of hoofed mammals, along with edentates and marsupials. Australia's mammal fauna at this time remains a mystery.

Eocene world
This world map shows land masses in Eocene time. Lines represent Equator, Tropics, and Polar Circles.

Eocene life
These fossil animals represent some of those that flourished in Eocene time. They are not shown to scale.

INVERTEBRATES
a Coelenterate
b Nematode worm

MAMMALS
c Bat
d Tillodont
e Primate
f Creodont
g Carnivore
h Condylarth
i Amblypod
j Sea cow
k Proboscidean
l Perissodactyl
m Artiodactyls
n Edentate
o Whale

BIRDS
p Ratite
q Shore bird

©DIAGRAM

221

Oligocene Epoch

The Oligocene or "few recent" (kinds of life) lasted from 36–26 million years ago. Australasia had hived off from Antarctica, and left it isolated by ocean. This cooled world climates everywhere. Grasses and temperate trees ousted tropical vegetation from large areas. Grazing and browsing mammals multiplied – beasts like horses, camels, and rhinoceroses; while brontotheres ranged over Asia and North America. Dogs, stoats, cats, pigs, and rat-like rodents were on the increase. Africa was home to mastodonts, creodonts, hyraxes, anthracotheres, and the ape-ancestor *Aegyptopithecus*. Meanwhile isolated South America produced sloths, armadillos, rodents resembling guinea pigs, elephant-like pyrotheres, and others. While these creatures flourished, old fashioned hoofed and flesh-eating mammals – the condylarths and creodonts – were on the wane. Meanwhile at sea, early whales died out, largely replaced by toothed whales.

Oligocene world
This world map shows land masses in Oligocene time. Lines represent Equator, Tropics, and Polar Circles.

Oligocene life
These fossil animals represent some of those that flourished in Oligocene time. They are not shown to scale.

INVERTEBRATES
a Crustacean

MAMMALS
b Primates
c Creodont
d Embrithopod
e Pyrothere
f Perissodactyls
g Artiodactyls

BIRDS
h Coraciiform
i Apodiform

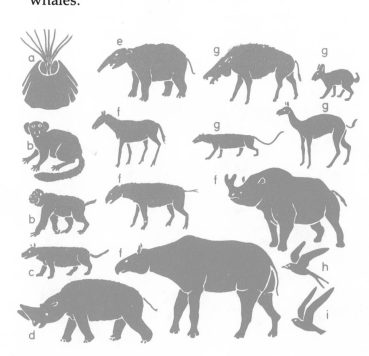

Miocene Epoch

The Miocene or "less recent" (with fewer modern creatures than the next epoch) lasted from 26–7 million years ago, longer than any other epoch. The world changed greatly now: ice covered Antarctica; the Mediterranean Sea dried up; India crashed into Asia; the Himalayas, Rockies, and Andes rose. But sea still isolated South America and Australia. Grasslands spread extensively and mammals reached their richest-ever variety. Many were hoofed grazers or browsers. Thus North America had horses, oreodonts, rhinoceroses, pronghorns, camels, protoceratids, and chalicotheres, with bear-dogs and sabre-toothed cats among the predators. Eurasia's "zoo" included early deer and giraffes, while African mammals included mastodonts, apes, and Old World monkeys. Great migrations saw elephants spread out from Africa to Eurasia and North America. Cats, giraffes, pigs, and cattle went the other way – from Eurasia to Africa. Horses found their way from North America into Eurasia. Meanwhile glyptodonts, armadillos, anteaters, New World monkeys, and horse-like litopterns evolved in isolated South America. Australia's Miocene marsupials and monotremes are little known.

Miocene world
This world map shows land masses in Miocene time. Lines represent Equator, Tropics, and Polar Circles.

Miocene life
These fossil organisms represent some of those that flourished in Miocene time. They are not shown to scale.

PLANTS
a   Grass

MAMMALS
b   Primates
c   Carnivores
d   Proboscidean
e   Litoptern
f   Notoungulate
g   Perissodactyls
h   Artiodactyls
i   Monotreme

BIRDS
j   Pelecaniform
k   Ratite

©DIAGRAM

223

Pliocene Epoch

Pliocene time
This world map shows land masses in Pliocene time. Lines represent Equator, Tropics, and Polar Circles.

The Pliocene ("more recent") of 7–2 million years ago ended the first, long, Tertiary Period of the Cenozoic Era. Continents had taken up their present-day positions, and land linked North and South America. Antarctica's ice cap and new ones in the northern hemisphere cooled lands and oceans. Vegetation was like today's. Grasslands replaced many forests, so grazing mammals spread at the expense of browsers. Cattle, sheep, antelopes, gazelles, and other bovids reached their peak in Old World lands. North American mammals included horses, camels, deer, pronghorns, peccaries, mastodonts, beavers, weasels, dogs, and sabre-toothed cats. Rhinoceroses and protoceratids died out in North America. But ground sloths and other mammals moved in from South America. Meanwhile, dogs, bears, horses, mastodonts, and others colonized South America from the north. Early elephants, antelopes, and the ancestors of man roamed Africa. But isolated Australia's only newcomers were rodents, rafting in on mats of vegetation drifting south from Indonesia.

Pliocene life
These fossil animals represent some of those that flourished in Pliocene time. A re-dating of this period may put some items earlier. They are not shown to scale.

MAMMALS
a Marsupial
b Primates
c Desmostylan
d Proboscidean
e Notoungulate
f Litoptern
g Perissodactyl
h Artiodactyls
i Edentate
j Rodent
k Lagomorph

BIRDS
l Falconiform

Pleistocene Epoch

The Pleistocene ("most recent") Epoch of the short Quaternary Period lasted from about 2 million to 10,000 years ago. Ice Age cold gripped northern lands as ice caps and glaciers waxed and waned. Advancing cold forced creatures south, though some returned in intervals of warmth. So much water lay locked up in ice that the level of the oceans fell. Horses, camels, deer, tapirs, mastodonts, mammoths, dogs, and sabre-toothed cats lived in North America (though horses and camels died out there). Beasts migrating into South America helped wipe out many of its native creatures. Meanwhile, monkeys, hyaenas, hippopotamuses, and straight-tusked elephants thrived in warm-phase Europe. Eurasia's cold-adapted beasts included woolly mammoth, woolly rhinoceros, cave bear, and cave lion – all now extinct. Australia was home to outsize marsupials including *Diprotodon*, a giant kangaroo, and a marsupial "lion". Spreading probably from Africa, mankind evolved efficient hunting skills. Maybe this explains the disappearance of most large mammals before the Pleistocene ended. About 10,000 years ago Ice Age cold gave way to the long warm phase we call the Holocene or Recent Epoch – the time we live in now.

Pleistocene life
These fossil animals represent some of those that flourished in Pleistocene time. They are not shown to scale.

MAMMALS
a Marsupials
b Primates
c Carnivores
d Proboscidean
e Perissodactyls
f Artiodactyls
g Edentate
h Rodent

BIRDS
i Ratite

©DIAGRAM

225

Chapter 10

FOSSIL HUNTING

What we know about past life we owe to the patient work of teams of experts – between them skilled at finding, retrieving, preserving, displaying, and explaining fossils. This last chapter shows would-be fossil-hunters briefly what most of these techniques involve (and incidentally how some help to bring us such benefits as coal and oil).

We end with a short survey of pioneer palaeontologists and a worldwide museum guide to fossil collections. Not all of these are open regularly to the public, so check access before a visit.

Top-hatted palaeontologists direct fossil excavations in this lithograph from *Geology of Sussex* (1827), by the pioneer British dinosaur hunter Gideon Mantell. From Tilgate Forest's Early Cretaceous sandstone rocks Mantell obtained the bones of fossil dinosaurs, crocodiles and plesiosaurs. (British Museum – Natural History.)

227

Finding fossils

Fossil hunters search where man or weather has exposed sedimentary rocks – particularly limestones, shales, and clays; sometimes sandstone too. Likely sites are sea cliffs, quarries, disused mines, road cuttings, building excavations, spoil heaps, streamsides with exposed bedrock, deserts, polar wastes, and mountainsides. Experienced collectors go armed with information obtained from local museums, guide books, and geological maps. They obtain landowners' consent; collect only in permitted areas (avoiding sites protected for their fossil rarities); and beware cliff falls, particularly after rain. Protective helmets may be needed, as well as old clothes and sturdy shoes, or rubber boots if working in soft mud.

Fossil hunters pace slowly, scanning gullies, cliffs, or piles of weathered rock. Some rocks bristle with fossils; others seemingly hold none. But patient

Rocks of the ages
This map shows where in the British Isles fossil hunters find fossil-bearing sedimentary rocks of different ages. Few fossils occur in igneous rocks such as basalt and granite, which are formed from upwelling molten rock, or in metamorphic rocks such as slate and marble, which are changed by tremendous heat or pressure.

- Upper Cenozoic rocks
- Lower Cenozoic rocks
- Mesozoic rocks
- Upper Palaeozoic rocks
- Lower Palaeozoic rocks
- Igneous and metamorphic rocks

searching and a practised eye could well reveal tell-tale shiny or discoloured shapes in rock. Whole fossils are a rarity. Collectors mostly find just scattered teeth or broken bits of fossil leaf or bone. Yet even fragments can betray the organisms they belonged to. For example, short columns made of disc-like plates could be the broken stems of crinoids, or "sea lilies". The broken plates of all echinoderms glint as they catch the light. Distinctive teeth help experts to identify fossil sharks and mammals. Deeply pitted bony plates come from crocodiles; less deeply pitted plates, from turtles. Turtle plates are thicker than those from certain fossil fishes. Fossil fishes survive mostly just as scales and individual bones. Our next two pages show how fossils small and large can be extracted from the rocks.

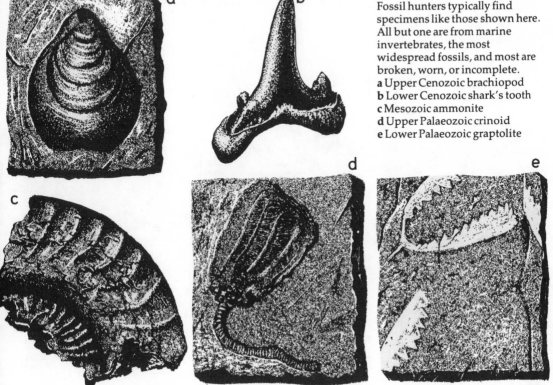

Fossil finds
Fossil hunters typically find specimens like those shown here. All but one are from marine invertebrates, the most widespread fossils, and most are broken, worn, or incomplete.
a Upper Cenozoic brachiopod
b Lower Cenozoic shark's tooth
c Mesozoic ammonite
d Upper Palaeozoic crinoid
e Lower Palaeozoic graptolite

Extracting fossils

Finding fossils is just the first stage of collecting. The fossil hunter must free fossils from their rocks, make a record of his finds, and transport these safely home.

You can go fossil collecting with just a geological hammer, hand lens, and old newspapers for wrapping finds. But collectors often take much more, inside a rucksack with extra space for specimens. A broad-bladed chisel called a bolster will split rocks along their bedding planes, revealing hidden fossils. Cold chisels help to cut them free. A flat-bladed trowel is helpful in soft rock. Old kitchen knives, brushes, even picks and spades, can have their uses. Sieves help you separate small teeth and bones from even smaller particles of clay or sand. The hand lens helps you identify tiny but important features. A steel rule serves for making measurements.

To remove a fossil from a rock, trim away as much rock as possible. Leave the rest for careful work at home. If your fossil breaks, mark joins and number

Fossil finder's toolkit
Dedicated fossil hunters may use all these items.
1 Rucksack
2 Geological map
3 Notebook
4 Marking pen
5 Compass
6 Hand lens
7 Geological hammer
8 Punch
9 Cold chisel
10 Bolster
11 Trowel
12 Kitchen knife
13 Toothbrush
14 Rule
15 Newspapers
16 Sticky tape
17 Paper tissues
18 Plastic bags
19 Boxes

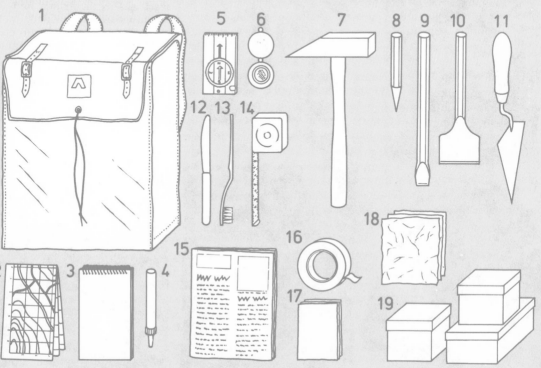

the pieces, using waterproof ink or a felt-tip pen. Wrap together fragments for later reassembly. Use a water- or solvent-based hardening solution to strengthen fragile fossils in soft clay or sand. When specimens have dried, wrap them in paper tissues, kitchen foil, moss, or sand, and place them in tins or boxes. Stronger specimens can go in polythene or linen bags.

Gather samples of loose rock and sieve them back at home for tiny fossil teeth, bones, or seeds. Number each item, then use a notebook to record its number, name, locality, and details of its parent rock. A sketch or photograph of rock layers at the site gives useful future reference.

Collecting big fossils, such as bones of dinosaurs, calls for special skills and teams of workers. Report such finds to the palaeontology department of a museum or university.

Fossil hunting
We show fossil hunters working near a cliff foot.
A This fossil hunter is finding fossils in lumps of rock. He cracks open ball-shaped lumps, and splits shale along its bedding plane. A helmet protects his head from falling stones.
B Another fossil hunter sketches and labels the nature, depth, and fossil content of exposed rock layers and their fossil beds.

1 Brown limestone
2 Fossil-oyster bed
3 Brown limestone
4 Hard sandstone: no fossils
5 Green shale: containing fossil nautiloids (**a**) and fossil trilobites (**b**)
6 Dark shale
7 Green shale
8 Talus: fallen rocks concealing lower rock beds

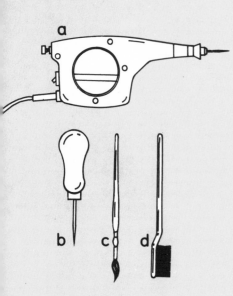

Cleaning and repairing fossils

Museums receive many fossil skeletons resembling unsorted pieces of a jig-saw puzzle stuck in rock. But there are ways to free each bone or bit of shell and clean it for display.

The first task is soaking, sawing, slicing, or otherwise removing protective packing wrapped around a fossil still embedded in its rock. Next, technicians may use special chemical solutions to harden exposed, fragile bones. Then they set to work with tools or chemicals, or both. People may chip away with hammer and chisel. But experts can work faster and more carefully with help from power tools, such as grinding burrs, or dental drills with rapidly revolving diamond cutting wheels. Vibrating tungsten points of pneumatic power pens, and gas jets firing an abrasive powder can cut through rock as if it were as soft as butter. Ultrasonic waves attack weaknesses in certain rocks. But sewing needles serve for cleaning tiny, fragile skulls.

Sometimes laboratory workers crush rock with a pestle and mortar, sieve the particles, and then inspect them with a microscope. All this separates and shows up microfossils.

Cleaning tools (above)
Of tools shown, the first two free fossils from hard rock; the others remove soft rock from fossils.
a Speed engraver, with a fast-vibrating point
b Awl
c Fine-bristle brush
d Toothbrush

Acid preparation (right)
Illustrations show four stages in using acid to reveal a fossil embedded in a rock.
1 Rock encloses almost all the fossil.
2 The rock is soaked in acid for 2- to 6-hour periods.
3 Each time the soaked rock has been removed from the acid, it is washed in deionized water for a day, completely dried, and then the exposed fossil is painted with a plastic glue.
4 The prepared specimen is left to soak in water for up to two weeks.

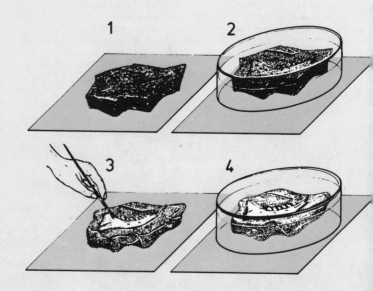

Chemicals have uses, too. Soaking limestone in dilute acetic or formic acid may remove the rock without dissolving fossils in it. Other acids attack rocks rich in iron and silica, while certain alkalis will break down shales and clays. Some chemicals are poisonous or burn the skin, so people handle these with special care.

Mending broken fossil bones and teeth is another job for the laboratory. Workers match two pieces at a time. They clean matching surfaces, then stick them together with a special adhesive. Rubber bands, metal clamps, or other aids hold the bits together until the adhesive dries.

Fossils rich in iron pyrites call for more than cleaning or repair. They need protection from decay. Damp air rots pyritic fossils, so museums store them in dry air. Treating already damaged specimens with ammonia arrests decay; and washing in fresh water prevents crumbling of pyritic fossils found at seashore sites. Even so, "pyrite disease" remains a major problem.

Inside a fish's skull
This drawing of an ancient fish's brain is based on one made by the Swedish palaeontologist Erik Stensiö in the 1920s. Stensiö used fine needles to remove the fossil skull, revealing rock formed in internal cavities that once contained brain, nerves, and blood vessels. Stensiö's discovery proved cephalaspids had been early fishes, not salamanders as some people once supposed.

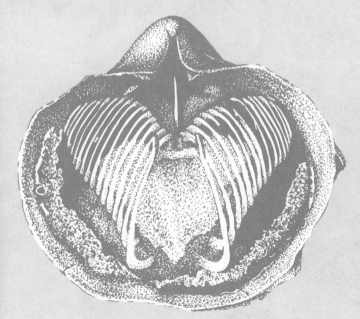

Inside a brachiopod
This much enlarged view shows the inside of one valve of a fossil brachiopod, after acid treatment. Acid has etched away unwanted rock, revealing spiral "rib-like" structures that once supported the creature's feeding system.

©DIAGRAM

Fossils for display

Sometimes, museum workers extract most of a fossil skeleton from rock. The museum might then decide to reconstruct the skeleton and mount it for display. This calls for expert knowledge of how bones or bits of shell fitted together in the creature when it was alive. Anatomists arrange bones in order, and judge their angles. Intelligent guesswork is needed for some bones of unfamiliar beasts. Then technicians can rebuild the skeleton in a life-like pose. They may set up the hip bones first, then most vertebrae, then skull, ribs, breastbone, limb bones, and the tail.

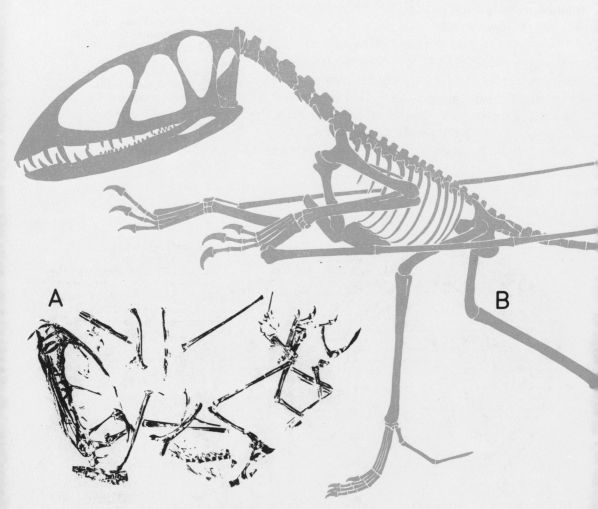

A

B

Different techniques serve for mounting specimens of different types and sizes. Workers hang heavy dinosaur bones by ropes from wooden scaffolding. They link the bones by means of angled metal pipes or rods, fastened to the bones with metal clips. Then they remove the scaffolding, leaving the skeleton exposed to view. This task can take a team of workers several months.

Museums also produce life-size, lightweight copies of important fossil skeletons. They make a plaster mould of every bone, then fill each mould with fibreglass. Museums exchange such casts to increase the variety of specimens they show.

From mounted skeletons, artists can draw or model fossil creatures as they would have looked in life. Important aids are the tell-tale bumps, grooves, and scars on bones, showing where muscles were attached. Experts studying such clues can work out shapes, weights, and volumes of beasts that lived hundreds of millions of years ago.

Dimorphodon restored
Illustrations picture three stages of a "resurrection" of a Jurassic pterosaur, *Dimorphodon macronyx.*
A *Dimorphodon*'s fossil bones as found scattered in Lower Lias rock near the southern English seaside town of Lyme Regis.
B Reconstruction of the whole fossil skeleton, about 1m (3ft 3in) long. Knowledge of anatomy enabled experts to infer shapes, sizes, and positions of any missing bones.
C Restoration of *Dimorphodon* in life-like pose.

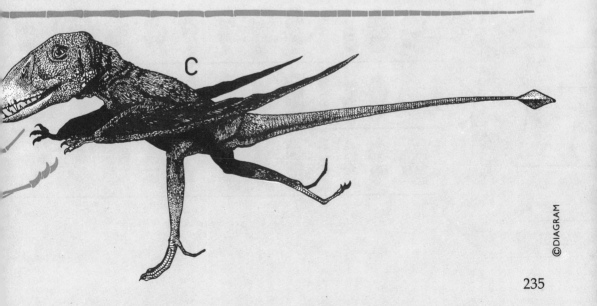

C

©DIAGRAM

235

Using fossils

Collecting and displaying fossils gives amateurs a fascinating glimpse of prehistoric life. For professional geologists, fossils mean much more. They offer clues to past climates, extinct communities, and the extent and depth of different rock formations. Indirectly, certain fossils even serve as guides to sedimentary rocks rich in useful minerals or fuels.

Macrofossils (big fossils) such as brachiopods, graptolites, and trilobites are among the chief guides to Palaeozic sedimentary rocks. Molluscs are their Mesozoic and Cenozoic counterparts. But for these latter eras, foraminiferans, ostracods, and other microfossils matter more.

Teams of geologists recover microfossils from sediments obtained as deep-drilled cores. In the laboratory, experts sieve and concentrate the samples. Then they identify these with a powerful microscope and "field guides" to microfossils. It is

Correlating rocks
Fossil foraminiferans showing rapid evolutionary change helped palaeontologists correlate these Jurassic rock strata in Montana. In three strata nine fossil forms stayed the same, but others vanished or appeared. Thus each layer has some fossil feature enabling experts to identify that stratum wherever it occurs.
1 Siltstone stratum
2 Limestone and shale
3 Shale
4 Limestone and shale
5 Not exposed
6 Shale
7 Limestone

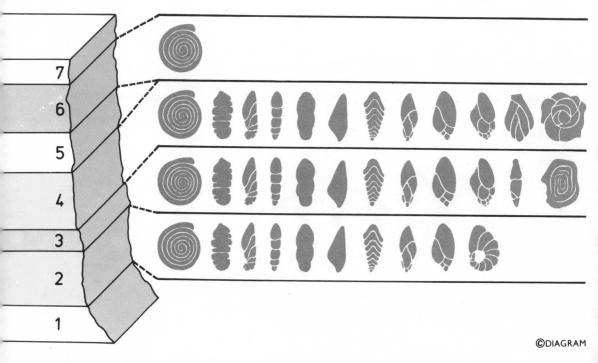

©DIAGRAM

236

possible to tell if a core came from a locality and depth likely to hold minerals or fossil fuels. All fossil fuels and certain other useful substances result from long-dead organisms, changed and concentrated in some way. For instance, coal forms slowly as a mass of plant material is fossilized, then buried under sediments and changed by heat and pressure. Petroleum and natural gas form when impermeable sea-bed sediments trap permeable rock containing the remains of billions of marine micro-organisms whose bodies once secreted oily droplets.

Limestone used for building blocks or in cement comprises fossil skeletons of algae and invertebrates that lived in shallow prehistoric seas.

Even some phosphate fertilizer comes from sea-bed layers enriched by phosphorus from fish bones and the shells of sea-bed creatures.

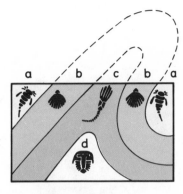

Fossils and folding (above)
Fossils of beasts that lived at different times help experts to date rocks that have been tilted or even overturned by folding.
a Devonian eurypterids
b Silurian brachiopods
c Ordovician crinoid
d Cambrian trilobite

Fossil fuel formation (left)
Block diagrams show parts played by prehistoric life forms in coal and oil formation.
1 Coal formation
a Rotting swamp plants form peat.
b Pressure of accumulating sediments turns peat to lignite.
c Increased pressure turns lignite to bituminous coal.
d Heat and deformation change bituminous coal to anthracite.
2 Oil formation
a Marine micro-organisms die and fall to the sea bed.
b Pressure cooking and bacteria act on micro-organisms to release hydrocarbons.
c Hydrocarbons migrate up through porous rock until trapped by impervious rocks here tilted by a risen salt dome.

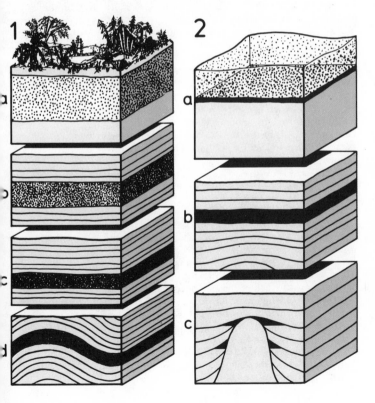

237

Famous fossil hunters 1

These four pages list major fossil hunters and others whose achievements added significantly to our understanding of prehistoric animals and plants. No list this brief could be complete; ours stresses pioneers, especially from the Western World. Outstanding personalities have also come from the USSR, Poland, China, India, and elsewhere.

Roy Chapman Andrews

Mary Anning

William Buckland

Agassiz, Jean Louis (1807–73) Swiss-born naturalist who made key studies of fossil fishes and showed that ice caps had covered much of Pleistocene Europe and North America.
Alberti, Friedrich August von (1795–1878) German geologist who in 1824 named the Triassic system from a threefold division of rocks found in Germany.
Ameghino, Florentino (1854–1911) Argentinian palaeontologist who described many South American fossil dinosaurs and mammals, largely from specimens collected by his brother Carlos.
Andrews, Roy Chapman (1884–1960) American leader of expeditions for the American Museum of Natural History. His Gobi Desert journeys of the 1920s produced the first-known dinosaur eggs.
Anning, Mary (1799–1847) British fossil collector from Lyme Regis, Dorset. She found the first British pterosaur and the first complete ichthyosaur and plesiosaur.
Bakker, Robert American palaeontologist who in the 1970s argued, controversially, that dinosaurs had been warm-blooded, and that birds are dinosaurs.
Barrande, Joachim (1799–1883) French pioneer researcher of Palaeozoic rocks and fossils. His 32-volume work included the first accounts of 4000 fossil species.
Beyrich, Heinrich Ernst (1815–96) German palaeontologist who coined the term Oligocene in 1854.
Brongniart, Adolphe (1801–76) French botanist who founded palaeobotany. In 1822 he published the first account of all known fossil plants.
Brongniart, Alexandre (1770–1847) French geologist who in 1829 named the Jurassic Period from its limestone rocks in the Jura Mountains. With Georges Cuvier he compiled a chronological sequence of Tertiary rocks in France.
Bronn, Heinrich Georg (1800–62) German palaeontologist and geologist who laid the basis for a chronological study of fossil organisms in Germany.
Broom, Robert (1866–1951) Scottish palaeontologist who made major discoveries about mammal-like reptiles and the origin of mammals.
Brown, Barnum (active 1890s and early 1900s) American fossil hunter who pioneered the Canadian "dinosaur rush", collecting bones from the Red Deer River Valley in 1910.
Buckland, William (1784–1856) British geologist who in 1824 described *Megalosaurus*, the first dinosaur to get a scientific name.
Buffon, Georges, Comte de (1707–88) French naturalist. He helped pioneer the idea that a succession of plants and animals dated back farther than theologians believed.

Hawkins' Iguanodon
This is a model of Benjamin Waterhouse Hawkins' restoration of *Iguanodon*. The full-size cement, stone and metal original still stands in a London park.

Conybeare, William Daniel (1787–1857) British geologist who in 1822 named the Carboniferous system. He also first described *Ichthyosaurus*.

Cope, Edward Drinker (1840–97) American zoologist, who described many American fossil fishes, reptiles, and mammals – especially from the newly opened West.

Cushman, Joseph (1881–1949) American palaeontologist who pioneered the use of foraminiferans as guides to the relative ages of certain rocks.

Cuvier, Georges (1769–1832) French anatomist and palaeontologist who pioneered the scientific study of fossil vertebrates. He believed groups of prehistoric beasts had perished from a series of natural catastrophes.

Dart, Raymond Arthur (1893–) Australian anatomist who in 1925 first described an *Australopithecus* fossil, from Botswana.

Darwin, Charles Robert (1809–82) British naturalist who proposed the theory of evolution by natural selection.

Deshayes, Gérard Paul (1797–1875) French conchologist, whose studies of fossil shells laid a basis for subdividing the Tertiary Period into epochs.

Desnoyers, Jules French geologist who in 1829 separated the Quaternary Period from the Tertiary Period. (Tertiary was a term coined in the mid 18th century.)

Douglass, Earl (1862–1931) American fossil collector. In 1909 he found a mass of embedded dinosaur bones in Utah (see p. 246 – Jensen, Utah: Dinosaur National Monument).

Dubois, Eugène (1858–1940) Dutch anatomist and palaeontologist who in 1891 made the first find of a *Homo erectus* fossil.

Ehrenberg, Christian (1795–1876) German naturalist who pioneered the study of microfossils. His *Mikrogeologie* (1854) depicted radiolarian skeletons and showed the role of microfossils in building limestones such as chalk.

Fischer von Waldheim, Gotthelf (1771–1853) German scientist who helped pioneer palaeontology in Russia.

Gilmore, Charles Whitney (1874–1945) American museum curator and expedition leader. He greatly enlarged the fossil reptile collection of the American National Museum of Natural History.

Hall, James (1811–98) Main founder of the "American school" of palaeontology. He wrote on the Palaeozoic invertebrates of New York State.

Halloy, Jean-Baptiste-Julien Omalius d' (1783–1875) French geologist who in 1822 named the Cretaceous Period.

Hawkins, Benjamin Waterhouse (1807–89) British sculptor who created the first life-size restorations of dinosaurs, completed in 1854.

Georges Cuvier

Charles Darwin

©DIAGRAM

Famous fossil hunters 2

Huene's Plateosaurus
Plateosaurus, probably a quadruped, appears bipedal in many restorations. Friedrich von Huene's discoveries added to our knowledge of this prosauropod dinosaur.

James Jensen with the shoulder blade of "Supersaurus"

Othniel Charles Marsh

Hooke, Robert (1635–1703) English scientist who gave the first descriptions of fossil wood, and foreshadowed the idea of fossils as clues to evolution.

Huene, Friedrich von (active early 1900s) German palaeontologist. In 1921 he found a fossil herd of *Plateosaurus* dinosaurs near Trossingen in Germany.

Hutton, James (1726–97) Scottish geologist who showed that natural agents still at work had produced geological changes over vast periods of time.

Hyatt, Alpheus (1838–1902) American geologist and palaeontologist who helped to classify ammonites.

Janensch, Werner (active early 1900s) German palaeontologist whose expeditions of 1909–12 recovered a wealth of dinosaurs from Tendaguru, now in Tanzania.

Jensen, James American fossil hunter who found two of the largest known dinosaurs ("Supersaurus" and "Ultrasaurus") in Colorado in the 1970s.

Lamarck, Jean-Baptiste de (1744–1829) French palaeontologist, pioneer in the scientific study of fossil invertebrates.

Lapworth, Charles (1842–1920) British geologist who, using graptolites, identified the Ordovician system in 1873.

Leakey, Louis Seymour (1903–72) British palaeontologist who worked in East Africa, with his wife Mary and son Richard. Their finds have added much to our knowledge of man's early evolution.

Leidy, Joseph (1823–91) American anatomist who pioneered the study of fossil vertebrates in North America. In 1856 he became the first to name an American dinosaur.

Lhuyd, Edward (1660–1709) Welsh natural historian who in 1699 produced the first book about British fossils.

Linnaeus, Carolus (1707–78) Swedish botanist who established a basis for classifying living things.

Lyell, Sir Charles (1797–1875) British geologist whose *Principles of Geology* helped found the modern science of geology. In the 1830s he named the Eocene, Miocene, Pliocene, and Pleistocene epochs.

Mantell, Gideon Algernon (1790–1852) British amateur geologist who in 1825 described *Iguanodon*, the second dinosaur to be named.

Marsh, Othniel Charles (1831–99) American palaeontologist whose expeditions discovered scores of fossil vertebrates in the West and Mid West. He described countless fossils including 17 dinosaur genera that still bear the names he gave them.

Murchison, Sir Roderick Impey (1792–1871) British geologist who established the sequence of early Palaeozoic rocks. He identified the Silurian and Permian systems.

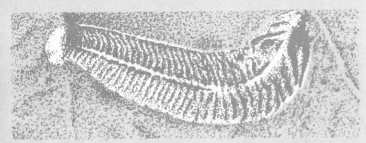

Sprigg's Spriggina
The name of the ancient sea worm *Spriggina* commemorates Sprigg's remarkable discoveries of Precambrian fossil animals in South Australia.

Nopcsa, Baron Franz (1877–1933) Palaeontologist born in what is now Romania; he discovered or described a number of dinosaurs.

Orbigny, Alcide Charles d' (1802–57) French naturalist whose study of fossil invertebrates revealed regional distributions of species in ancient seas.

Osborn, Henry Fairfield (1857–1935) American palaeontologist and expedition leader who organized major fossil hunts in Colorado, Wyoming, and Central Asia. He wrote over 600 scientific papers, and became president of the American Museum of Natural History.

Owen, Sir Richard (1804–92) British anatomist and palaeontologist who coined the name "Dinosauria". He wrote the first work on general palaeontology in English, and became the first director of the British Museum (Natural History).

Sir Richard Owen

Phillips, John (1800–74) British geologist who in 1840 named the Mesozoic and "Kainozoic" (Cenozoic) eras.

Romer, Alfred Sherwood (1894–1973) American palaeontologist who contributed new ideas about how fishes, amphibians, and reptiles evolved. He wrote major textbooks on fossil and living vertebrates.

Scheuchzer, Johann Jakob (1672–1733) Swiss botanist who wrote one of the first books to picture fossil plants.

Schimper, Wilhelm Philipp (1808–80) German palaeontologist who in 1874 named the Palaeocene Epoch.

Schlotheim, Ernst von (1764–1832) German palaeontologist who helped pioneer the use of fossils as clues to the relative ages of rock strata.

Sedgwick, Adam (1785–1873) British geologist who named the Palaeozoic Era, in 1838, and the Cambrian and Devonian systems.

Alfred Sherwood Romer

Seward, Sir Albert Charles (1863–1941) British palaeobotanist who wrote key books on fossil plants.

Smith, William (1769–1839) British geologist known as the father of English geology. He identified layers of sedimentary rock by their fossils, and in 1815 made the first geological map of England and Wales.

Sprigg, R.C. Australian geologist who in 1947 discovered major Precambrian fossils in South Australia.

Sternberg, Kasper von (1761–1838) Czech palaeobotanist who related the classification of fossil plants to that for living plants.

Walcott, Charles (1850–1927) American palaeontologist who made important studies of trilobites.

Woodward, Arthur Smith (1864–1944) British palaeontologist who made a major catalogue of fossil fishes, with many kinds reclassified.

Woodward, John (1665–1728) English palaeontologist who compiled one of the first classifications of fossils.

Zittel, Karl Alfred von (1830–1904) German scientist who compiled major handbooks of fossils, and a key history of palaeontology.

William Smith

©DIAGRAM

241

Museum displays 1

Thousands of museums show or store fossils or models of prehistoric animals and plants. In some Western countries the chief town in almost every state or county depicts local prehistoric life. These six pages give just a brief selection of interesting collections worldwide. Items appear in alphabetical order of country and then city.

ARGENTINA
Buenos Aires: Argentine Natural Science Museum Features a fine collection of fossil vertebrates from Argentina.
La Plata: Museum of La Plata University Includes fossil dinosaurs from Argentina.
AUSTRALIA
Adelaide: South Australian Museum Contains fossil mammals, reptiles (especially crocodiles), fishes, amphibians, etc.
Brisbane: Queensland Museum Includes fossil mammals, reptiles, fishes, and some fossil bird remains. Little on show before 1986.
Melbourne: Geological Museum Includes a big collection of fossil graptolites from Victoria.
Melbourne: Museum of Victoria Has mostly mammal fossils.
Perth: Western Australian Museum Includes fossil invertebrates and vertebrates (mostly mammals).
Sydney: Australian Museum Includes mostly Devonian fishes and mammals. Exhibition galleries of its Department of Palaeontology show the story of life on Earth.
AUSTRIA
Vienna: Natural History Museum Its 30,000 specimens include mounted skeletons of a moa, Austrian Ice Age mammals, and remains of an armoured dinosaur.
BELGIUM
Brussels: Royal Institute of Natural Sciences This has the world's best and biggest collection of *Iguanodon* skeletons. Other exhibits include mosasaurs and invertebrate Mesozoic and Cenozoic fossils.
CANADA
Calgary, Alberta: Zoological Gardens An outdoor park shows 50 full-size models of dinosaurs.
Drumheller, Alberta: Tyrrell Museum of Palaeontology On show are major fossils and restorations of Late Cretaceous dinosaurs from Alberta.
Edmonton, Alberta: Provincial Museum of Alberta Includes a partial dinosaur skeleton and life-size models of several other dinosaur genera.
Ottawa, Ontario: National Museum of Natural Sciences A "life through the ages" sequence includes mounted examples of western Canada's Late Cretaceous dinosaurs, and a reconstructed prehistoric forest.
Toronto, Ontario: Royal Ontario Museum Canada's largest public museum is rich in North American vertebrate fossils. The Dinosaur Gallery includes dinosaurs, mosasaurs, an ichthyosaur, a plesiosaur, and an early crocodile.

A giant ammonoid in the South Australian Museum

A giant moa in the Natural History Museum, Vienna

CHINA

Beijing (Peking): Beijing Natural History Museum On show are many Chinese fossils. The five dinosaur skeletons include an immense reconstructed *Shantungosaurus*. Museums in some other major Chinese cities also have significant fossil displays.

CZECHOSLOVAKIA

Prague: Národní Museum The national museum includes a number of fossils found in the 19th century. Many were the first of their kind to be discovered.

FRANCE

Paris: National Museum of Natural History This has the largest fossil collection in France. It includes bones or casts of dinosaurs from various continents. (Regional museums also feature fossil displays.)

GERMANY (EAST)

East Berlin: Natural History Museum, Humboldt University Its fossil collection has important dinosaur skeletons from Late Jurassic Tanzania; the *Brachiosaurus* is the world's largest mounted dinosaur.

GERMANY (WEST)

Darmstadt: Hesse State Museum Unusual exhibits include an American mastodont skeleton.

Frankfurt am Main: Senckenberg Natural History Museum On show is a wealth of vertebrate fossils, including ichthyosaurs, plesiosaurs, and dinosaurs from various parts of the world.

Munich: Bavarian State Institute for Palaeontology and Historical Geography This includes the first known skeleton of the Jurassic dinosaur *Compsognathus*.

Stuttgart: State Museum for Natural History Its collection includes important dinosaur fossils.

Tübingen: Institute and Museum for Geology and Palaeontology Fossils on show include dinosaur skeletons and casts.

GREECE

Athens: Department of Geology and Palaeontology, University of Athens This has important fossil Pliocene mammals.

INDIA

Calcutta: Geology Museum, Indian Statistical Institute The chief exhibit is the only mounted skeleton of the early sauropod *Barapasaurus*. (Other museums also feature fossils.)

ITALY

Bologna: G. Capellini Museum has a large dinosaur cast.

Genoa: Civic Museum of Natural History The collection includes fossil plants, invertebrates, and vertebrates.

Milan: Civic Museum of Natural History Fossils feature in its collection.

Padua: Museum of the Institute of Geology This museum includes fossil plants.

Rome: Museum of Palaeontology, Institute of Geology and Palaeontology This has fossil Quaternary mammals, footprints, and invertebrates.

JAPAN

Osaka: Museum of Natural History Fossil plants, invertebrates, and vertebrates contribute to a history in nature theme.

Tokyo: National Science Museum On show is a significant fossil collection, including a dinosaur display.

The mounted *Brachiosaurus* in an East Berlin museum

An ichthyosaur in Stuttgart's Natural History Museum

An *Allosaurus* in Osaka's Museum of Natural History

© DIAGRAM

243

Museum displays 2

KENYA
Nairobi: Kenya National Museum Includes remains of fossil man from East Africa.

MEXICO
Mexico City: Natural History Museum Includes a cast of the sauropod dinosaur *Diplodocus*.

MONGOLIA
Ulan-Bator: State Central Museum Features an imposing collection of dinosaurs from the Gobi Desert.

MOROCCO
Rabat: Museum of Earth Sciences This is planned to house fossils including the largest known skeleton of the sauropod dinosaur *Cetiosaurus*.

NIGER
Niamey: National Museum Fossils include *Ouranosaurus*, a big sail-backed iguanodontid dinosaur.

POLAND
Warsaw: Palaeobiology Institute, Academy of Sciences Has a major collection of Mongolian dinosaurs, but seldom on show.

SOUTH AFRICA
Cape Town: South African Museum This has important fossil reptiles from Permian and Triassic southern Africa.

SPAIN
Madrid: Natural Science Museum Includes a cast of the sauropod dinosaur *Diplodocus*.

SWEDEN
Uppsala: Palaeontological Museum, Uppsala University This has fossils of Chinese dinosaurs including a sauropod.

UNITED KINGDOM
Cambridge: Sedgwick Museum, Cambridge University Fossils include some local dinosaur remains.

Cardiff: National Museum of Wales Has a large fossil collection, including Palaeozoic Welsh invertebrates.

Cheddar, Somerset: Cheddar Caves Museum Features Pleistocene remains, including a Palaeolithic human burial.

Clitheroe, Lancs: Clitheroe Castle Museum Has a significant collection of Carboniferous fossils.

Dorchester: Dorset County Museum Includes dinosaur footprints.

Edinburgh: Royal Scottish Museum Has a major national collection of early fossils including invertebrates (notably crinoids and eurypterids), vertebrates, and plants.

Elgin: Elgin Museum Includes world-famous fossils of Scotland's Early Mesozoic reptiles.

Glasgow: Hunterian Museum Includes a *Triceratops* skull and paintings of Scottish Mesozoic reptiles.

Glasgow: Victoria Park Its Fossil Grove building houses remarkable fossil Carboniferous tree stumps.

Huddersfield: Tolson Memorial Museum Includes a section with geological exhibits.

Ipswich: Ipswich Museum Local invertebrate and vertebrate fossils, including bones, teeth, and tracks of dinosaurs.

Keighley, West Yorks: Cliffe Castle Has a geological gallery.

Leicester: Leicestershire Museum and Art Gallery Its fossils include remains of Precambrian invertebrates.

Ouranosaurus in the National Museum, Niamey

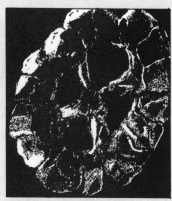

Coccolith in the Sedgwick Museum, Cambridge

A fossil tree stump in Glasgow's Victoria Park

London: British Museum (Natural History) One of the world's great natural history museums, with more than three million fossils on show or stored, plus many casts. Displays represent most major forms of prehistoric life.

London: Crystal Palace Park Has the world's oldest display of full-size model dinosaurs, completed in 1854.

London: Geological Museum Permanent exhibitions include "British Fossils" and "Britain Before Man".

Lyme Regis: Lyme Regis Museum *Ichthyosaurus* and other local Jurassic fossils figure in this small collection.

Maidstone: Maidstone Museum Includes some *Iguanodon* bones with a related display.

Manchester: Manchester Museum Fossils and minerals figure in its total collection of over eight million items.

Newcastle upon Tyne: Hancock Museum One of England's finest natural history museums. It includes geological items.

Oxford: University Museum The five thousand fossil exhibits include unique dinosaur specimens.

Peterborough: City of Peterborough Museum and Art Gallery Includes an exhibition of local geology.

Portsmouth: Cumberland House Natural Science Museum and Aquarium Has a full-size restoration of the dinosaur *Iguanodon*.

Sandown: Museum of Isle of Wight Geology Houses local fossils, including part of the oldest-known bone-headed dinosaur, *Yaverlandia*.

Sunderland: Sunderland Museum Has a sizable collection of Permian fossils, including fishes and the lizard-like gliding reptile *Weigeltisaurus*.

Tenby, Dyfed: Tenby Museum Has a significant geological collection on show.

Torquay: Natural History Society Museum On show are Prehistoric finds from Kent's Cavern.

York: Yorkshire Museum Extinct and fossil birds figure in this museum's collection.

UNITED STATES OF AMERICA

Amherst, Massachusetts: Amherst College Displays a major collection of dinosaur footprints.

Austin, Texas: Texas Memorial Museum On show are Late Palaeozoic and Mesozoic reptiles (including sea turtle, mosasaur, and dinosaurs), and Pleistocene mammals.

Berkeley, California: University of California Museum of Paleontology Includes Triassic and Jurassic reptiles.

Boulder, Colorado: University Natural History Museum On show are Jurassic dinosaur fossils.

Buffalo, New York: Buffalo Museum of Science Fossils include dinosaur bones, footprints, eggs, skin impressions, and other items.

Cambridge, Massachusetts: Museum of Comparative Zoology, Harvard University Displays a major collection of fossil vertebrates. The museum includes fossil fishes, the best North American collection of South America's Early Mesozoic amphibians and reptiles, and North American dinosaurs.

Canyon, Texas: Panhandle Plains Museum On show are local Triassic reptiles.

© DIAGRAM

The London specimen of *Archaeopteryx*

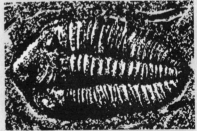

A reconstruction of *Weigeltisaurus* from Sunderland Museum

A trilobite in Harvard's Museum of Comparative Zoology

Museum displays 3

Dunkleosteus's huge head on show in Cleveland, Ohio

A bird from the La Brea tar pits of Los Angeles

Chicago, Illinois: Field Museum of Natural History A major museum that includes fossil plants, invertebrates, and vertebrates from South America and the Western United States – for example mounted dinosaurs.

Cincinnati, Ohio: University of Cincinnati Fossils include some dinosaur remains.

Cleveland, Ohio: Natural History Museum This is famous for *Dunkleosteus* and other Devonian fossil fishes, the only mounted skeleton of the sauropod dinosaur *Haplocanthosaurus*, and Pleistocene mammals including a mastodont.

Denver, Colorado: Denver Museum of Natural History Its "Succession of Life" displays feature dinosaurs, marine reptiles, and 50 million years of mammal evolution.

Durham, North Carolina: North Carolina Museum The fossil collection includes some dinosaur bones.

East Lansing, Michigan: The Museum, Michigan State University A Hall of Life includes fossils and wall paintings from successive geological eras.

Flagstaff, Arizona: Museum of Northern Arizona Items include fossils of a small, early, armoured ornithopod dinosaur.

Fort Worth, Texas: Fort Worth Museum of Science and History Exhibits include mounted dinosaurs.

Hays, Kansas: Sternberg Memorial Museum Toothed birds are among its display of local Cretaceous fossils.

Houston, Texas: Houston Museum of Natural Science Large exhibits include much of a *Diplodocus* skeleton.

Jensen, Utah: Dinosaur National Monument Comprises more than 200,000 acres of fossil-rich canyons. The Carnegie Quarry is now a covered Visitor Center where you can view technicians freeing thousands of dinosaur bones from rock.

Laramie, Wyoming: W.H. Reed Museum Exhibits include part of a fossil sauropod, *Apatosaurus*.

Lawrence, Kansas: University of Kansas Museum of Natural History The collection includes Mesozoic fossils.

Lincoln, Nebraska: University of Nebraska State Museum This has a good display of fossil mammals.

Los Angeles, California: Los Angeles County Museum of Natural History Holds a big collection of Cretaceous fossil vertebrates (e.g. *Pteranodon, Tylosaurus*) and the world's largest collection of later Pleistocene fossil vertebrates. At its La Brea Center viewers can see experts excavating prehistoric bones from tar pits.

Newark, Delaware: University of Delaware Fossils include some sauropod dinosaur remains.

New Haven, Connecticut: Peabody Museum of Natural History, Yale University This has a major collection of fossil vertebrates – especially American dinosaurs and early mammals.

New York City, New York: American Museum of Natural History This great collection has a rich display of mounted skeletons – including fishes, amphibians, reptiles, and mammals. No other museum contains so many dinosaurs. Exhibits include fossil eggs, tracks, and skin imprints.

Norman, Oklahoma: Stovall Museum, University of Oklahoma Fossils include a big, flesh-eating dinosaur from Oklahoma.

Palaeoscincus in the American Museum of Natural History

A fossil forest
Logs literally turned to stone lie jumbled in Rainbow Forest. This is one of the six fossil forest areas in Arizona's Painted Desert.

A reconstruction of the early fossil bat in a Princeton Museum

Painted Desert Arizona: Petrified Forest National Park Has the world's largest concentration of petrified wood – six "forests" of 150-million-year-old giant conifer logs, transformed to agate and chalcedony.

Peoria, Illinois: Lakeside Museum and Art Center Fossils include sauropod dinosaur remains.

Philadelphia, Pennsylvania: Academy of Natural Sciences This has some of the first fossil dinosaurs found in North America.

Pittsburgh, Pennsylvania: Carnegie Museum of Natural History On show are major fossil displays, including a Mesozoic Hall with some of the world's best-preserved, mounted specimens of Late Jurassic dinosaurs, as well as mosasaurs and a marine turtle.

Princeton, New Jersey: Museum of Natural History, Princeton University Fossils include Late Cretaceous dinosaurs and one of the first-known fossil bats.

St Paul, Minnesota: The Science Museum of Minnesota Contains local Cretaceous fossils, including champsosaurs.

Salt Lake City, Utah: Utah Museum of Natural History Fossils include dinosaurs from the famous Cleveland-Lloyd Quarry, also pterosaur tracks.

San Francisco, California: California Academy of Science Fossil items include single dinosaur bones.

Scranton, Pennsylvania: Everart Museum Its fossils include some dinosaur bones.

Vernal, Utah: Utah Natural History State Museum Exhibits include a *Diplodocus* skeleton.

Washington, D.C.: National Museum of Natural History, Smithsonian Institution This major museum has fossils representing most phases of prehistoric life. The Department of Paleobiology holds one of the world's largest collections of type specimens (the first fossils of their kind to get a description and name).

USSR

Leningrad: Central Geological and Prospecting Museum Fossil exhibits include an Asian hadrosaurid dinosaur.

Leningrad: Museum of Zoology Unique exhibits include a mammoth that had been preserved by permafrost.

Moscow: Palaeontological Museum Its impressive displays include five skeletons of the big, Mongolian flesh-eating dinosaur *Tarbosaurus*.

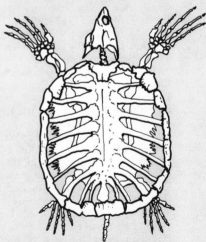

Archelon skeleton in the National Museum of Natural History, Washington, D.C.

FURTHER READING

General

Archer, M. and Clayton, G. (editors) *Vertebrate Zoogeography & Evolution in Australasia (Animals in Space & Time)* Hesperian Press, Australia 1984

Colbert, E. H. *Evolution of the Vertebrates* Wiley, 1980

McFarland, W. N.; Pough, F. H.; Cade, T. J.; and Heiser, J. B. *Vertebrate Life* Macmillan, USA 1979; Collier Macmillan, UK 1979

Moody, R. *Prehistoric World* Chartwell Books, 1980

Romer, A. S. *Vertebrate Palaeontology* University of Chicago Press, 1966

Sheehan, A. (editor) *The Prehistoric World* Galley Press, 1983

Steel, R. and Harvey, A. P. (editors) *The Encyclopaedia of Prehistoric Life* Mitchell Beazley, UK 1979; McGraw-Hill, USA 1979

Swinnerton, H. H. *Fossils* Collins, 1960

Note: News of major fossil discoveries appears in serious newspapers and in science magazines such as *Nature*, but many finds appear in specialist journals such as the *Journal of Paleontology*; *Paleontology*; and *Paleobiology*.

Chapter 1: Fossil Clues to Prehistoric Life

Colbert, E. H. *Wandering Lands and Animals* Dutton, USA 1973; Hutchinson, UK 1974

Fortey, R. *Fossils: The Key to the Past* Heinemann, 1982

Raup, D. M. and Stanley, S. M. *Principles of Paleontology* W. H. Freeman and Co, USA 1978

Chapter 2: Fossil Plants

Stewart, W. N. *Paleobotany and the Evolution of Plants* Cambridge University Press, 1983

Taylor, T. N. *Paleobotany* McGraw-Hill, USA 1981

Chapter 3: Fossil Invertebrates

Lehmann, U. and Hillmer, G. *Fossil Invertebrates* Cambridge University Press, 1983

Moore, R. C. et al. *Invertebrate Fossils* McGraw-Hill, USA 1952

Chapter 4: Fossil Fishes

Schultze, H.-P. *Handbook of Paleoichthyology* Fischer, West Germany 1978– (10 volumes planned)

Chapter 5: Fossil Amphibians;

Chapter 6: Fossil Reptiles

Charig, A. J. *A New Look at the Dinosaurs* Heinemann, UK 1979; Smith Publishers, USA 1979

Kemp, T. *Mammal-like Reptiles and the Origins of Mammals* Academic Press, 1982

Kuhn, O. (editor) *Encyclopedia of Paleoherpetology* Fischer, West Germany, 1969– (19 volumes planned)

Lambert, D. with the Diagram Group *Collins Guide to Dinosaurs* Collins, UK 1983; and as *A Field Guide to Dinosaurs* Avon Books, USA 1983

Swinton, W. E. *Fossil Amphibians and Reptiles* British Museum (Natural History), 1973

Chapter 7: Fossil Birds

Feduccia, A. *The Age of Birds* Harvard University Press, 1980

Swinton, W. E. *Fossil Birds* British Museum (Natural History), 1975

Chapter 8: Fossil Mammals

Kurtén, B. *The Age of Mammals* Weidenfeld and Nicolson, UK 1971; Columbia University Press, USA 1972

Clapham, F. M. (editor) *The Rise of Man* Sampson Low, 1976

Gribbin, J. and Cherfas, J. *The Monkey Puzzle*, Bodley Head, UK 1982; McGraw-Hill, USA 1983

Halstead, L. B. *The Evolution of the Mammals* Book Club Associates, 1979

Chapter 9: Records in the Rocks

Dott, Jr., R. H. et al. *Evolution of the Earth* McGraw-Hill, USA 1981

Spinar, Z. V. and Burian, Z. *Life Before Man* Thames and Hudson, UK 1972; McGraw-Hill, USA 1972

Chapter 10: Fossil Hunting

Hamilton, R. and Insole, A. N. *Finding Fossils* Penguin Books, 1977

Macfall, R. P. and Wollin, J. *Fossils for Amateurs: A Guide to Collecting and Preparing Invertebrate Fossils* Van Nostrand, USA 1972

Murray, M. *Hunting for Fossils* Macmillan, USA 1974

Rixon, A. E. *Fossil Animal Remains: Their Preparation and Their Conservation* Humanities Press, USA 1976; Athlone Press, UK 1976

INDEX

A

Aardvarks 27, 150, 168
Acanthodes 79
Acanthodians (Acanthodii) 72, 75, 78–79, 207–208, 210, 213
Acanthodiformes 79
Acer trilobatum 43
Acritarchs 49
Acropora 53
Acteosaurus 114
Actinopoda 49
Actinopterygians (Actinopterygii) 75, 84–85
Adapis 158
Aeger 67
Aegialornis 146–147
Aegyptopithecus 160, 222
Aellopos 83
Aeluroids 164, 166
Aepyornis 142–143
Aepyornithiformes 139
Aëtosaurs 116–117
Agnathans (Agnatha) 72, 75–77, 207–208
Agrichoeres 189
Agrichoerus 189
Aistopods (Aistopoda) 90, 96
Albatrosses 139, 144
Alcyonaria 52
Alecidiniformes 139, 147
Algae 18, 22–23, 30–32, 34–36, 43, 204, 237
Allodesmus 167
Allosaurus 122, 217, 243
Allotheria 148, 150–152
Alticamelus 188
Amblypods (Amblypoda) 150, 168, 170–171, 178, 220–221
Ammonites 18–19, 28, 57, 209, 217, 219–220, 229, 240
Ammonoids (Ammonoidea) 56–57, 213
Amphibians (Amphibia) classification of 88, 90
  evolution of 18, 23, 90–91
  in geological periods 208–214, 216
  types of 88–97
Amphichelyids 104
Amphineura 47
Ampyx 63
Amynodontids 184
Anapsids (Anapsida) 98, 100–101
Anaspida 77
Anchiceratops 219
Ancylopoda 182
Andrewsarchus 169
Andrias scheuchzerii 88
Anetoceras 57
Angiosperms (Angiospermae) 22, 42–43
Animals (Animalia), classification of 20–21, 23, 30

Ankylosaurs 128–129
Ankylosaurus 128–129
Annelid worms (Annelida) 23, 47, 58–59, 204–205, 207
Anomodonts (Anomodontia) 130, 132
Anoplotheres (Anoplotheroidea) 187–189
Anoplotherium 189
Anseriformes 139, 144
Anteaters 27, 152–153, 196, 223
Antelopes 194–195, 224
Anthozoa 52
Anthracosaurs 92–94, 213
Anthracotheres (Anthracotherioidea) 186–187, 221–222
Anthropoids 158
Antiarchi 81
Ants 15
Anura 88, 90, 97, 214
Apes 158–161, 223
Aphaneramma 94
Apoda 88, 90, 97
Apodiformes 139, 147, 222
Apodops 97
Apterygiformes 139
Arachnids (Arachnida) 65, 209
Araeoscelids (Araeoscelidea) 100, 106–107
Araeoscelis 107
Araneae 65
Araucaria 41
Archaeocetes 198–199
Archaeocyathids 28, 50–51
Archaeocyathines (Archaeocyatha) 47, 51
Archaeomeryx 191
Archaeopsittacus 146
Archaeopterygiformes 139–140
Archaeopteryx 136, 139–141, 216, 245
Archaeornithes 136, 139
Archaeosperma 41
Archaeotherium 186
Archaeotrogon 147
Archelon 104–105, 247
Archosaurs (Archosauria) 28, 100–101, 116, 118, 215
Arctoids 166–167
Arctyonids 169
Argentavis 146
Armadillos 196–197, 222–223
Arsinoitherium 172–173
Arthrodira 81
Arthropleura 60
Arthropods (Arthropoda) 18, 47, 54, 58, 60–67, 205–208, 210, 212, 214, 216, 218
Artiodactyls (Artiodactyla) 150, 186–195, 221–225
Aschelminthes 47, 58
Asiamerica 218
Askeptosaurus 113
Aspergillus 31

Asses 181
Asteroidea 71
Asteroxylon 39, 208
Asterozoa 47
Astrapotheres (Astrapotheria) 150, 178–179
Astrapotherium 179
Athrolycosa 65
Auks 144
Aulophyseter 199
Australopithecus 160, 239
Aves, see Birds
Aysheaia 60

B

Bacillariophyceae 37
Bacteria 22–23, 30–31
Bactritoids (Bactritoidea) 56–57
Badgers 166
Balanus 67
Baluchitherium 184
Bandicoots 150, 154
Barnacles 66–67
Barylambda 171
Barytheres (Barytherioidea) 174
Basilosaurus 198
Batoidea 83
Batrachosaurs (Batrachosauria) 90, 92–93
Bats 18, 24, 150, 156–157, 221, 247
Bauria 134
Bauriamorphs 134
Baurusuchus 118–119
Bear-dogs 223
Bears 164, 166–167, 224–225
Beavers 200–201, 224
Bees 60, 218
Belemnites (Belemnitida) 56–57
Bennettitaleans (Bennettitales) 41, 214, 216
Birds (Aves)
  classification of 136, 139
  evolution of 18, 23, 138–139
  in geological periods 216, 218–225
  types of 136–147
Birkenia 77
Bison 195
Bivalves 25, 28, 54, 69, 206, 220
Bladder wrack 34–35
Blastoids 47, 70
Blastomeryx 192–193
Blastozoa 47, 70
Blue-green algae 22–23, 30–31, 34, 36, 204
Bony fishes 72, 75, 80, 84–87, 106, 212–213
Borhyaenids 155
Bothriocidaris 71
Bothriodon 187
Bothriolepis 81

Botryocrinus 70
Bovids 224
Bovoids (Bovoidea) 187, 192, 194–195
Brachiopods (Brachiopoda) 18, 47,
    68–69, 229, 233, 236–237
Brachiosaurus 124–125, 243
Branchiotremes (Branchiotremata) 23,
    47, 70–72, 206
Brittle stars 71
Brontotheres (Brontotherioidea)
    182–183, 222
Brontotherium 183
Brown algae 34–35
Bryophytes (Bryophyta) 34–35, 43
Bryozoans (Bryozoa) 47, 68–69, 210
Buchloe 42
Buffalo grass 42
Butterflies 60, 218

C

Cacops 94
Caecilians 88, 97, 220
Cainotheres (Cainotherioidea) 187–189
Cainotherium 189
Calamites 39
Calcispongea 51
Calymene 62
Calyptoptomatids 56
Cambrian Period, description of 17, 205
    fossils from 237
    life in 51–52, 54, 56–57, 60, 62, 66,
        68, 70–71
Cameloids (Cameloidea) 87–189
Camels 186–189, 222–225
Canadia 59
Canids (Canidae) 166–167
Canoids 166–167
Caprimulgiformes 139, 146
Caprimulgus 146
Captorhinomorphs 102
Carboniferous Period, amphibians in
    90–91, 93–94, 96
    description of 17, 210–211
    fossils from 15
    invertebrates in 53–54, 60–61, 65, 71
    plants in 38–41
    reptiles in 102
Carnivores (Carnivora) 131, 133–134,
    150–151, 154, 162–167, 178,
    220–223, 225
Carnosaurs 122–124, 128, 130, 217
Carodnia 179
Carpoids 47, 70
Castoridae 201
Castoroides 201
Casuariformes 139
Catarrhines (Catarrhini) 159–161
Cats 164–166, 222–225
Cattle 186–187, 190, 192, 194, 224
Cave bears 167, 225
Cave lions 225
Cavies 200
Cavioids (Caviomorpha) 201

Cenozoic Era, description of 17, 220, 224
    fossils from 228–229, 236
    life in 43, 47, 75, 91, 101, 139, 142,
        151, 166, 168, 172, 174, 182, 187,
        201
Centipedes 60–61, 211
Cephalaspids (Cephalaspida) 76–77, 233
Cephalaspis 77
Cephalocarids 63
Cephalochordates (Cephalochordata) 23,
    74
Cephalopods 54, 57
Ceramium 35
Ceratites 57
Ceratomorpha 182
Ceratopsians 128–129
Cervoids (Cervoidea) 187, 192–193
Cetacea 150, 198–199
Cetotherium 199
Chaetetopsis 51
Chalicotheres 182–183, 223
Champsosaurus 113
Charadriiformes 139, 144
Charnia 52
Charophyta 35
Cheirolepis 86
Cheirotherium 12
Chelicerates (Chelicerata) 23, 47, 64–65
Chelonians (Chelonia) 100, 104–105,
    214, 218
Chevrotain 186, 191
Chimaeriformes 83
Chiropterans (Chiroptera) 150, 156–157
Chitinozoans 49
Chitons 47, 54
Chlorophyta 35
Chondrichthyans (Chondrichthyes) 72,
    75, 82–83, 208, 210, 212, 216, 218
Chondrosteans (Chondrostei) 84–85
Chordates (Chordata) 20–22, 72, 74
Choristodera 113
Chrysophyta 36
Ciconiiformes 139, 145
Ciliates (Ciliata) 47–49
Cirripedes 66
Civets 164–166
Cladoselache 82
Cladoselachiformes 82
Cladoxylales 40
Cladoxylon 40
Clams 54
Classification system 20–23
Climatiformes 79
Climatius 79
Club mosses 22, 39, 43, 210
Cnidarians (Cnidaria) 47, 52–53
Coat-of-mail shell 54
Coatis 166
Coccolithophyceae 37
Coccoliths 37, 244
Coccolithus 37
Cockroaches 211
Coelacanths (Coelacanthini) 28, 86–87

Coelenterates 22, 52–53, 204–206, 221
Coelodonta 184
Coelolepida 77
Coeloptychium 51
Coelurosaurs 123, 138
Coelurus 123, 217
Coenopteridales 40
Coleoids (Coleoidea) 56–57
Coliiformes 139, 146
Colius 146
Columbiformes 139, 146
Colymboides 144
Compsognathus 122, 138, 243
Conchifera 47
Condylarths (Condylarthra) 168–169,
    178, 182, 198, 220–222
Coniconchs 56
Coniferales 41
Conifers 40–41, 212, 214, 216, 247
Conocardium 55
Conodonts (Conodontophorida) 47, 68
Continents, formation of 202, 204–224
Conularia 52
Cooksonia 39–40, 43
Coraciiformes 139, 147, 222
Corals 28, 47, 52–53, 204, 207,
    209–210
Cordaitales 41
Cordaites 41
Cormorants 144
Coryphodon 171
Cotylosaurs (Cotylosauria) 100, 102–104,
    106, 112, 130, 210, 212, 214
Crabs 60, 66–67
Cranes 139, 144
Cranioceras 193
Crataraster 71
Creodonts (Creodonta) 150, 162–164,
    220–222
Cretaceous Period, description of 17, 28,
    218–219
    fishes in 83–84, 87
    invertebrates in 51, 55, 57, 66–67, 69,
        71
    mammals in 152, 154, 156–157, 163,
        169
    plants in 37, 42
    reptiles in 104, 108–109, 113–114,
        119, 121, 123, 126–127, 129
Cricetops 200
Crinoids (Crinoidea) 24–25, 47, 70, 210,
    229, 237, 244
Crocodiles 101, 116, 118–119, 138, 215,
    229, 242
Crocodilians (Crocodilia) 100, 112–113,
    116–119, 214, 216, 218
Crossopterygii 87
Crustaceans (Crustacea) 23, 47, 66–67
Cryptodires (Cryptodira) 104
Cryptoprora 49
Cuckoos 139, 146
Cuculiformes 139, 146
Cyanophyta 31

Cycads (Cycadales) 41, 214, 216, 219
Cyclocorallia 53
Cyclostomata 75
Cyclotosaurus 95
Cymbella 37
Cymbospondylus 111
Cynodictis 166
Cynodonts 134
Cynognathus 134–135
Cypridea 66
Cyrtobactrites 57
Cystoids 70

D

Daphoenodon 167
Daspletosaurus 123
Decapoda 67
Deer 187, 190–194, 223–225
Deinonychosaurs 123
Deinonychus 123
Deinosuchus 118–119
Deinotheres (Deinotherioidea) 174–175
Deinotherium 175
Deltatheridians 162–163
Deltatheridium 163
Demospongea 51
Dendrocystites 70
Dentalium 55
Dermopterans (Dermoptera) 150, 156–157, 220
Desmostylans (Desmostylia) 150, 172–173, 224
Demostylus 173
Devonian Period, amphibians in 91–92
  description of 17, 208
  fishes in 76–77, 79–80, 82–84, 86–87
  invertebrates in 57, 61, 63
  plants in 35, 39, 40–41
Diadectes 93
Diapsids (Diapsida) 98, 100–101, 103, 112
Diarthrognathus 135
Diatoms 37
Diatryma 139, 143, 220
Diatrymiformes 139
Dichobune 186
Dichobunoids (Dichobunoidea) 186–187
Dichograptus 71
Dickinsonia 59
Dicots (Dicotyledonae) 42
Dicynodonts 132–133
Didymaspis 77
Dimetrodon 130–131
Dimorphodon 121, 235
Dinichthys 80–81
Dinictis 165
Dinilysia 114
Dinocephalians 132
Dinocerata 168
Dinophyceae 36–37
Dinornis 142–143
Dinornithiformes 139

Dinosaurs 18, 28, 100–101, 116, 122–129, 138, 151, 214–219, 231, 238–247
Diplocaulus 96
Diplodocids 125
Diplodocus 125, 244, 246–247
Diplograptus 71
Diplorhina 75
Dipnoi 87
Diprotodon 155, 225
Diprotodonts (Diprotodonta) 150, 154
Dipterus 87
Dissorophids 97
Docodonts (Docodonta) 150, 152–153
Dodo 146
Dogs 164–167, 222, 224–225
Dolichosaurids 114
Dolphins 24, 111
Doves 146
Dromasaurs 132
Dromiceiomimus 123
Dryopithecines 160
Dryopithecus 160
Ducks 139, 144
Duckweeds 42
Dunkleosteus 80–81, 246
Dynamopterus 146

E

Echidna 153
Echinoderms (Echinodermata) 23, 47, 70–71, 75, 206–207, 218, 229
Echinozoa 47, 71
Edaphosaurs (Edaphosauria) 130–131
Edaphosaurus 130–131
Edentates (Edentata) 150, 196–197, 221, 224–225
Elaphas falconeri 176
Elaphas trogontherii 177
Elasmobranchii 75
Elasmosaurus 109
Elasmotherium 184
Elephant birds 139, 142–143
Elephantoidea 174
Elephants (Elephantinae) 150, 172, 174–177, 221, 223–225
Embolomeri 93
Embrithopods (Embrithopoda) 27, 150, 172–173, 222
Emus 139, 142
Encrinurus 63
Entelodontoids (Entelodontoidea) 186–187
Eocardia 201
Eocene Epoch, birds in 15, 144–147
  description of 17, 221
  invertebrates in 49, 53, 58
  mammals in 156–158, 162–163, 165–166, 168–171, 173–174, 178–181, 184, 186–187, 189, 191, 197, 199
  reptiles in 113
Eogyrinus 93

Eosuchians (Eosuchia) 100, 107, 112–114, 116, 212, 214, 218
Eotheria 148, 150, 152, 214
Epigaulus 201
Equoidea 182
Equus 180–181
Errantia 59
Eryma 67
Eryon 67
Eryops 94
Erythrosuchus 117
Eucalyptus 42
Eucaryotes, see Eukaryotes
Euelephantoids (Euelephantoidea) 174–175
Eukaryotes (Eukaryota) 20–23, 30
Eupantotheres (Eupantotheria) 150, 152–154, 156
Euparkeria 12, 117, 138
Euramerica 218
Euryapsids (Euryapsida) 98, 100–101, 106–107, 214
Eurypterids (Eurypterida) 64–65, 76, 237, 244
Eurypterus 64
Eusuchians 118–119
Eusthenopteron 87, 91–92
Eutheria 150, 156–157

F

Fabrosaurus 126–127
Falconiformes 139, 146, 224
Felids (Felidae) 165–166
Felis leo spelaea 165
Feloids 164
Ferns 22, 40–41, 43, 208–209, 211, 214, 216, 219
Fishes, classification of 72, 75
  evolution of 18, 23
  fossils of 229, 233
  in geological periods 206–208, 210–213, 216–218, 220
  life of 24–25
  types of 72–87
Fissipeds 164–166
Flatworms 22, 54
Fleshy-finned fishes 75, 86–87, 213
Flowering plants 42–43
Foraminiferans (Foraminifera) 49, 220, 236, 239
Fossil fuels 237
Fossil hunters 228–230, 238–241
Frogs 88, 90, 97, 215, 219
Fucus 34–35
Fungi 22–23, 30–31, 216
Fusulina 48–49
Fusulines (Fusulinida) 49

G

Galepus 132
Galliformes 139, 146
Gallinuloides 146
Gannets 144

Gastropods (Gastropoda) 54–55, 206, 220
Gaviiformes 139, 144
Gazella 194
Gazelles 194, 224
Geese 144
Gemuendina 80
Genet 165
Genetta 165
Geological periods 16–17, 204–225
Gephyrostegoidea 93
Gephyrostegus 93
Geranopterus 147
Gerrothorax 95
Ginkgoes 41, 214
Ginkgoales 41
Giraffes 186–187, 192–194, 223
Giraffids 193
Glossopteris 26–27
Glyptodon 196–197
Glyptodonts 196–197, 223
Goats 194–195
Gomphotherioidea 174
Gondwanaland 207–208, 214, 216, 218
Gonioteuthis 57
Gontiatites 57
Gonyaulacysta 36–37
Gordius 58
Gorgonopsians 134
Graminales 42
Graptolites (Graptolithina) 70–71, 206–207, 209, 229, 236, 242
Grasses 42, 223–224
Grebes 139, 144
Green algae 34–35, 204
Grouse 139
Gruiformes 139, 144
Gulls 144, 219
Gymnosperms 22, 40–43, 210, 214

H
Hadrosaurids 127, 247
Halysites 53
Hares 200–201
Hawks 139, 146
Hedgehogs 156
Hegetotheres 179
Helaletes 184
Helaletids 184
Hemichordates 72
Hemicyclaspis 76–77
Henodus 107
Herbivores 122, 124, 126, 130–133, 151–152, 157, 164, 168–171, 173, 182–183, 219–220
Herons 145
Hesperornis 140, 218
Hesperornithiformes 139–140
Hippomorpha 182
Hippopotamuses (Hippopotomoids) 186–187, 225
Holocene Epoch 17, 225
Holocephali 75

Holosteans (Holostei) 84–85
Homalozoa 47, 70
Hominids 160–161
Hominoids (Hominoidea) 160–161
Homo erectus 161, 239
Homo habilis 160–161
Homo sapiens 161
Homoeosaurus 115
Horses 150, 180–182, 186, 221–225
Horseshoe crabs 63–65
Horsetails 22, 39, 43, 209–212, 214
Hyaenas (Hyaenidae) 164–166, 225
Hyaenodon 163
Hyaenodonts (Hyaenodontids) 162–163
Hyalospongea 51
Hybodus 83
Hydrozoans (Hydrozoa) 52
Hylobatids 160
Hylonomus 102
Hypertragulids 191
Hypsilophodon 126
Hypsognathus 103
Hyracodon 184
Hyracodontids 184
Hyracotherium 180–181
Hyraxes (Hyracoidea) 150, 172, 222
Hystricidae 201

I
Ibises 145
Icaronycteris 157
Icarosaurus 113
Ichthyornis 140, 218
Ichthyornithiformes 139–140
Ichthyosaurs (Ichthyosauria) 24, 100, 108, 110–111, 215–217, 238, 242–243
Ichthyosaurus 111, 239, 245
Ichthyostega 92
Ichthyostegids (Ichthyostegalia) 90, 92
Ictidosaurs 135
Iguanodon 126–127, 242, 245
Iguanodontids 127, 244
Inarticulata 69
Insectivores 27, 156–158, 161–162, 164, 168, 218, 221
Insects (Insecta) 18, 23, 46–47, 60–61, 209, 211, 217
Invertebrates, classification of 21–23, 47
evolution of 22–23, 47
fossils of 237
in geological periods 204–208, 210–214, 216–218, 220–222
types of 44–71
Irregulares 51
Ischnacanthiformes 79
Ischnacanthus 79
Ischyodus 83

J
Jawless fishes 72, 75–77
Jellyfishes 47, 52–53, 204–205

Jurassic Period, amphibians in 94, 97
birds in 140
description of 17, 216–217
fishes in 83–85
fossils from 235–236, 238
invertebrates in 55, 57, 66–67, 69
mammals in 152–153
plants in 37, 41
reptiles in 108–111, 115, 119, 121, 123–126, 128–129, 135

K
Kakabekia 31, 204
Kangaroos 154–155, 225
Karaurus 97
Kiaeraspis 77
Kingfishers 139, 147
Kinkajous 166
Kiwis 139
Koalas 150, 154
Kronosaurus 109

L
Labyrinthodonts (Labyrinthodontia) 88, 90, 92–95, 97, 208, 210, 212, 214
Lagomorphs (Lagomorpha) 150, 200–201, 224
Lambeosaurus 127
Lamellibranchia 55
Lampshells 68, 206–207
Lancelets 74, 205
Lanius 147
Latzelia 61
Laurasia 207–208, 214, 218
Lemurs (Lemuroidea) 27, 158, 161, 221
Lepidodendron 38–39
Lepidosaurs (Lepidosauria) 100–101, 112, 114
Lepidotes 85
Lepospondyls (Lepospondyli) 88, 90, 96, 212
Leptolepis 85
Leptomeryx 191
Lepus 201
Lingula 68–69
Lissamphibia 97
Litopterns (Litopterna) 150, 178–179, 223–224
Liverworts 34–35, 43
Lizards 100–101, 112, 114, 215
Llamas 187–189
Lobsters 66–67
Lonchodomas 63
Lunaspis 81
Lungfishes 28, 86–87
Lycaenops 134
Lycophyta 38–39, 43
Lycosuchus 134
Lystrosaurus 26, 133

M
Macrauchenia 179
Macraucheniid 179

Macropoma 87
Magnolia 42
Maidenhair trees 41
Malacostracans (Malacostraca) 66–67
Mammal-like reptiles 130–135
Mammals (Mammalia), classification of
  20–21, 23, 148, 150
  evolution of 18, 150–151
  in geological periods 214–215,
    218–225
  types of 148–201
Mammoths 174, 176–177, 225
Mammuthus imperator 177
Mammuthus primigenius 177
Man 158–161, 224
Maple 42–43
Marattiales 40
Marchantia 35
Marsupials (Marsupiala) 150, 154–156,
  221, 224–225
Marsupicarnivores (Marsupicarnivora)
  150, 154–155
Mastodon 175
Mastodontoidea 174
Mastodonts 162, 174–175, 222–225
Medullosa 40
Medusas 52
Megalictis 167
Megaloceros 148, 193
Meganeura 61
Megasecoptera 61
Megatherium 197
Megazostrodon 153
Megistotherium 162
Merostomata 65
Merychippus 180–181
Merycodus 194–195
Merycoidodon 189
Merycoidodonts (Merycoidodontoidea)
  187–189
Mesembriportax 195
Mesohippus 180
Mesonychids (Mesonychidae) 169, 198
Mesosaurs (Mesosauria) 100, 102–103
Mesosaurus 103
Mesosuchians 118–119
Mesozoic Era, description of 16–17,
  214–215, 219
  extinctions in 28
  fossils from 228, 236, 241
  life in 40–41, 43, 47, 56–57, 75, 91,
    98, 101, 112, 116, 122, 151
  (also see Cretaceous, Jurassic, Triassic
    Periods)
Metacheiromys 197
Metamynodon 184
Metatheria 150
Metazoans (Metazoa) 46
Metriorhynchus 118–119
Miacids (Miacidae) 164–166
Miacis 165
Microsaurs (Microsauria) 90, 96
Millepora 52

Millerosaurus 103
Millipedes 47, 58, 60–61, 207
Miocene Epoch, birds in 144, 146–147
  description of 17, 223
  mammals in 159–160, 162–163, 165,
    167, 173, 175, 178–179, 181,
    183–184, 187, 189–191, 193–195,
    199
  plants in 42–43
Miohippus 180–181
Miotapirus 184
Mississippian Period 17, 210–211
Moas 139, 142–143, 242
Moeritheres (Moeritherioidea) 174
Moeritherium 174
Mole rats 200–201
Molluscs (Mollusca) 19, 23, 25, 28, 47,
  54–57, 205–208, 210, 214, 216, 218,
  220, 236
Monera 23, 31
Monkey puzzle tree 41
Monkeys 158–161, 223, 225
Monocots (Monocotyledonae) 42
Monoplacophora 54
Monorhina 75
Monotremes (Monotremata) 150,
  152–153, 156, 223
Morganucodon 135, 153
Moropus 183
Mosasaurs 114, 242, 245, 247
Moschops 132–133
Mosses 35, 43
Moulds 14, 31
Mousebirds 139, 146
Multisparsa 69
Multituberculates (Multituberculata) 150,
  152–153, 220–221
Museums 231–235, 241–247
Musk-oxen 194
Mussels 55
Mustelids (Mustelidae) 166–167
Mylodon 164
Myomorphs (Myomorpha) 200–201
Myotragus 195
Myriapods (Myriapodia) 23, 47, 60–61
Mysticetes 199
Mytilus 55

N
Nautiloids (Nautiloidea) 56–57, 231
Nautilus 56
Neanis 147
Necrolemur 158
Nectrideans (Nectridea) 90, 96
Nematode worms 23, 221
Nematomorpha 58
Neognathous birds 139
Neopilina 28
Neornithes 136, 139
Newts 88, 90
Nightjars 139, 146
Nostoc 31

Nothosaurs (Nothosauria) 100, 106–107,
  215
Nothosaurus 107
Notoprogonia 178
Notostylops 178
Notoungulates (Notoungulata) 150,
  178–179, 223–224
Nummulites 49

O
Octocorallia 52
Octopuses 54, 56–57
Odobenidae 166
Odontocetes 199
Odontognathous birds (Odontoholcae)
  136, 139
Ogygoptynx 146
Okapis 193
Oligocene Epoch, birds in 144, 146–147
  description of 17, 222
  invertebrates in 49, 67
  mammals in 159–160, 166, 172–175,
    179–181, 183–184, 186, 189, 199
  plants in 35
Oligokyphus 135
Onychophorans (Onychophora) 47, 60
Ophiacodon 131
Ophiacodonts (Ophiacodontia) 130–131
Ophiderpeton 96
Ophthalmosaurus 111
Opossum rats 27, 154
Opossums 219
Ordovician Period, description of 17, 206
  life in 35, 51–53, 55, 57, 62–63,
    68–71
Oreodonts 187–189, 223
Oreopithecids 160
Ornithischian dinosaurs (Ornithischia)
  100, 112, 126–129, 214–216, 218
Ornithomimus 122
Ornithopods 126–128, 246
Orohippus 181
Orthoceras 57
Osteichthyes 72, 75, 84
Osteoborus 167
Osteodontornis 144–145
Osteolepis 86
Ostracoderms 76–77
Ostracods (Ostracoda) 66, 236
Ostrich dinosaurs 219
Ostriches 139, 142
Otariidae 166
Otters 166
Ouranosaurus 126–127, 244
Owls 139, 146
Oxyaenids 162–163

P
Pachycephalosaurids 127
Pachycephalosaurus 126–127, 219
Pachydyptes 144
Pachyrukhos 179
Pakicetus 198

Palaeanodonts 196–197
Palaeinachus 67
Palaeocastor 200
Palaeocene Epoch, amphibians in 97
    birds in 146
    description of 17, 220
    fishes in 83
    invertebrates in 49
    mammals in 153, 157–158, 168–171,
    179–180, 201
Palaeodonts (Palaedonta) 186–187
Palaeognathous birds 139
Palaeolimulus 65
Palaeoloxodon 177
Palaeomerycids 193
Palaeoniscum 85
Palaeophonus 65
Palaeoreas 195
Palaeotherium 181
Palaeotragus 193
Palaeozoic Era, description of 16–17,
    205, 213
    extinctions in 28
    fossils from 228–229, 236
    life in 43, 47, 71, 74, 90, 92, 100–101,
    112
    (also see Cambrian, Carboniferous,
    Devonian, Ordovician, Permian,
    Silurian Periods)
Paliguana 114
Pandas 166
Pangaea 26–27, 208, 212
Pangolins 150, 196
Pantodonts (Pantodonta) 168, 170–171
Pantolambda 171
Pantotheres (Pantotheria) 150, 152–153
Pantylus 96
Paraceratherium 184–185
Paracyclotaurus 95
Paramys 200–201
Parazoans (Parazoa) 46, 50–51
Pareiasaurs 103
Parrots 139, 146
Passeriformes 139, 147
Patriofelis 163
Paucituberculates (Paucituberculata) 150,
    154
Peccaries 186–187, 224
Pecorans (Pecora) 187
Pelecaniformes 139, 144–145, 223
Pelicans 139, 144
Peloneustes 109
Pelycosaurs (Pelycosauria) 28, 100,
    130–132, 134, 212
Penguins 139, 144
Pennsylvanian Period 15, 17, 102, 131,
    210–211
Peramelines (Peramelina) 150, 154
Percrocuta 165
Perissodactyls (Perissodactyla) 150,
    180–183, 220–225

Permian Period, amphibians in 91, 93–95
    description of 17, 212–213
    fishes in 78–79, 83, 85
    fossils from 26–27
    invertebrates in 53, 55–56, 58, 65
    plants in 40–41
    reptiles in 100, 102–103, 106–107,
    112–113, 130–134
Petalichthyida 81
Petalonamae 49
Petrels 144
Phaeophyta 35
Phenacodus 168–169
Phiomia 175
Phiomorpha 201
Phlaocyan 167
Phocidae 166
Pholidota 150, 196
Phoronids (Phoronida) 47, 68–69
Phoronis 69
Phorusrhacids 142
Phorusrhacos longissimus 143
Phthinosuchians (Phthinosuchia) 130,
    132
Phthinosuchus 132
Phylloceras 57
Phyllolepida 81
Phyllolepis 81
Phytosaurs 116–118
Piciformes 139, 147
Pigeons 139, 146
Pigs 186–187, 222
Pikas 200
Pinnipeds 164, 166–167
Placental mammals 156–158
Placoderms (Placodermi) 72, 75, 80–82,
    84, 207–208
Placodonts (Placodontia) 24, 100,
    106–107, 215
Placodus 106–107
Plagiosaurs 94–95
Planetetherium 157
Plants (Plantae), classification of
    21–23, 30, 32, 34, 43
    evolution of 18–19, 43
    in geological periods 207–212, 214,
    216, 218–219, 223
    types of 32–43
Plateosaurus 125, 240
Platybelodon 174–175
Platygonus 187
Platypus 151, 156
Platyrrhini 159, 161
Pleistocene Epoch, birds in 146–147
    description of 17, 225
    mammals in 148, 155, 165, 167, 175,
    177–179, 184, 187, 193, 195, 197,
    201
Plesiadapis 158
Plesiadapoidea 158, 161
Plesiosaurs 100–101, 108–109, 216–219
Plesiosaurus 98
Pleuracanthiformes 83

Pleurocystites 70
Pleurodires (Pleurodira) 104
Pleurotomaria 55
Pliocene Epoch, birds in 146
    description of 17, 224
    mammals in 155, 165, 173, 175, 178,
    181, 187, 191, 193, 195, 197, 201
    protophytes in 37
Pliohippus 181
Pliomera 63
Pliosaurs (Pliosauroidea) 108–109
Plopterids 145
Podiceps 144
Podicipediformes 139, 144
Podocnemis 104
Poëbrotherium 189
Pollen 19, 42
Polychaete worms (Polychaeta) 58–59
Polyplacophora 54
Polypodiophyta 40, 43
Polytholosia 51
Pongids 160
Porana oeningensis 43
Porcupines 200–201
Porifera 47, 50–51
Precambrian time, description of 16–17,
    204
    life in 31, 49, 52, 54, 59
Presbyornis 15, 144
Primates 18, 20–21, 150, 158–161,
    220–225
Proboscideans (Proboscidea) 150, 172,
    174–177, 221, 223–225
Procaryotes, see Prokaryotes
Procellariiformes 139, 144
Procolophonians 103
Procolophonoids 102–103
Procyonids (Procyonidae) 166–167
Proganochelyds (Proganochelydia) 104
Proganochelys 104
Prokaryotes (Prokaryota) 22–23, 30, 34
Pronghorns 187, 194–195, 223–224
Prosauropods 124–125
Prosimians 158
Prosqualodon 199
Proterosuchians 116–117
Proterotheriids 179
Proterozoic Era 17, 47
Proteutherians 157
Protists (Protista) 22–23, 30, 34, 36–37,
    46
Protoceratids 191, 223–224
Protogyrinus 93
Protophytes (Protophyta) 22, 36–37
Protosiren 173
Protosuchians 118–119
Protosuchus 118–119
Prototheria 148, 150, 152
Protozoans 22, 46, 48–49
Protrogomorphs (Protrogomorpha)
    200–201
Protungulatum 169
Protypotherium 178

Prozeuglodon 199
Prymnesium 36
Przewalski's horse 181
Psaronius 40
Pseudosuchians 116–117, 120, 122, 138
Psittaciformes 139, 146
Pteranodon 121, 219, 246
Pteraspida 77
Pteraspis 77
Pteridospermales 40
Pterobranchs (Pterobranchia) 70–71
Pterocorallia 52
Pterodactyloids 120–121
Pterosaurs (Pterosauria) 24, 100–101,
    112, 116, 120–121, 138–139,
    215–219, 235, 238, 247
Pterygotus 64–65
Pthinosuchus 132
Ptyctodontida 81
Puffinus 144
Pyrotheres (Pyrotheria) 27, 150,
    178–179, 222
Pyrotherium 179
Pyrrophyta 36

Q
Quaternary Period 17, 161, 225, 239, 243
Quetzalcoatlus 121, 218

R
Rabbits 150, 200–201
Racoons 166
Radiolaria 49, 239
Rails 139, 144
Ramapithecus 160
Ranales 42
Raphus 146
Ratites 142, 221, 223, 225
Rats 200–201
Ray-fins 75, 84–86
Red algae 34–35
Reptiles (Reptilia), classification of 98,
    100
    evolution of 18, 23, 91, 100–101
    in geological periods 210–212,
    214–219
    types of 98–135
Rhabdopleura 71
Rhachitomes 94
Rhamphodopsis 81
Rhamphorhynchoids 120–121
Rheas 139
Rheiformes 139
Rhenanida 80
Rhinoceroses 180, 182, 184–185,
    221–225
Rhinocerotids (Rhinocerotoidea) 182, 184
Rhipidistians (Rhipidistia) 86–87, 90–91,
    213
Rhizopods (Rhizopoda) 47–49
Rhodophyta 35
Rhynchocephalians (Rhynchocephalia)
    100, 112, 114–115, 214

Rhyncosaurids 115
Rhyncosaurs 115, 215
Rhyniella 61
Rhyniophytes (Rhyniophyta) 38–39, 43,
    207
Rodents (Rodentia) 150, 200–201,
    220–221, 224–225
Rollers 139, 147
Rostroconchs (Rostroconchia) 55
Rotaliida 49
Roundworms 47, 58
Rudists 55
Rugose corals 52
Ruminants (Ruminantia) 187–195
Rutiodon 117

S
Sabre-tooth cats 165, 223–225
Saghatherium 172
Salamanders 88, 90, 97, 219
Sarcopterygians (Sarcopterygii) 75, 86,
    208, 218
Saurischian dinosaurs (Saurischia) 100,
    112, 122–126, 214, 216, 218
Sauropodomorphs 124
Sauropods 124–126, 244, 246–247
Sauropterygia 100, 107–109
Scale trees 209, 211–212
Scaphonyx 115
Scaphopoda 55
Scelidosaurus 128–129
Scenella 54
Sciuromorphs (Sciuromorpha) 200–201
Sciurus 201
Scleractinia 53
Sclerospongea 51
Scolecodonts 58
Scorpions (Scorpionida) 47, 64–65, 207
    211
Scutosaurus 103
Scyphozoans (Scyphozoa) 52
Sea anemones 47, 52–53
Sea cows 150, 172–173, 221
Sea cucumbers 71
Sea lilies 47, 70, 207, 229
Sea scorpions 64–65, 207, 213
Sea urchins 25, 47, 70–71, 220
Seals 164, 166–167
Seaweeds 34–35
Seaworms 60
Sebecosuchians 118–119
Sedentaria 59
Seed-ferns 40
Segmented worms 47, 58–59
Selachii 83
Serpula 59
Seymouria 93
Seymouriamorpha 93
Shantungosaurus 126–127, 243
Sharks 72, 75, 80, 82–83, 208, 220, 229
Sharovipteryx 120
Shearwaters 144
Sheep 168, 186, 190, 194

Shellfish 24–25
Shrews 156
Shrimps 66–67
Silicoflagellates 37
Silurian Period, description of 17, 207
    fishes in 76–80
    invertebrates in 52–53, 59, 63–65, 70
    plants in 37, 39
Siphonia 51
Sirenians (Sirenia) 150, 172–173
Sivatherium 193
Sloths 196–197, 222, 224
Smilodon 164–165
Snails 46–47, 54–55, 211
Snakes 100–101, 112, 114, 219
Sordes 121
Spalacotherium 135
Sphenacodonts (Sphenacodontia)
    130–131
Spenisciformes 139, 144
Sphenophyta 38–39, 43
Spiders 47, 60, 64–65
Spiny anteaters 27, 152–153
Spiny fishes 72, 78–79
Sponges 22, 46–47, 50–51, 204–205
Spores 19, 31, 34, 38–40
Spriggina 59, 241
Squamata 100, 112, 114, 218
Squids 54, 56–57
Squirrels 200–201
Stagonolepis 117
Starfishes 70–71, 205
Stauropteris 40
Stegoceras 126–127
Stegodon 175
Stegodonts (Stegodontoidea) 174
Stegosaurs 128–129, 216
Stegosaurus 128–129
Stenomylus 188
Stenonychosaurus 123
Stenopterygius 110
Stensiö, Erich 72, 233
Stephanoceras 57
Stephanochara 35
Stereospondyls 94–95
Sthenurus 155
Stictonetta 144
Stoats 222
Stonefly 61
Stonewort 35
Stony corals 52–53
Storks 139
Streptelasma 52
Strigiformes 139, 146
Stromatolites 28
Struthioniformes 139
Stylinodon 157
Subungulates 172–175
Suina 186
Suoids (Suoidea) 186–187
Swans 144
Swifts 139, 147

Symmetrodonts (Symmetrodonta) 150, 152
Synapsids (Synapsida) 98, 100–101, 130
Syndyoceras 190
Synthetoceras 191

T
Tabellaecyathus 51
Tabulate corals (Tabulata) 28, 53
Taeniodonts (Taeniodontia) 150, 156
Taeniolabis 153
Tanystrophaeus 107
Tapirids 184
Tapirs (Tapiroidea) 180, 182, 184, 221, 225
Tarsiers (Tarsioidea) 158, 161, 221
Teleosts (Teleostei) 84–85
Temnospondyls (Temnospondyli) 90, 92, 94, 97
Tentaculates 68–69, 206, 216, 220
Teratornis 147
Terns 144
Terror cranes 142–143
Tertiary Period, description of 17, 220–224
    (also see Eocene, Miocene, Oligocene, Palaeocene, Pliocene Epochs)
Thalattosauria 113
Thaumatosaurus 109
Thecodont (Thecodontia) 12, 100–101, 112, 116–118, 122, 126, 138, 215
Thelodus 77
Therapsids (Therapsida) 100, 130, 132–135, 212, 214
Therians (Theria) 148, 150–152
Theriodonts (Theriodontia) 130, 134–135
Therocephalians 134
Theropods 122–123
Thoatherium 179
Thyestes 77
Thylacoleo 155
Thylacosmilus 155
Tillodonts (Tillodontia) 150, 156–157, 221
Tinamiformes 139
Tintinnopsis 49
Titanosuchus 133
Titanotheres 182–183
Toads 88, 90, 97
Tortoises 100, 104, 215
Toxodon 178–179
Toxodonts 178
Traguloids (Traguloidea) 187, 190–192, 194
Tragulus 191
Tree ferns 32, 209, 211–212, 216
Trees 38, 41–42, 218, 245
Tremataspis 77
Trematops 91
Triadobatrachus 97

Triassic Period, amphibians in 91, 94–97
    description of 17, 214–215
    fishes in 84
    fossils from 12, 26
    invertebrates in 51–52, 55, 57, 66–68
    mammals in 152–153
    plants in 37
    reptiles in 100, 102–104, 106–108, 111, 113–115, 117, 119–120, 124–126, 130, 133–135
Triceratops 128–129, 244
Triconodonts (Triconodonta) 150, 152–153
Trigonostylopoids (Trigonostylopoidea) 150, 178–179
Trigonostylops 179
Trilobites (Trilobitomorpha) 18, 24–25, 47, 62–63, 205–207, 209, 213, 231, 236–237, 245
Trimerorhachis 94
Trimerus 63
Tritylodonts 135
Trogosus 157
Tubulidentata 150, 168
Turtles 100, 104, 219, 229, 245, 247
Tusk shells 55
Tylopods (Tylopoda) 187–189
Tylosaurus 114, 246
Typotheres 178
Tyrannosaurus 122–123, 219

U
Uintatheres 27, 168, 170–171, 221
Uintatherium 170–171
Ulva 35
Ungulates 18, 168–195, 221
Urochordates 23
Urodeles (Urodela) 88, 90, 97
Ursids (Ursidae) 166–167
Ursus spelaeus 167

V
Vascular plants 38–39
Velvet worms 47, 60, 205
Venyukovia 133
Venyukoviamorphs 132–133
Vertebrates (Vertebrata)
    classification of 20–22
    evolution of 18
    types 72–201
Viverridae 165–166

W
Walruses 164, 166
Weasels 166, 224
Weigeltisaurus 113, 245
Whales 18, 150, 198–199, 221–222
Williamsonia 41
Wolves 166–167
Wombats 154
Woodpeckers 139, 147
Woolly mammoth 176–177, 225

Woolly rhinoceros 184, 225
Worms 15, 22–23, 25, 46–47, 58–59, 204–205, 221

X
Xenacanthus 83
Xenungulates (Xenungulata) 150, 178–179
Xiphactinus 84
Xiphosura 65

Y
Yews 214
Youngina 112–113
Younginiformes 113

Z
Zalambdalestes 157
Zamia 41
Zebras 181
Zeuglodon 198
Zoantharia 52–53